For Calum and Kirstie

Human Performance in General Aviation

Edited by
DAVID O'HARE
University of Otago
Dunedin, New Zealand

Routledge
Taylor & Francis Group

LONDON AND NEW YORK

First published 1999 by Ashgate Publishing

2 Park Square, Milton Park, Abingdon, Oxon OX14 4RN
711 Third Avenue, New York, NY 10017, USA

Routledge is an imprint of the Taylor & Francis Group, an informa business

First issued in paperback 2016

British Library Cataloguing in Publication Data
Human performance in general aviation
   1.Aeronautics - Human factors
   I.O'Hare, David
   629.1'3

Library of Congress Cataloging-in-Publication Data
Human performance in general aviation / edited by David O'Hare.
     p. cm.
  Includes index.
ISBN 978-0-291-39852-9
   1.Aeronautics--Human factors. 2.Private flying. 3.Airplanes--Piloting--Human factors. I. O'Hare, David.

  TL553.6 .H864 1999
  629.132'52--dc21

                                        99-45177

ISBN 978-0-291-39852-9 (hbk)
ISBN 978-1-138-25608-8 (pbk)

Transfered to Digital Printing in 2011

Contents

List of Contributors

Prue Anderson is a commercial pilot and has an Honours degree in Aviation Science from the University of Newcastle. She is currently pursuing a Master of Business Administration degree at Macquarie University. Her research interests are in the area of team skills and innovative teaching techniques and she has published several articles in the area. Prue holds the position of Distance Education Manager with the Department of Aviation Studies at the University of Western Sydney, Macarthur and her current research project is to investigate teaching strategies and methods to overcome the isolation of studying as a distance education student.

Richard Batt is an Air Safety Investigator with the Bureau of Air Safety Investigation, Australia. As a human performance specialist his responsibilities include accident and incident investigation and air safety research. He has a B.Sc. in cognitive psychology from Flinders University, Australia, and has worked in research in cognitive psychophysiology at the University of Adelaide, Australia. He is currently completing a PhD in the Department of Psychology, University of Otago, New Zealand, working on a project investigating the role of aircraft accident and incident case histories in aviation safety and training. He is a member of the International Society of Air Safety Investigators and the Australian Aviation Psychology Association.

Dennis Beringer is presently a Research Engineering Psychologist at the Civil Aeromedical Institute, Oklahoma City, and manages the flight simulation facility for the Human Factors Laboratory. There he conducts flight simulation studies investigating a variety of issues involving human performance in the general aviation cockpit, including benefits of cockpit innovations (integrated instruments, multi-function displays) and pilot responses to automation (autopilot failures). He is also currently liaison to the AGATE (Advanced General Aviation Transport Experiments) project for CAMI in the primary flight and multifunction displays groups. His previous work in academics as director of two human factors research labs involved advanced multi-axis manual controls, cockpit displays of traffic information, integrated status displays, and pilot/computer interface technologies. He continues to be involved as consulting editor and reviewer for several journals. Consulting activities have included both flight deck and air traffic control system designs. Dr Beringer has been a pilot since 1969.

Don Harris is the Senior Research Fellow in Human Factors Engineering in the Human Factors Technology Group at the College of Aeronautics, Cranfield University, where he has worked for the last 12 years. His principal research interests lie in assessing the training effectiveness of low fidelity flight simulators; the analysis of primary flight control configurations in the cockpit; the assessment of aircraft handling qualities; and the effects of low blood alcohol levels on pilot performance. His teaching duties cover all aspects of human factors in aviation. He is also a human factors accident investigator for the British Division of Army Aviation. Don has also recently been actively involved in the European consultation process to develop human factors guidelines for flight deck certification. Outside of his immediate aviation interests, Don was the chairman of the first and second International Conferences on Engineering Psychology and Cognitive Ergonomics and is a member of the editorial boards of the International Journal of Cognitive Ergonomics and the International Journal of Cognition, Technology and Work. Before moving into a new house he was also an enthusiastic glider pilot!

Irene Henley is an Associate Professor and foundation Head of the Department of Aviation Studies at the University of Western Sydney, Macarthur in Australia. Prior to this appointment, she taught in the Department of Aviation at the University of Newcastle. A former Transport Canada Flight Training Standards Inspector, Dr Henley holds a Canadian Airline Transport Pilot Licence, Class I Flight Instructor Rating with aerobatic and float endorsements, an Australian Grade 1 Flight Instructor Rating, and is a Civil Aviation Safety Authority Approved Testing Officer. She has over twenty years of teaching experience as well as over twenty years experience in civil aviation. Her research interests include flight instructor training and evaluation, student pilot and flight instructor stress, and innovative and effective teaching strategies in aviation education

Ruth M. Heron is a specialist in aviation cognitive ergonomics, currently living in Vancouver, British Columbia, Canada. She obtained the M.Sc. in 1972 and the Ph.D. in 1975 from the University of Calgary in Alberta, emphasising perception, cognitive science, and methodology. Subsequently, as a professor at Queen's University in Kingston, Ontario, she taught ergonomics, carried out laboratory research in measurement, and acted as a consultant in transportation ergonomics. Much of her later activity in applied ergonomics has been through her role as Principal Ergonomist at the Transportation Development Centre in Montreal, where she designed, developed, and directed large-scale R&D projects in transportation technology. Those in aviation include such areas as flight simulator development; pilot training, workload, and fatigue; flight deck technology and automation; and ATC work station design, shift work, fatigue, skill, selection, and training. Other professional activity includes membership on various editorial boards and human factors committees, and roles in various advisory capacities. Notably, she has been adviser to Transport Canada's member of the ICAO Human Factors Committee; recently she was appointed to the membership of the Advisory Board for the Institute for Human Factors and Interface Technology at Simon Fraser University in Burnaby, B. C. Dr Heron has published and presented papers in numerous countries throughout the world. She is a member of the Association of Aviation Psychologists and of the Human Factors Association of Canada, a Fellow of the Ergonomics Society in the U.K., and a Certificant of the Board of Certified Professional Ergonomists. Until recently, she flew a Cessna Aerobat out of Burlington, WA.

David R. Hunter is a research psychologist with the Federal Aviation Administration. He manages an extensive programme of research aimed at improving aviation safety through the development and marketing of interventions to improve the decision-making of pilots. Previously, he has worked on the design of new military helicopters while with the US Army Research Institute for the Behavioral and Social Sciences, and has also conducted research on pilot and air traffic controller selection for both the United States Air Force and the Royal Air Force. Dr Hunter is a former military pilot who flew in combat in Vietnam and has been both a military and civilian flight instructor.

Kirsten Kite is a librarian and information specialist who has provided consulting services in many areas of human factors, including warning compliance, vehicle conspicuity, tractor safety, and truck ingress-egress systems. She is co-author, with Alan Stokes, of *Flight Stress: Stress, Fatigue, and Performance in Aviation*, and of *Display Technology: Human Factors Concepts*.

Michael D. Nendick is a lecturer in the Department of Aviation & Technology at the University of Newcastle, Australia. Mike has an aviation background that includes service as an Royal New Zealand Air Force Navigator and civilian air traffic controller. He also holds a private pilot licence. Prior to taking up his current position Mike undertook MSc studies at Massey University on human factors aspects of GPS use. At the University of Newcastle he teaches advanced navigation and aviation human factors. Mike has published on human factors and GPS issues and contributed the chapter "Human Factors & GPS" to the recent Australian Civil Aviation Safety Authority book on GPS. He is currently commencing doctoral research in the aeronautical decision making field.

David O'Hare is a senior lecturer in the Department of Psychology at the University of Otago, Dunedin, New Zealand. He obtained his PhD in 1978 from the University of Exeter in England. He lectured at the University of Lancaster before moving to New Zealand in 1982. He has a wide range of interests in aviation psychology and cognitive ergonomics including decision making, human error analysis, injury prevention, attention management and training. He has worked as a consultant for the Civil Aviation Authority (New Zealand), the Transport Accident Investigation Commission and the Federal Aviation Administration. He is a full member of the International Society of Air Safety Investigators and the Human Factors and Ergonomics Society. He has held pilot licences in New Zealand and the United Kingdom as well as a variety of gliding qualifications. He is co-author (with Stanley Roscoe) of *Flightdeck Performance: The Human Factor* published in 1990.

Stanley N. Roscoe was a transport pilot in the Troop Carrier Command during WW II. He earned his PhD in aviation engineering psychology at the University of Illinois in 1950 and pioneered the application of human engineering principles and man-in-the-loop simulation in aircraft system design during the 1950s and '60s. He is retired from Hughes Aircraft Company (1977) where he was Manager of the Display Systems department and Senior Scientist; from the University of Illinois at Urbana-Champaign (1979) where he was associate director for research of the Institute of Aviation and head of the Aviation Research Laboratory; and from New Mexico State University (1986) where he was head of the Behavioral Engineering Laboratory. He is still occasionally active as professor emeritus of psychology at the University of Illinois at Urbana-Champaign and New Mexico State University; as president of ILLIANA Aviation Sciences Limited, and as vice-president of Aero Innovation Inc. He is the primary author of the WOMBAT Situational Awareness and Stress Tolerance Tests.

Alan Stokes is director of the Cognitive Systems Engineering professional programme at Rensselaer Polytechnic Institute, where he conducts research on emotion and cognition in the Minds and Machines Laboratory. He was previously at the University of Illinois Aviation Research Laboratory, where he investigated stress and decision making, and at Florida Institute of Technology, where he was chair of the Human Factors Programme and director of the Space Coast Centre for Human Factors Research. He is the author of numerous books and other publications in the human factors field, and is currently working on a synthesis of applied human factors and evolutionary psychology.

Mark Wiggins is a senior lecturer in human factors at the University of Western Sydney, Macarthur. A member of the Australian Psychological Society and the Human Factors and Ergonomics Society, he has a Masters Degree in Psychology from the University of Otago and undergraduate degrees from the University of New England. Mark is a private pilot and glider pilot and his research interests include decision-making, computer-based training, system safety and research design and methodology in the aviation environment. He is currently pursuing a PhD in psychology at the University of Otago.

Dmitri Zotov graduated from the RAF College with the Queen's Medal in 1961. He served in the RAF for 18 years and was involved with the development flight trials of the HS Nimrod at Boscombe Down; after the successful completion of these trials he was awarded the MBE. Subsequently he worked in Operational Research. After returning to New Zealand he flew professionally for several years and then spent seven years as an Inspector of Air Accidents. Subsequently he joined the staff of Massey University to lecture in Air Safety Investigation. His principal areas of interest have been the investigation of human factors in aircraft accidents, and the systemic factors which lead to accidents. He has a Masters degree in Aviation and is a member of the Royal Aeronautical Society.

Acknowledgements

The manuscript for this book was produced in the Department of Psychology at the University of Otago. I am very grateful for the support of the Department and of the University. I would like to thank Professor Jeff Miller for his kindness and encouragement. I am greatly indebted to Debbie Park for all her expertise and hard work on countless versions of this manuscript. I am also grateful to Norma Bartlett for her careful inspection of the manuscript and invaluable advice about formatting. I should also like to extend my appreciation to Adele Arnold for her work on the index.

This project would not have been undertaken without the support and encouragement of John Hindley at Ashgate. Obviously, an edited collection such as this would not be possible without the individual talents and enthusiasm of the contributors. I was fortunate to secure the commitment of such an outstanding group, every one of whom has an involvement in both the practical aspect of general aviation and in the research process. Thank you all.

David O'Hare

Foreword

Captain Dr A. Gordon Vette has become a leading figure in aviation safety since the publication of his pioneering inquiry ('Impact Erebus') into the events surrounding the crash of an Air New Zealand DC-10 on the slopes of Mount Erebus in 1979. Vette's thorough analysis of the perceptual conditions confronting the crew and of the organisational factors which led to the events of this particular flight have become recognised as the forerunner to the concerns with systemic failure which are now widely held in aviation safety. He has had distinguished careers as a military pilot, a civil flight instructor and as an air transport pilot, accumulating 21,000 hours of flight time on aircraft up to and including the B747 where he was Senior Flight Instructor for Air New Zealand. He was awarded the Johnston Memorial Trophy by the Guild of Air Pilots and Air Navigators for his rescue of a missing aircraft in the Pacific. This event was the subject of a major Hollywood film ('Mercy Mission'). He has continued to be actively involved in flight testing and in research on applications of synthetic vision, virtual reality and enhanced ground proximity warning systems for enhancing pilot situational awareness. He is Chairperson of the Advisory Board of ICARUS, the New Zealand confidential incident reporting scheme and is a passionate advocate of human factors and flight safety across the aviation spectrum. He was recently awarded an honoury doctorate in engineering by the University of Glasgow.

It is a very great honour to be asked to write the foreword to this book, particularly as David O'Hare, Stan Roscoe, and Dmitri Zotov have given substantial support to my work on flight safety enhancement, post Erebus, and my own doctoral studies in this area. We are well served to have such veterans continuing their research and supervising the work of students in the flight safety field. This excellent collaborative text provides a welcome blend of high academic achievement with practical experience. This team draws the problems of the sadly neglected general aviation area into sharp focus and offers sensible cost effective ways of making improvements. I found myself reading each chapter and saying "yes, yes," knowing that the real bonus of harmonious collaboration is that the total contribution of the team will always exceed the sum of the individual parts.

We need the thrust of getting more aviation practitioners to gain higher academic qualifications, and collaborating to produce cutting edge research texts, such as this one, and in the process gaining attractive qualifications for their selection. onto some of the more powerful and effective industry committees.

There are many such groups looking for very active members. One worth mentioning is the fast growing Aviation Study Group with a network of very well qualified members world wide, including the president of ICAO, and chaired by the extremely energetic Dr James Vant of Linacre College at Oxford University. This is one way we can help reduce the viscosity of the treacle of resistance to change through which we have to swim in order to gain even the most cost-effective changes. In this regard I am pleased that my own flight safety research trust fund, funded from proceeds from my book and video Impact Erebus and hopefully from my book Impact Erebus II (in preparation), has assisted numerous graduate and post graduate candidates, including some contributors to this book. Stan Roscoe refers to some of the difficulties to be overcome in noting the lost lessons from the past. "Many of the findings from research and from research on selection, transfer of training, and display and control designs have far wider applicability to GA than the industry has realised". So true.

In light of the above, it is not hard for me to believe that it is 30 years since I flew some of the early HUD (head-up display) trials onto Santa Catalina island with Don Bateman, chief engineer of Sunstrand now Allied Signal and participated in some of the early work on ground proximity warning devices. Don's work is legend and was recently recognised by the Guild of Air Pilots; he knows all about the treacle. I have enjoyed working with him again in recent years in my study of enhanced situational awareness systems, synthetic VFR, collision avoidance devices, etc. The more resources we can get to help such innovators and the various supporting committees, the sooner they will be able to complete the objective of having cost-effective protection in these areas right down to the smaller GA aircraft. I have also been privileged to work with Dr Paul Moller (Moller flying car *www.moller.com*) for about 15 years on his engineering design team and as his test pilot for the VTOL flying car project. It is intended that this highly innovative VTOL GA vehicle will be capable of zero zero weather capability eventually. This vehicle has eight low pollution, small frontal area, highly reliable, multi-fuel capable rotary engines producing 165Hp and weighing only 65 lbs which should be of great interest to the GA fraternity.

It is important, as David O'Hare and his collaborators clearly illustrate, that we understand that GA is the breeding ground for the entire aviation industry. We cannot afford to accept it as the weakest link. And any money and effort spent there is sound insurance for us all. This includes work on the design of GA aircraft systems discussed in several chapters (e.g., Beringer, Chapter 10 and Heron and Nendick, Chapter 9) in the present book, as well as techniques for pilot pre-selection which have been little used in general aviation. Hopefully these will have a prescriptive remedial ability to reduce wastage rates. Like Stan Roscoe (see Chapter 2), I enjoyed flying the Ercoupe, marvelled at its docile handling and was surprised when they ceased manufacture. This design would certainly have prevented many GA pilots from being killed. GA accidents represent a major loss of resources, therefore accident investigations are very important and generally cost effective. However, inaccurate results send the wrong signals. Systemic error is common and frequently not recognised by some investigators.

As an active flight instructor and examiner I endorse all of the material in Don Harris's chapter on 'Flying Light Aircraft' (Chapter 6) and am a strong advocate of the "follow me through" technique. I find it seldom distracts the student and I use it almost continuously on new exercises. It is my own experience, particularly when teaching aerobatics, that the follow through technique can be beneficially further modified by briefing in advance to expect momentary bursts of "instructor overrides", initially perhaps even full control throw and of high rate diminishing to taps on the controls at the precise instant required, when verbal explanation would be too slow. It is important that control still remains with the student who will have felt the correctly timed tactile feedback along with the enhanced performance result. All of this uses the open and closed loop kinetic and kinaesthetic feedback effects referred to in this chapter to produce best results in the shortest time.

For flight training on the aircraft I use the conventional thrust/power for rate of descent and elevators for airspeed control on base leg and only on the first few non-precision approaches. I switch very quickly to stabilised precision approaches (this is consistent with the research on command guidance and flight path prediction described by Roscoe in Chapter 2) where on final approach when configured to land we select and hold an acceptable glide slope and aiming point. From this moment on we must use thrust/power to control speed and flight controls to control trajectory or glide slope, if our selected slope is steep we may choose to hold constant speed of 1.3Vs + increment for gust etc. Or, as normal, use the constant bleed technique from that point, bleeding the speed at the same rate as we are closing with the threshold to arrive at the threshold at 1.3Vs + inc. This teaches glide slope or trajectory control using flight controls just as the autopilot does, and use of thrust to control speed, just as the auto throttle does. It is also the way it is actually done by experienced pilots flying true stabilised precision approaches where glide slope is "inviolate" and is the technique you use when flying HUD approaches. Most importantly, students learn, almost intuitively, from the beginning to judge their glide slope control by flare path or runway perspective, and how the aim point remains stationary with all points expanding uniformly from it as the threshold is approached. Also apparent is the angle of convergence of the sides of the runway meeting at the visual horizon and remaining constant to flarepoint. This results in shorter time to solo and a better end product.

The book contains numerous valuable suggestions for the aviation practitioner such as the use of reflective journals and problem-based learning discussed by Henley, Anderson, and Wiggins in Chapter 5, the innovative uses of computer-based training discussed by Wiggins (Chapter 7) and O'Hare and Batt (Chapter 8). Hunter (Chapter 3) documents the swift rise in pilot computer access capability for CBT etc. but points out that all alternate means of safety training should be explored and must be marketed as a product competing for pilot time and resources so we must design a product and delivery system accordingly

The nature of stress and coping in general aviation is discussed in detail by Stokes and Kite (Chapter 4). I can remember times when, under considerable stress and without the valuable lessons of this text, I have perceived the very mismatch they talk about between the demands of a situation and my perceived ability to meet the demands. I have had to resort to going with the 'gut feeling', with good results, not realising that the old reptilian limbic system was probably responding with emotional processing of information that the cortex and speech centres do not have. Finally I would like to see a much greater effort put into confidential reporting systems from the GA sector. This too is an extremely important area of endeavour as it allows early remedial action

Just reading this book will significantly educate and serve all within the general aviation system from private pilots to instructors and small business operators. This book must go out to leading persons in influential areas such as aero clubs, flying schools, accident investigators, manufacturers, regulators and legislators etc. I thank David and his team for this outstanding contribution to general aviation.

Gordon Vette

Part 1
Introduction

1   Introduction to Human Performance in General Aviation

David O'Hare

Introduction

The development of aviation clearly constitutes one of the great defining characteristics of the twentieth century. From the pre-World War I flying machine to the long-range, high-capacity, advanced-technology commercial jet of the 1990s lies a gap of a mere 95 years. Over 1,000 million fare-paying passengers were transported by commercial aircraft in 1996, returning an operating profit of US$12 billion to the airlines that carried them (ICAO, 1997). Unfortunately, 1,135 of those passengers lost their lives in 23 fatal scheduled air transport crashes Nevertheless, set against the numbers carried, such figures lend credence to the view that commercial air transport in the late 20th century is an exceedingly safe mode of transportation.

Of course it was not always this way. The accident rate in commercial transport has declined throughout its history. Advances in technology, especially the development of the reliable turbojet engine, and other developments in materials, navigation systems, and flying practices have all contributed. At the close-out of the twentieth century, increasing efforts are being devoted to tackling the most difficult problem of all—the contribution of human error to failures in the aviation system. This is generally alleged to contribute to 75-80% of air transport crashes.

However, if one thinks of aviation as a truly complex system involving the designers and manufacturers of all the equipment involved (aircraft, ATC, etc.), the maintainance of the components in the system, the organisation and management of all the processes involved, the selection and training of operational personnel (pilots flight attendants, controllers etc.) and the government regulators, then it could reasonably be pointed out that, unless one truly believes in 'Acts of God', almost 100% of crashes are due to some kind of human performance failure, somewhere in the system. For some reason, it is rarely pointed out that 100% of successful outcomes are also due to human performance.

The main focus for human performance analysis has been on commercial air transport. In view of the huge numbers of passengers carried, and the global economic significance of the air transport industry, this is perfectly understandable.

As we turn the corner into the new millennium this will surely continue to be the case. It is possible to argue, however, that it is time that a proportion of this effort be re-directed towards other areas of aviation. There are at least two principal reasons for this. Firstly, the size and significance of the problems elsewhere in the aviation system are such as to warrant increased attention. Secondly, as the use of the term 'aviation system' implies, there is a fundamental inter-relatedness between different participants in the aviation enterprise. For example, small, general aviation (GA) aircraft and large commercial transports are often obliged to share the same airspace. The capabilities and performance of each affect the overall system outcome. At worst, a number of commercial air transport crashes have been attributed to airspace incursions by small, GA aircraft. The collision between an Aeromexico DC-9 airliner and a Piper PA-28 at 6500 feet over Cerritos in the LAX Terminal Control Area on August 31, 1986 (National Transportation Safety Board, 1998) provides a terrible example of the dangers.

In some respects, the overall safety of the system is determined by the performance of the weakest link. This may often turn out to be the single crew, low-technology, less rigorously trained GA operator rather than the multi-crew, high-technology, expensively trained air transport participant. The challenges facing the former are summed up in this comment from an ASRS report (Callback, June 1998) submitted by an airline captain: "This Bonanza is a lot harder to fly than the B-757 I drive at work". None of this should be taken to imply that all GA operators are incompetent, badly trained 'accidents-waiting-to-happen'. Far from it. Most GA operators are careful, responsible, and highly competent. However, the fact remains that the size of the accident problem in GA is considerably greater than that in the air transport sector, and the potential for unintended interactions between the two communities is ever present.

The General Aviation (GA) Problem

General Aviation is defined as "civil aviation other than scheduled and non-scheduled commercial air transport" (ICAO, 1997, p. 9). In practice, GA includes the private pilot flying simple, light, single engine aircraft such as the Cessna 172 or Piper PA-28, training organisations preparing pilots for private and commercial exams, agricultural and aerial work, air rescue services, business flights, and a host of other operations flown by both fixed-wing and rotary-wing aircraft. Of these, business and recreational flying accounts for the largest proportion of flying hours as shown in Figure 1.1

There are over half-a-million active GA private pilot licence holders alone, worldwide flying an estimated 337,910 aircraft used mostly in GA. As Figure 1.2 shows, both the total numbers of hours flown and the numbers of fatalities occurring in air crashes are greater in the GA sector than in the commercial air transport sector. In addition, the GA sector is responsible for a great many more departures and miles flown than the air carriers (Olcott, 1997). As David Hunter (Chapter 3) points out, very little is known about the characteristics of GA pilots.

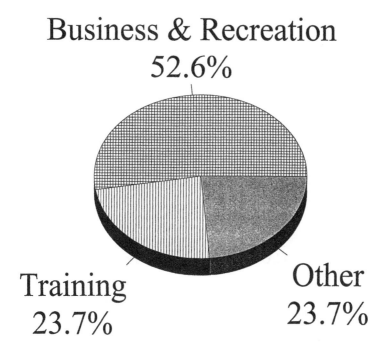

Business & Recreation
52.6%

Training
23.7%

Other
23.7%

Figure 1.1   Main sources of GA activity
Source: Year in Review, ICAO Journal, 1997

The more that is known about this group, the better we will be able to target effective improvements in regulation, education and training. Hunter reminds us of another very important point which is that virtually every pilot has at one time or another been a GA pilot, and that attitudes and behaviours developed at this stage are likely to influence performance further down the track.

This point is also emphasised by Irene Henley, Prue Anderson and Mark Wiggins (Chapter 4). They address a number of issues which arise in developing human factors training in GA with a particular focus on the role of the flight instructor. They also discuss in detail two innovative strategies, the use of reflective journals and problem-based learning, which can improve the learning process.

Whilst politicians and regulators focus almost exclusively on the toll arising from air transport crashes, the losses—both human and economic are as great, or greater in GA. Lives lost in concert, as in major disasters, are attention getting and newsworthy, whereas a similar toll spread thinly in ones and twos is easily ignored. This stark fact has resulted in public tolerance of high motor vehicle fatality rates and an apparent intolerance of 'disasters' involving multiple fatalities

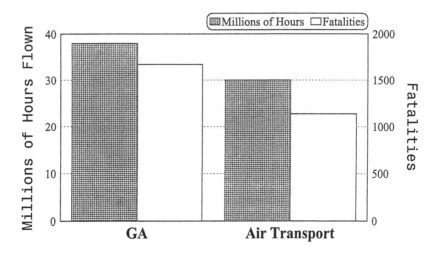

Figure 1.2   Total flight activity levels and fatality levels for GA and Air Transport Operations
Source: Year in Review, ICAO Journal, 1997

such as airliner crashes. It has also been claimed that the general public is willing to tolerate risk levels in voluntary activities (such as skiing or recreational aviation) that are approximately a thousand times greater than those from equally beneficial involuntary activities (Starr, 1969). If this is indeed true, and if it is taken as a prescriptive guide, then the almost exclusive focus on air transport safety might seem warranted. However, doubts have been raised about the validity of this assumption (Fischhoff, Lichtenstein, Slovic, Derby, and Keeney, 1981) and whilst a significant proportion of GA can be categorised as recreational (see Figure 1.1), a much greater proportion is undertaken for more utilitarian purposes.

It seems safer to assume that all loss of life is inherently undesirable and to the same degree. Aviation safety should be concerned with managing the risks of aviation to produce the fewest possible losses. This does not imply that the interests of the fare-paying passenger should be downgraded in any way, but that other participants in the aviation system are deserving of better efforts to understand and improve human performance and to make use of technological advances in all aviation systems which can lead to improved reliability and safety. In some areas of GA, such as sport and recreational aviation for example, so little attention has been paid to the scientific analysis of human performance that significant advances in understanding and consequent safety improvements might be relatively easy to achieve and would be extremely cost effective. In contrast, efforts to improve the already very low crash rate in the air transport sector will require significant advances in our understanding of human performance and improvements are likely to be extremely costly to implement throughout the industry.

Here Today ... Here Tomorrow

As noted above, the accident rates for commercial air transport have shown significant improvements since the beginnings of the commercial air transport industry in the 1930s. A particularly dramatic fall in accident rates has occurred since the introduction of jet airliners into commercial service in the 1960s. The reliability of jet engines compared to reciprocating engines was both directly (fewer engine failures) and indirectly (fewer powerplant problems meant fewer problematic situations for crew to manage) responsible for improved levels of safety. The rate of improvement in safety levels has of course decreased substantially in recent years (Nagel, 1988). GA aircraft are of course still predominantly of the piston-engine variety.

Safety levels in GA, defined in terms of accidents per 100,000 flight hours, can best be described as static. Figure 1.3 shows the GA accident rates for three countries—Australia, New Zealand and the U.S.A. between 1988 and 1994. Whilst the U.S rates are consistently a third lower than those in Australia and New Zealand, all three countries show completely static rates over this period. Unlike commercial air transport, there have been no major changes in GA aircraft technology during this time. Indeed, most of the aircraft which are flown in GA are essentially 1940s designs with cockpit displays and controls which have shown little change in decades. There are encouraging signs of significant changes on the way in GA design technology and these are discussed in some detail by Dennis Beringer (Chapter 10).

The impact of one new technological innovation, namely the development of cheap, portable GPS navigation units is now beginning to be felt throughout GA. The problems created by the proliferation of less than optimally designed GPS systems are discussed by Ruth Heron and Mike Nendick (Chapter 9). They consider both the basic ergonomics of the GPS system as well as problems arising from pilot training and flight preparation.

On the other hand, there has been an increasing recognition of the importance of human factors in pilot performance during this period, and some form of human factors training is now included in the pilot licencing syllabus in several countries. Unfortunately, it is not possible to draw any conclusions about the merits or otherwise of current forms of human factors training from these accident data. The normal caution about accidents being such rare events that underlying changes cannot be easily inferred is clearly applicable in this case. Also, since GA covers such a wide range of activities, aircraft, and pilots, the overall data may obscure individual trends in different areas. It is important therefore, to carefully examine the performance of different groups (e.g., rotary-wing versus fixed-wing, recreational versus aerial work etc.) so that similarities and differences can be properly determined.

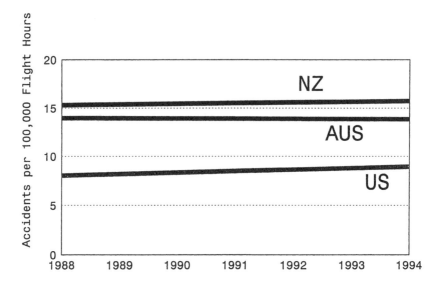

Figure 1.3   GA accident rates for New Zealand, Australia, and the U.S.A.
*Sources: (NZ) O'Hare, Chalmers, & Bagnall, 1996, (AUS) The
Parliament of the Commonwealth of Australia (1995) (USA) NTSB
web page (http://www.ntsb.gov/aviation/aviation.htm/)*

    As the International Civil Aviation Organization (ICAO) has pointed out: "general aviation accidents constitute a major loss of resources. As a consequence, substantial benefits are to be gained from accident prevention programmes aimed at this group" (ICAO, 1984, p. 9). However, as David O'Hare (Chapter 4) points out, a safety programme is not solely about accident prevention. Losses in terms of personal injury and material damage can be reduced by interventions targeted at events occurring both during, and after, the crash sequence itself. The design of such interventions depends on the development of a sound understanding of the risk factors for injury and damage in aircraft crashes.

    An important set of risk factors for GA accidents are related to weather conditions which are beyond the capabilities of the pilot, the aircraft, or both. In the United States alone "an average of 17 people a month are killed in aviation accidents attributable to weather" (AVflash Sunday, August 2, 1998). One potential solution lies in the provision of improved weather information to pilots. Dennis Beringer (Chapter 10) predicts that these data will be routinely available in the GA cockpit in the future. The other, complementary approach to the problem, is through training pilots to be better able to deal with the vagaries of weather. An innovative approach to the development of computer-assisted learning (CAL) systems for weather-related decision making is described by Mark Wiggins (Chapter 7).

Enhancing Safety in GA

In the past two decades there has been a major shift in our conceptualisation of accident causation. Several major air transport disasters provided the impetus for far-reaching analyses which looked beyond the accident flight to factors located deep within the organisations that operated, designed, maintained, and regulated the aircraft. The first of these, the loss of an Air New Zealand DC-10 on the slopes of Antarctica in 1979 led to an unprecedently far reaching public inquiry (Mahon, 1984) as well as an in-depth investigation by another Air New Zealand pilot (Vette, 1983). This accident and its aftermath is now considered a turning-point in the analysis of air transport accidents (Maurino, Reason, Johnston and Lee, 1995). Dmitri Zotov (Chapter 12) touches on the question of organisational pressures in his discussion of accident investigation in GA. The investigation of GA accidents requires the same thoroughness and painstaking attention to detail required in the investigation of an air transport accident. In the end though, the accident investigation process can only lead to safety improvements if the lessons learnt are acted upon.

Several chapters offer suggestions and techniques for improving the level of individual performance. Alan Stokes and Kirsten Kite (Chapter 9) provide a detailed discussion on the critical question of stress. In addition to outlining a number of operational recommendations for coping with stress, Stokes and Kite suggest that training pilots in decision making might be extended by 'packaging vicarious experiences…into Event-Based Simulations' (p. 72). David O'Hare and Richard Batt (Chapter 8) develop a similar proposal for using simulation to develop the experiences that pilots formerly acquired through a lengthy and often haphazard process of 'seasoning'. Don Harris (Chapter 6) looks at the development of the psychomotor skills required to fly the aircraft. The basic instructional techniques have evolved over the past 80 years or so to a high degree of proficiency. Improvements in instructional effectiveness are now more likely to be achieved by the appropriate use of low-cost flight simulators. These chapters all point the way to the more innovative use of currently existing technologies (e.g., cheap, PC-based flight simulation).

Conclusion

The future for GA looks bright. Technological developments may bring the dream of an 'everyman's' aircraft a little closer. The 21st century GA cockpit may actually be quite different from that of the preceding century. New ways of looking at training and innovative uses of low-end computer technology may bring substantial improvements in training effectiveness. A better understanding of the GA pilot and careful analysis of the risks and hazards of GA may ultimately contribute to improved safety levels. Each of these points is illuminated in greater detail in the chapters that follow.

Whilst the individual chapters in this book may be read in any order the reader desires, I would strongly recommend that Stanley Roscoe's (Chapter 2) succinct overview of aviation psychology be read first. Roscoe identifies the central issue of "why the general aviation community has willingly accepted the fact that flying is more dangerous than it needs to be" (p. 18). Each of the contributors to this book has asked themselves much the same question and the chapters that follow represent their efforts to provide the answers upon which a safer, more productive, and more enjoyable general aviation may be based.

References

Fischhoff, B., Lichtenstein, S., Slovic, P., Derby, S. L., & Keeney, R. L. (1981). *Acceptable risk*. Cambridge: Cambridge University Press.

International Civil Aviation Organization (1997). Year in review: Annual civil aviation report. *ICAO Journal, 52*(6). Whole issue.

International Civil Aviation Organization (1984). *Accident prevention manual* (Doc 9422-AN/923). Montreal: International Civil Aviation Organization.

Mahon, P. (1984). *Verdict on Erebus*. Auckland: William Collins.

Maurino, D., Reason, J., Johnston, N., and Lee, R. (1995). *Beyond aviation human factors*. Aldershot: Ashgate.

Nagel, D. C. (1988). Human error in aviation operations. In E. L. Wiener, & D. C. Nagel (Eds.*), Human factors in aviation* (pp. 263-303). San Diego: Academic Press.

National Transportation Safety Board. (1998). *Aircraft accident database* (http://www.ntsb.gov/Aviation). (Report DCA86AA041A).

O'Hare, D., Chalmers, D., & Bagnall, P. (1996). *A preliminary study of risk factors for fatal and non-fatal injuries in New Zealand aircraft accidents. Final report to the Civil Aviation Authority of New Zealand*. Dunedin: University of Otago Injury Prevention Research Unit.

Olcott, J.W. (1997). General aviation. In *McGraw Hill encyclopedia of science and technology, Vol. 7* (8th ed.). New York: McGraw-Hill.

Starr, C. (1969). Social benefit versus technological risk. *Science, 165*, 1232-1238.

The Parliament of the Commonwealth of Australia (1995). *Plane safe inquiry into aviation safety: The commuter and general aviation sectors*. Canberra: Australian Government Publishing Service.

Vette, G. (1983*). Impact Erebus*. Auckland: Hodder and Stoughton.

2 Forgotten Lessons in Aviation Human Factors

Stanley N. Roscoe

Introduction

Aviation psychology emerged from World War II as a new scientific discipline—a clearly recognisable branch of experimental psychology (Fitts, 1947; Chapanis, Garner, & Morgan, 1949). Several notable psychologists had been involved in pilot selection during World War I, but most of the tests they devised would now be considered little more than practical jokes on eager young pilot wannabes (Koonce, 1984). Furthermore, prior to World War II psychologists had no memorable involvement in pilot training or the experimental study of human factors in aviation equipment design, the other two subfields of aviation psychology.

Though the motivation for the scientific study of aviation human factors derived from military needs in World War II, it soon became evident that human factors deserved similar attention in the field of general aviation. While much of the support for such work in universities around the world continued to come from the military, in the United States at least the Civil Aeronautics Administration supported a programme that produced a stream of publications known as the "gray cover reports," in contrast to the U.S. Army Air Forces' series known as the "blue cover reports" (Roscoe, 1997).

Many universities contributed to these studies, but the laboratories that were most productive of experimental research and professional aviation psychologists in the 1940s and '50s were those of Donald Broadbent and Christopher Poulton at Cambridge in the UK and of Alexander Williams at Illinois, Alphonse Chapanis at Johns Hopkins, Paul Fitts at Ohio State, Leonard Mead at Tufts, J. C. R. Licklider at MIT, and Neil Warren at USC in the U.S.A. (Roscoe, 1997). Collectively these aviation psychologists were concerned with anything affecting the performance of pilots, air traffic controllers, and aviation maintenance personnel.

The laboratory of Alex Williams at Illinois was the most representative of the entire field of aviation human factors (Roscoe, 1994). In addition to the scientific study and application of ergonomic principles in the design of aviation displays and controls, as practised in the other university laboratories, Williams' lab pioneered the quantitative measurement of the transfer of training in airplane simulators to pilot performance in airplanes and the development of air-traffic-control simulators. These early studies, and those to follow at Illinois, provided the basis for modern training programs in general aviation as well as commercial and military aviation.

Regrettably, no sector of aviation has capitalised fully on the principles of cost-effective training transfer or the ergonomic design of aviation displays and controls established during the early days of aviation human factors research. Nor has either civil or military aviation taken full advantage of the safety, productivity, and accompanying economic benefits to be gained from the objective measurement of human aptitude for future performance in airplanes, control towers, and maintenance centres. Many of the findings from research on selection, transfer of training, and display and control design have far wider applicability to general aviation than the industry has realised.

Aptitude Prediction

Predicting how well an individual will perform in pilot training is a critical consideration not only for the school but also for the applicant. Before admitting the applicant, the school needs to know in advance how much time and instructor attention will likely be required, and the trainee needs the same information and, in addition, what his or her chances are of eventually being a safe pilot. It is costly for both to undertake training that will not result in criterion performance levels or, worse yet, will result in certification of a pilot who will be unable to exercise good judgement in an emergency.

As so often happens with pilots, and with air traffic controllers for that matter, the individual may have all the skills and knowledge normally required but be unable to put them together in a complex, confusing flight situation. In any case, the ultimate purpose of selection tests is not merely to predict who will succeed and fail in training, although that alone would be a highly profitable product of a sharply discriminating test. The ultimate objective is to predict who will be able to maintain situational awareness and not mix up task priorities when in command of an airplane 10, 20, or even 30 years in the future.

In predictions of this type, traditional basic abilities such as reaction times and manual dexterity—even hand flying skill-are of little concern. Diagnosing the situation while not losing control of the routine requires a different kind of ability. While it is true that reasonable levels of intelligence and motor skill are necessary to fly an airplane safely, these abilities are not sufficient. Superior airmanship evidently requires a high degree of situational awareness, the ability to accept information from several sources concurrently and to order priorities and allocate attention sensibly under severe operational stress (O'Hare, 1997).

To avoid confounding basic aptitude with the effect of prior training in specific tasks such as flying airplanes, the elements that comprise a good selection test must be unlike any real-world activities—unlike operating computers, unlike controlling any specific vehicle. Also, the individual tasks must be sufficiently simple to allow their mastery in a short practice period before combining them in the test situation. Situational complexity can be achieved by the manner in which the individually simple subtasks are combined in an adaptive scenario involving multiple sources of information and multiple response alternatives (Roscoe, Corl, & LaRoche, 1997).

The requirements and constraints just described evolved from a research programme at the University of Illinois Institute of Aviation in the 1970s. During this period I mentored several master's theses and doctoral dissertations by graduate students studying the concurrent performance of two or more tasks with shifting priorities requiring time-sharing and frequent reallocation of attention. These computer-based tests proved to have unusually high predictive validities for success in pilot training (e.g., Damos, 1972, 1977; Gopher & North, 1974; Jacobs, 1976; North, 1977; North & Gopher, 1974; Roscoe & North, 1980). Use of such tests can be expected to offer handsome rewards in general aviation.

Training Transfer

The use of airplane simulators for pilot training is based on the observation that training transfer is directly related to the similarity of the device to an airplane. The consequence has been the specification of training device requirements solely on the basis of engineering criteria. However, the proper criterion is the actual time saved in airborne training for each incremental investment in ground training. Furthermore, research has shown that innovations in training strategies, in some cases involving relatively low fidelity simulation with intentional departures from reality, can yield high transfer (Lintern & Roscoe, 1980; Lintern, Roscoe, & Sivier, 1990).

During the 1970s it became evident, from operational experience as well as research, that complex cockpit motion systems have no measurable training benefit (Jacobs & Roscoe, 1980; O'Hare & Roscoe, 1990; Waag, 1981). There is no manoeuvre, procedure, or flight scenario that cannot be taught with high positive transfer in a fixed-base simulator. This does not mean that moving cockpits have not performed a valuable service. Their face fidelity and the joy of flying them played a big role in gaining pilot acceptance of simulators for testing as well as training, but they can no longer be justified economically for any training application.

Simulation of the outside visual world presents a far more complex set of design tradeoffs than does cockpit motion. There is ample evidence that visual systems, even quite simple ones (Roscoe, 1980; Payne et al., 1954), can produce large benefits. By the use of command guidance and flight-path prediction symbology superposed on a simple simulated contact view, low-flight-time students can be taught landings and other ground-referenced manoeuvres with little trial and error or instructor intervention (Lintern, Roscoe, Koonce, & Segal, 1990; Lintern, Roscoe, & Sivier, 1990).

Typically such computer-animated visual scenes are projected via a lens system that causes all the light rays to enter the pilot's eyes in parallel, the intention being to create a three-dimensional depth effect. However, contrary to common belief, visual systems that present collimated virtual images do not cause the eyes to focus at optical infinity (Hull, Gill, & Roscoe, 1982; Iavecchia, Iavecchia, & Roscoe, 1988; Randle, Roscoe, & Petitt, 1980). Instead, focus lapses inward toward the individual pilot's dark focus distance—the resting or tonic state of visual accommodation in the absence of visible texture—which is at about arm's length for young, healthy eyeballs.

No one clearly understands why parallel light rays do not necessarily draw eye focus to optical infinity, as going theory would predict. However, the fact that they do not causes a scene to appear shrunken; a simulated airport runway will seem smaller, farther away, and closer to the horizon than it should (Randle, Roscoe, & Petitt, 1980; Roscoe, 1985). As a consequence, pilots flying approaches in simulators with collimated visual systems tend to come in high and fast and touch down long and hard (Palmer & Cronn, 1973). It has been argued that increasing the simulated field of view and scenic detail will solve the problem, but that is wishful marketing.

A better solution is to project the animated scene on the interior of a spherical-section screen at a distance of several feet from the pilot's eyes. Although some magnification of the scene may still be needed (Roscoe, 1984; Roscoe, Hasler, & Dougherty, 1966), all pilots with normal vision will focus the screen at a distance close to the near limit of optical infinity, and simulated objects will appear at their intended locations. The momentary field of view need not fill the entire screen; rather, a limited field of view should be directed automatically to the area of interest for the manoeuvre being performed (Westra, 1982).

Although not as effective as a projection screen several feet from the pilot's eyes, a monitor viewed from a distance of two feet or more with the scene slightly magnified also allows more accurate judgements of distance and angular position than does a collimated virtual image. The major limitation of a directly viewed monitor immediately above the instrument panel is the severely restricted horizontal field of view. Nevertheless, the high transfer of *ab initio* training from simulator to airplane demonstrated at the University of Illinois was achieved with such an arrangement (Lintern, Roscoe, Koonce, & Segal, 1990).

In the magical world of computers, only our imaginations limit the introduction of "unreal-worldly" creations to facilitate learning. By adding force vectors to forward, downward, and side views of the simulated airplane, the real-time dynamic effects of control inputs can be directly seen and quickly understood (see Figure 2.1). Another innovation is to provide a computer-generated map display in the cockpit of the simulator. By means of a touch-sensitive screen, all manner of navigation problems and plans can be set up and practised, including ground-referenced manoeuvres as well as enroute and terminal area procedures and communications.

If, in an ideal world, pilot certification were based only on competence, the flight hours required could be reduced to no more than the amount needed to demonstrate the required proficiency, and the question of how much credit can be allowed for training in any ground-based device would become purely academic. Competition in the marketplace would provide the incentive to develop teaching systems that train better pilots faster and cheaper. The cost-conscious operator would use a given device until its hourly cost is greater than the cost of the flight time it saves, plus a little more when the weather is bad.

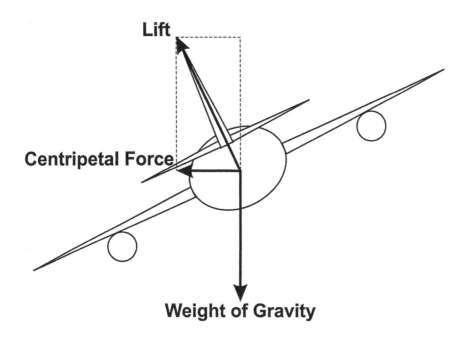

Figure 2.1   In a view of an airplane from the rear, force vectors that respond to simulator control inputs allow the student to practice lateral and vertical manoeuvres and quickly understand the dynamics of flight control. Similar enhancements can be applied to downward and side views to show the effects of changes in thrust. Adding guidance symbology to induce correct responses early in the training sequence minimises trial and error and further speeds the learning process

Ergonomic Cockpits

From the earliest blind flying experiments by Lt. Jimmie Doolittle with the original gyro-stabilised Sperry horizon there was controversy over whether the airplane symbol or the horizon bar should move relative to the fixed display coordinates (Roscoe, Johnson, & Williges, 1980). Those with "common sense" argued that the horizon bar should move to maintain "congruency" with the real horizon (Poppen, 1936), and that's the way Sperry built it. But Doolittle and other early-day instrument fliers had trouble seeing it that way and remembering that the fixed "little airplane" was what they were supposed to control and not the longer moving horizon bar.

All pilots do learn to control flight attitude by reference to the artificial horizon, and by the time they qualify for an instrument rating they have long since learned the correct responses to the display's indications. But they still see the horizon bar as the part of the display that moves and not the little airplane symbol, and in the perceptual confusion of a sudden, unanticipated entry into an unusual attitude, there is a strong tendency to control the part of the display that is moving, not the part that is fixed. They naturally expect the moving part to move in the same direction as the movement of the controls, as in steering a car or a bicycle.

The time has past when it would be reasonable to consider reversing the control/display relationships in flight attitude indicators. In fact, the reversed relationship is not the best. A much easier change results in an even better display, one that would be trivial to implement in modern planes with "glass cockpits" and more difficult but not unreasonable with electromechanical instruments. A flight path predictor can be added to the conventional moving horizon display by allowing the airplane symbol to move in immediate response to control inputs and in the same, expected direction. It sounds simple, but the effect is magical.

A flight path predictor display of this type was tested extensively in both airplanes and flight simulators at the University of Illinois starting in the 1970s (Roscoe, Johnson, & Williges, 1980). In addition to responding immediately to control inputs, the flight path prediction was improved by having the airplane symbol move laterally from the display centre in proportion to the rate of turn, thereby creating a superior flight-director presentation as well. This type of flight path predictor was then integrated in the computer-animated visual system of a primary training simulator, resulting in immediate improvement in landing performance (Lintern, Roscoe, & Sivier, 1990).

The experiments at Illinois have covered a wide range of operationally realistic contact and instrument flight tasks. Not only do beginning students learn to land airplanes and fly by instruments more quickly, but also their terminal performance is far more precise. The latter is also true for experienced instrument pilots who have no trouble taking advantage of the flight path predictor without having to unlearn their over-learned responses to the moving horizon display. And because pilots of highly automated planes do little hand flying these days, they will welcome the assistance of a flight path predictor when they do have to take control.

The perceptual-motor problems encountered by pilots during the evolution of aircraft instrumentation have received serious experimental attention during the half century since World War II. The beauty of the research is that we now have several well-established display design principles that have application far beyond the immediate settings of the individual experiments. The cumulative benefit of the application of the various principles is not merely the additive sum of their individual benefits; it is more akin to their mathematical product, as suggested by the findings shown below (adapted from Lintern, Roscoe, & Sivier, 1990).

The curves in Figure 2.2 represent the performances of independent groups of *ab initio* pilot trainees learning to land a flight simulator. The unaugmented pictorial display that served as a control condition presented a simplified view of an airport scene that was responsive to the changing attitude and flight path of the simulated airplane. The augmented pictorial display included, in addition, the integrated pictorial presentation of flight path guidance and prediction symbology with direction of motion compatibility and optimum scaling. The augmented symbolic display embodied the same principles except that they were incorporated in a symbolic format.

The display principles that make integrated pictorial aircraft attitude and flight path displays so effective are also applicable to downward-looking map-type navigation displays with integrated traffic and terrain information and integrated altitude, vertical flight path, thrust, and speed displays (see Figure 2.3). Unfortunately, these principles have not been applied consistently in modern jet airplanes with "glass cockpits" or in general aviation planes. If they were, huge reductions in training requirements would follow, with greater safety and productivity and a sudden competitive advantage for the first to adopt this scientifically based approach.

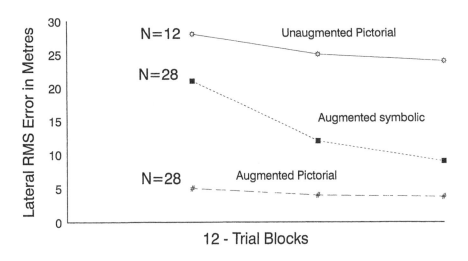

Figure 2.2   **Performances of independent groups of *ab initio* pilot trainees learning to land a flight simulator**

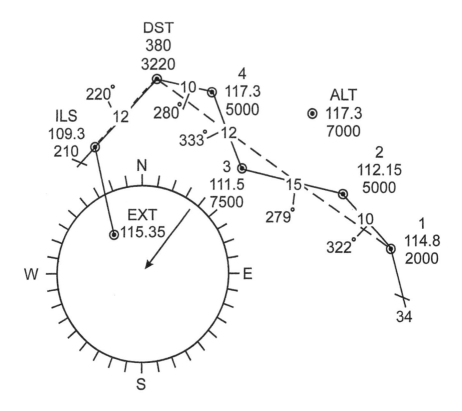

Figure 2.3   A typical flight plan set up via a touch-sensitive map display and keypad, including departure and arrival airports, numbered waypoints and associated radio frequencies; route segment courses, distances, and flight altitudes; ILS capture point (DST); final approach course (echoed by arrow pointing to centre of compass rose); and missed approach waypoint (EXT)

The Defence of Manhood

For many years I have pondered the mystery of why the general aviation community has willingly accepted the fact that flying is more dangerous than it needs to be. Why are the standards for admission to pilot training not based on demonstrated aptitude? Why are pilots who have not been trained to fly an airplane safely in and out of a cloud licensed to kill themselves and their families when they inadvertently or, worse yet, intentionally enter one? And why are flight attitude indicators still allowed to induce horizon control reversals that cause about 100 fatalities a year in the United States alone (Bryan, Stonecipher, & Aron, 1954)?

In the 1940s a new general aviation airplane was introduced that wouldn't stall or spin or make uncoordinated turns and almost never bounced on a landing. It was called the Ercoupe, and it was literally laughed out of business by flight instructors. It was simply too easy to learn to fly and too safe to be a challenge. Who wanted an airplane you couldn't slip into the wind to compensate for drift? And who wanted a plane that wouldn't stall or spin, for Pete's sake? Only those who weren't men enough to learn how to make coordinated turns or intentionally miscoordinated approaches would be caught flying one. What was there left to teach?

The story of the Ercoupe is well known, at least by old timers, and while I certainly don't think that all general aviation planes should have the flight characteristics of an Ercoupe, there was and still is a place in general aviation for planes that do. In a like manner, there is a place for objective tests of pilot aptitude to assure a minimum standard, for flight training devices with high procedural fidelity but also with intentional departures from literal fidelity to speed the learning of correct responses, and for flight displays in general aviation airplanes that follow experimentally validated principles of ergonomic design.

References

Damos, D. L. (1972). *Cross-adaptive measurement of residual attention to predict pilot performance* (TR ARL-72-25/AFOSR-72-12). Savoy, IL: University of Illinois at Urbana-Champaign, Aviation Research Laboratory.

Damos, D. L. (1977). *Development and transfer of timesharing skills* (TR ARL-77-11/AFOSR-77-10). Doctoral dissertation, University of Illinois at Urbana-Champaign.

Chapanis, A., Garner, W. R., & Morgan, C. T. (1949). *Applied experimental psychology*. New York: Wiley.

Fitts, P. M. (1947). *Psychological research on equipment design (Research Report 19)*. Washington, DC: US Army Air Forces Aviation Psychology Program.

Gopher, D., & North, R. A. (1974). The measurement of attention capacity through concurrent task performance with individual difficulty levels and shifting priorities. *Proceedings of the Human Factors Society—18th Annual Meeting*. Santa Monica, CA: Human Factors Society.

Hull, J. C., Gill, R. T., & Roscoe, S. N. (1982). Locus of the stimulus to visual accommodation: Where in the world or where in the eye? *Human Factors, 24*, 311-319.

Iavecchia, J. H., Iavecchia, H. P., & Roscoe, S. N. (1988). Eye accommodation to head-up virtual images. *Human Factors, 30*, 689-702.

Jacobs, R. S., & Roscoe, S. N. (1980). Simulator cockpit motion and the transfer of flight training. In S. N. Roscoe (Ed.), *Aviation psychology* (pp. 204-216). Ames: Iowa State University Press.

Koonce, J. M. (1984). A brief history of aviation psychology. *Human Factors, 26*, 499-508.

Lintern, G., & Roscoe, S. N. (l980). Visual cue augmentation in contact flight simulation. In S. N. Roscoe (Ed.), *Aviation psychology* (pp. 227-238). Ames: Iowa State University Press.

Lintern, G., Roscoe, S. N., Koonce, J. M., & Segal, L. D. (1990). Transfer of landing skills in beginning flight training. *Human Factors, 32,* 319-327.

Lintern, G., Roscoe, S. N., & Sivier, J. E. (1990). Display principles, control dynamics, and environmental factors in pilot training and transfer. *Human Factors, 32,* 299-317.

North, R. A., & Gopher, D. (1974). Basic attention measures as predictors of success in flight training. *Proceedings of the Human Factors Society—18th Annual Meeting.* Santa Monica, CA: Human Factors Society.

O'Hare, D. (1997). Cognitive ability determinants of elite pilot performance. *Human Factors, 39,* 540-552.

O'Hare, D., & Roscoe, S. N. (1990). *Flightdeck performance: The human factor* (pp. 61-91). Ames: Iowa State University Press.

Palmer, E., & Cronn, F. W. (1973). Touchdown performance with a computer graphics night visual attachment. *Proceedings of the AIAA Visual and Motion Simulation Conference* (pp. 1-6). New York: American Institute of Aeronautics and Astronautics.

Payne, T. A., Dougherty, D. J., Hasler, S. G., Skeen, J. R., Brown, E. L., & Williams, A. C., Jr. (1954). *Improving landing performance using a contact landing trainer* (Technical Report TR SPECDEVCEN 71-16-11, AD121200). Port Washington, NY: Office of Naval Research, Special Devices Center.

Poppen, J. R. (1936). Equilibratory functions in instrument flying. *Journal of Aviation Medicine, 6,* 148-160.

Randle, R. J., Roscoe, S. N., & Petitt, J. (1980). *Effects of magnification and visual accommodation on aimpoint estimation in simulated landings with real and virtual image displays* (Technical Paper NASA-TP-1635). Washington, DC: National Aeronautics and Space Administration.

Roscoe, S. N. (1980). Transfer and cost effectiveness of ground-based flight trainers. In S. N. Roscoe (Ed.), *Aviation psychology* (pp. 194-203). Ames: Iowa State University Press.

Roscoe, S. N., Johnson, S. L., & Williges, R. C. (1980). Display motion relationships. In S. N. Roscoe (Ed.), *Aviation psychology* (pp. 68-81). Ames: Iowa State University Press.

Roscoe, S. N. (1984). Judgments of size and distance with imaging displays. *Human Factors, 26,* 617-629.

Roscoe, S. N. (1985). Bigness is in the eye of the beholder. *Human Factors, 27,* 615-636.

Roscoe, S. N. (1994). Alexander Coxe Williams, Jr., 1914-1962. In H. L. Taylor (Ed.), *Division 21 members who made distinguished contributions to engineering psychology* (pp. 68-93). Washington, DC: Division 21 of the American Psychological Association.

Roscoe, S. N. (1997). The adolescence of engineering psychology. *Human Factors History Monograph Series, 1,* 1-16.

Roscoe, S. N., Corl, L., & LaRoche, J. (1997). *Predicting human performance.* Pierrefonds, QC, Canada: Helio Press.

Roscoe, S. N., Hasler, S. G., & Dougherty, D. J. (1966). Flight by periscope: Making takeoffs and landings; the influence of image magnification, practice, and various conditions of flight. *Human Factors, 8,* 13-40.

Waag, W. L. (1981). *Training effectiveness of visual and motion simulation* (Technical Report AFHRL-TR-79-72). Brooks Air Force Base, TX: Air Force Human Resources Laboratory.

Westra, D. P. (1982). *Simulator design features for carrier landing: II. In-simulator transfer of training* (Technical Report NAVTRAEQUIPCEN 81-C-0105-1). Orlando: Naval Training Equipment Center.

Part 2
The General Aviation Pilot

3　The General Aviation Pilot: Variety is the Spice of Flight

David R. Hunter

Introduction

A student takes her first solo cross-country flight on the way to a private pilot certificate. An engineer flies his family of four to an isolated airport on the ocean for a weekend of relaxation. A busy sales representative hops into his light twin for a whirlwind tour of manufacturing facilities and possible sales. An agricultural application pilot lines up his turbine powered aircraft using global positioning satellite (GPS navigation) for a run across a Kansas wheat field. At a busy metropolitan trauma centre, an aeromedical evacuation pilot sets her helicopter down on the landing pad as a team of technicians rush forward to receive the victim of another traffic accident. A pilot contacts oceanic control on his way across the North Atlantic in a corporate jet that rivals the capabilities of most airliners. Each of these examples, from the weekend pilot in a simple two-place trainer to the corporate pilot in a glass-cockpit jet, is a part of the world of general aviation.

As may be seen from those examples, variety is the word that best seems to describe general aviation—both the pilots and their aircraft. And, while variety may well be the spice of life, it certainly makes it difficult to characterise the general aviation pilot. By definition (at least in the Federal Aviation Administration), general aviation (and hence general aviation pilots. includes everything that is neither military nor air carrier. This definition by exclusion means that general aviation encompasses single engine sport aircraft, high performance twin-engine aircraft, and the corporate jets, whose sophistication (and cost) often exceeds that of the air carrier aircraft. The range of pilot characteristics is correspondingly broad, ranging from recreational pilots with only a hundred hours of experience to flight instructors with thousands of hours of experience in light aircraft, and the occasional retired airline pilot with tens of thousands of hours.

It is important that this variety be recognised in planning research, developing training, and designing new technologies which target general aviation. Clearly, there are vast differences between the capabilities and needs of the weekend recreational pilot flying from an uncontrolled airfield and the corporate pilot who operates in an air carrier-like environment. Even within the private pilot category there are substantial differences among pilots that I will touch upon in more detail later using data from some recent surveys of pilots. To make this a more manageable task, for the remainder of this chapter I shall primarily focus on the private pilot certificate holders that we generally associate with the term general aviation and shall exclude (with only a few digressions) the general aviation pilots who fly for a living. This does not mean of course that all holders of commercial pilot certificates are earning their living as a pilot. From our studies it seems that in many cases, commercial pilot certificate holders obtained the higher rating to increase their pilot skills and, possibly, to enhance their image, without any plans for exercising the privileges that a commercial certificate offers beyond that conferred by a private certificate.

We should also note that, except for military pilots, all pilots are considered general aviation pilots at some point in their career. The award of the initial certificate as a private pilot, while it may only be a brief plateau on the way to a commercial career, puts every new pilot in the general aviation category. Thus, we may also distinguish between pilots who are merely transitioning through the general aviation category and those whom we might consider as permanent residents. Certainly, the permanent residents are the focus of most of our training and safety awareness efforts, yet we cannot afford to ignore those who are only passing through, since the skills, habits, and attitudes that they develop during this phase may stay with them for the rest of their aviation career. For better or worse, much of what is learned in a Cessna 152 carries over to the flight deck of an A320 or B777.

Determining just who is a general aviation pilot is an important undertaking; because, to understand the factors which affect human performance in general aviation, we must first understand the pilots who comprise this sector of aviation. Although the characteristics of pilots who are involved in accidents are routinely tabulated (c.f., NTSB, 1989), such information is generally lacking for the much larger group of pilots who have not experienced an accident. Some data are available from the periodic medical examinations required of active pilots; however, these data are generally limited to simple measures such as total and recent flight time. To gain a better understanding of the characteristics of the pilot population, in particular general aviation pilots, the FAA conducted a large-scale, nationwide survey of pilots (Hunter, 1995). Questionnaires were mailed to 20,000 active pilots and almost 7,000 responses were returned. Of that number, there were 2,548 private pilots, 2,845 commercial pilots, and 1,218 airline transport pilots. Even though about half of the 2,845 commercial certificate holders had never flown for hire, and hence are probably general aviation pilots, I will limit the data reported below to the 2,548 private certificate holders, as they are without question general aviation pilots.

First, let us examine some basic demographic characteristics of this group (see Table 3.1). The source of initial training is interesting as recent research (Jensen, personal communication, 1997) suggests that there are enduring effects of initial training that may last throughout a pilot's flying career. Specifically, there is some evidence that pilots training at well-established pilot training schools (those certificated under Federal Aviation Regulation Part 141) or in a university setting perform better than pilots. who received their initial training at other locations, even ten years or more after completion of the training. If these tentative findings are corroborated by other studies, then source of training could prove an important variable in understanding differential performance by pilots and, possibly, their differential risk potential.

The substantial proportion of private pilots (39%) who hold an instrument rating may reflect the increased emphasis on obtaining instrument rating and lowering of total time requirements by the FAA. Increasing the proportion of the pilot population who hold an instrument rating is a worthy goal, since encounters with weather specifically continued Visual Flight Rules (VFR) flight into Instrument Meteorological Conditions (IMC) have historically proven to be the largest single cause of fatal general aviation accidents. (NTSB, 1989). Possession of an instrument rating, and maintenance of proficiency, may shield against this type of accident.

This movement toward increased qualifications of the general aviation pilots may also be influenced by factors external to aviation itself. An examination of the educational characteristics of this group (Figure 3.1) shows that almost two-thirds have college degrees, and 14% have completed graduate or professional training leading to a doctorate. These data suggest that as a group, general aviation pilots. are well able to deal with complex technical training, such as that which an instrument rating entails. Further, these data should also be taken into consideration when designing new training programs, so that they adequately engage and challenge participants. It may be a mistake, however, to assume that these individuals all come to the aviation training setting with well-developed, or even roughly equivalent general problem solving skills. Research on expertise (c.f., Ericsson & Smith, 1991) suggests that expertise gained in one domain does not necessarily transfer to other domains. Rather, there may be an overall negative effect, as an expert in a particular domain may have an inflated, inaccurate appraisal of his or her skill levels in another domain.

Another interesting measure of pilot qualification is number of hours of flying experience. Typically, experience is equated with expertise, such that pilots. with high numbers of hours (usually over 10,000) are considered to be experts, while pilots with relatively few hours are considered novices, or perhaps simply competent. From the distribution of total flight time given in Figure 3.2, we may see that, according to that definition, there are very few expert general aviation pilots. However, the notion of equating flight hours with expertise has been called into question by many studies (c.f., Giffin & Rockwell, 1984; Wickens, Stokes, Barnett, & Davis, 1987; Barnett, 1989; Stokes, Kemper, & Marsh, 1992; Wiggins & O'Hare, 1995) that have generally found little relationship between measures of pilot performance (for example decision optimality) and total flight experience. Recent research by Jensen and his associates at The Ohio State University (Jensen, 1997; Kochan, et al., 1997; Guilkey, et al., in press) has also demonstrated that total flight time may be a very fallible measure of pilot expertise. Jensen's studies show that substantial performance differences may be observed among groups of high- and low-time general aviation pilots., and that in many cases pilots with as few as 1,000 total flight hours may well exhibit decision skills superior to pilots with ten times as much experience.

Table 3.1   U.S. general aviation pilot characteristics (N = 2,548)

Source of training for certificate.	
Civilian (Part 141) school	19%
CFI at a fixed-base operator	48%
CFI working with a flying club	12%
Independent CFI	18%
Instrument rating -Yes	39%
Multi-engine rating - Yes	11%
Sex	
Male	96%
Female	4%

Source: Hunter, 1995

More detailed information on both total and recent (last 6 and 12 months) flying experience is given in Table 3.2. From this Table we may see that half of the private pilots in this survey reported flying 30 hours a year, or less; roughly 2.5 hours per month. Clearly, skill maintenance, particularly for pilots flying more complex aircraft in demanding environments, becomes problematic at that level of frequency of flight. Lack of practice of flight skills, particularly procedural and decision-making skills, acts to curtail human performance. Since the factors that constrain flight practice for general aviation pilots are typically associated with lack of resources (time and money), human factors researchers are challenged to develop novel approaches to skill maintenance that may be effected outside of the aircraft. Later chapters will address this challenge in more detail; however, the solutions that are proposed must take into account the characteristics and preferences of the targeted population. A wonderful training programme that no one uses because it costs too much, requires too much time, or is boring, is a useless exercise. Moreover, it siphons resources from other more effective programs.

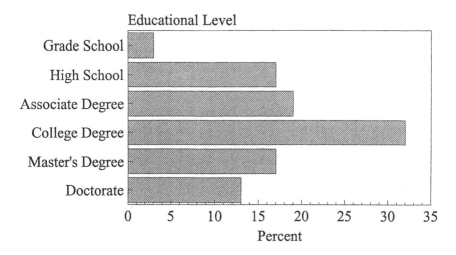

Associate Degree = 2 years of college
College Degree = 4 year, Bachelor's Degree

Figure 3.1   Educational levels of general aviation pilots
Source: Hunter, 1995

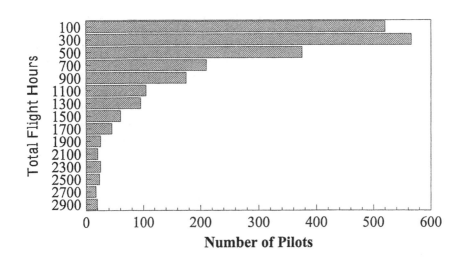

Figure 3.2   Total flight time for private pilots
Source: Hunter, 1995

Involvement in Hazardous Events

Far more often than they are involved in accidents, pilots experience hazardous events. These are events that do not result in an accident or incident. However, they might have done so had the situation changed even slightly. For example running low on fuel is a hazardous event. If there is an airport nearby then no accident occurs. But, had the headwinds been slightly higher, had the line up for takeoff taken slightly longer, or had the pilot drifted slightly from course during the flight, then a hazardous event might have easily turned into a potentially serious accident. Encounters with hazardous events provide one possible measure of the extent to which pilots are placing themselves at risk for an accident. Multiple encounters with hazardous events may indicate a lack of skill or knowledge (for example, in the computation of expected fuel consumption) or it may indicate a differential willingness to accept risk. Even among the relatively low-time pilots in this sample, running so low on fuel that they were seriously concerned about making it to an airport is not an uncommon event. Rather, 20% reported that they had experienced such an event at least once. Even more seriously, 6% reported that they had experienced an engine failure due to fuel starvation.

As noted earlier, continued VFR flight into IMC is the single largest cause of fatal general aviation accidents (NTSB, 1989). It is disturbing, therefore, but perhaps not particularly surprising, that 23% of these pilots reported having flown into instrument conditions while under VFR. While it is apparent from these data and from the accident statistics that this is a significant problem, as in the case of fuel events, it is not apparent from these data why general aviation pilots get into these situations, or precisely what may be done to reduce their occurrence. Findings like these have stirred action however, and later I will describe some of the efforts currently underway to address these problems.

Inadvertent stalls, followed by a spin, also are frequently cited as a cause of general aviation accidents. This stall/spin accident typically occurs during takeoff or landing operations when there is insufficient altitude to allow the pilot to make a successful recovery. It is interesting, therefore, that 6% of the pilots in this survey reported having inadvertently stalled an aircraft at some time. Like the data on fuel and VFR into IMC events, the incidence of inadvertent stall events suggests that brushes with potentially catastrophic events are not uncommon in general aviation. Rather, for the most part the brush with disaster is simply that, and the pilot and aircraft escape unscathed. Nothing is bent and no one is injured and, except in a very few cases, no one outside the aircraft knows what happened or how close they came to an accident. Hopefully, the pilot involved is aware of the potential severity of the situation and appreciates just how close they came to getting top billing in an accident report; however, there are, as yet, no firm data to support or refute that supposition and it remains an interesting question for future research.

Table 3.2   Flight experience (N = 2,548)

	Median	Mean	Standard Deviation
Total Time - Last 6 months	12	22	34
Total Time - Last 12 months	30	50	68
Total Time - Career	445	819	1293
Day Time - Last 6 months	11	24	152
Day Time - Last 12 months	27	46	95
Day Time - Career	396	777	1664
Night Time - Last 6 months	0	3	13
Night Time - Last 12 months	0	5	18
Night Time - Career	22	108	644
Under Hood - Last 6 months	0	2	5
Under Hood - Last 12 months	0	4	9
Under Hood - Career	20	41	67
Actual Instrument - Last 6 months	0	2	7
Actual Instrument - Last 12 months	0	4	16
Actual Instrument - Career	2	60	316
Student - Last 6 months	0	1	5
Student - Last 12 months	0	3	13
Student - Career	64	95	863
Landings - Last 6 months	40	61	109
Landings - Last 12 months	16	29	43

Source: Hunter, 1995

So, why do GA pilots frequently have these encounters with critical events? Is it a lack of knowledge? Do they lack the skills required to keep away from trouble? Are they willing to stretch the envelope of both pilot and aircraft and tolerate more risk than is prudent? In the absence of definitive research answers to these questions, a pragmatic approach seems appropriate—to do what seems reasonable to remediate to some degree the problems, while simultaneously pursuing research to develop a more definitive understanding of the problems and the answers. The earlier attempt to improve decision making through an understanding of hazardous attitudes and their antidotes (Buch & Diehl, 1982; Diehl & Lester, 1987) is certainly one example of this approach. While the review of accidents by Jensen and Benel (1977) established the importance of decision making to accident involvement, the development of the attitude awareness training that was produced as a possible solution to defective decision making was simply a pragmatic approach that applied a reasonable solution based upon a general understanding of the problem. The fact that several studies of this training showed that it worked (Buch & Diehl, 1982; Diehl, 1991) showed that even if it was not great science, it was nevertheless good engineering. In accident prevention, as in many other fields, a workable partial solution than can be applied immediately is often worth far more than a potentially complete solution that may be available only in the distant future. Engineers and pilots agree, "a bird in the hand is worth two in the bush."

Setting Personal Minimums

Jensen and his associates at The Ohio State University (Kirkbride, Jensen, Chubb, & Hunter, 1996) have taken a similar pragmatic approach in their development of a recent programme aimed at improving the decision making of general aviation pilots. This approach focuses on the pre-take off decision making of pilots. A study by McElhattton and Drew (1993) utilising data from the Aviation Safety Reporting System (ASRS) showed that the majority of errors that led to accidents and incidents occurred during the preflight phase. These errors were associated with decisions regarding the route of flight, weight and balance, fuel loading, and simply whether to go or not. It seems reasonable to presume, therefore, that eliminating these preflight decision errors could result in a reduction in the number of accidents and incidents. Accordingly, a training programme (Developing a Personal Minimums Checklist) that focused upon preflight decisions was developed.

A key component of this preflight decision making training was the concept of setting and using personal minimums. Personal minimums are minimum operating conditions that are more conservative than those required by government regulation or company or club policy. As an example, the typical weather requirements for VFR flight are 3 miles visibility and a minimum ceiling of 1,000 feet above ground level (AGL). While a pilot may not engage in VFR flight when the conditions are less than those values, there is no positive requirement for the pilot to fly whenever the conditions exceed those values. Pilots are free to set more conservative minimums than those required by regulation and, as shown in Table 3.4, many do so. As shown in that table, about half of the pilots surveyed would not engage in a local (near the airport) flight during the day, if the visibility were only 3 miles, and only 15% of the pilots would fly with a ceiling of only 1,000 feet. When considering flights away from the airport (cross-country) or at night, the pilots become even more conservative. The modal minimum conditions for a night cross-country flight were 15 miles visibility and 5,000 feet ceiling.

Clearly, prudence dictates that the legal conditions established for the entire gamut of pilots. are not always safe conditions for a particular pilot with a given level of experience and competence.

The personal minimums training programme utilises a videotape presentation depicting a relatively inexperienced pilot conducting a cross-country flight through terrain with which he has had no previous experience. At a key point in the video presentation, the tape is stopped and the instructor engages the participants in a discussion of the hazards this pilot is facing. The participants are shown how to categorise hazards into four principal areas corresponding to the acronym PAVE (Pilot, Aircraft, enVironment, and External). Later, after the participants have seen how the pilot in the video might have used a Personal Minimums Checklist to avoid his difficulties, the instructor assists the participants in completing their own Personal Minimums Checklists.

Placing the personal minimums in a checklist format is another key feature of this training. While pilots may espouse prudent personal minimums, unless they abide by those self-imposed restrictions no safety advantage is gained. By putting the minimums in the format of a checklist designed to be used before each flight, the likelihood of compliance is increased. In addition, other items (such as those shown in Table 3.5) which may be overlooked during the preflight decisions may also be included in such a checklist, thus ensuring that they are also considered. This checklist, then, becomes a personal standard operating procedure, tailored for a specific pilot, that sets the minimum operating conditions for that individual. It is believed that because each pilot creates his or her own personal minimums checklist tailored for their individual characteristics, that they will be more likely to use the checklist during the preflight planning process and that this use will have a positive effect on safety. Both these suppositions must, of course, be confirmed through appropriate evaluations. A pragmatic solution may suffice for the short term, but must eventually be verified.

Attitudes and Opinions

Attitudes influence behavior and general aviation pilots have a considerable range of attitudes about flying. These attitudes are reflected in the diversity of agreement they expressed regarding a number of flying issues. This diversity may also be interpreted as an indicator of the differential tolerance for risk or as an indicator that pilots. perceive different levels of risk in an activity. For example, in the first question from Table 3.6 although 61% of the pilots. strongly disagree with the statement, "I would duck below minimums to get home" 39% did not subscribe to that complete rejection of the idea of ducking below minimums. One interpretation of these results is that there are some circumstances under which those pilots. *would* duck below minimums to get home. Therefore, there are some circumstances under which these pilots would consciously expose themselves to a potentially high level of risk (not to mention violating regulations). In addition, it seems likely that some of the 61% who strongly disagreed with ducking under minimums might still do so under certain circumstances. Thus the proportion who might violate the minimums is even higher than the 39%.

Table 3.3   Involvement in hazardous events (N = 2,548)

Number of aircraft accidents	
None	91%
1	8%
2	1%
3 or more	
Run low on fuel	
None	80%
1	16%
2	3%
3 or more	1%
Inadvertent stalls	
None	94%
1	5%
2 or more	1%
Disoriented (lost)	
None	83%
1	14%
2 or more	3%
Flown VFR into IMC	
None	77%
1	15%
2	6%
3 or more	3%

Source: Hunter, 1995

Table 3.4   Personal minimums (N = 2,548)

	Local Day	Local Night	Cross-Country Day	Cross-Country Night
Visibility				
1 mile	4%	1%	1%	1%
2 miles	5%	1%	1%	0%
3 miles	45%	11%	18%	6%
4 miles	6%	2%	3%	1%
5 miles	30%	33%	37%	20%
6 miles	3%	6%	6%	4%
8 miles	2%	6%	7%	6%
10 miles	5%	26%	20%	28%
15 miles	1%	5%	8%	35%
Ceiling				
1,000 feet	15%	2%	3%	1%
1,500 feet	24%	5%	5%	1%
2,000 feet	29%	16%	14%	5%
3,000 feet	25%	33%	38%	18%
4,000 feet	4%	13%	16%	13%
5,000 feet	3%	30%	24%	62%

Source: Hunter, 1995

A similar analysis might be made of the responses to the statement, "Most accidents are beyond the pilot's control." The 16% of the pilots. who do not reject this idea (by disagreeing or strongly disagreeing) seemingly have a very different idea of the risks and sources of risks inherent in aviation.

Table 3.5   Usual practices—always (or never) do these things (N = 2,548)

	Local flights	Cross-Country flights
Get weather briefing before takeoff	36%	89%
Top off / check fuel tanks	94%	98%
Compute weight & balance	20%	41%
Perform complete pre-flight	95%	98%
Use checklist for landing & takeoff	79%	83%
Compute expected fuel consumption	47%	82%
File a flight plan	4%	42%
Request weather updates	4%	24%
Verify fuel consumption in flight	38%	59%
Never Fly VFR above clouds	76%	62%
Never fly below 1,000 AGL under clouds	70%	75%
Never fly below 500 AGL below clouds	94%	94%

Source: Hunter, 1995

The data in Table 3.6 indicate that over half of these general aviation pilots felt they were capable of instrument flight, yet only 39% had an instrument rating and the median number of hood or actual instrument hours over the last year was zero. These data suggest that the pilots may be seriously overestimating their capability to fly in instrument conditions. A self-rating of knowledge and proficiency completed by pilots attending FAA sponsored safety seminars yielded somewhat more conservative estimates of proficiency. For that group, 35% of the pilots rated their instrument flying knowledge and proficiency as very good or good. (Hunter, 1997) Although the comparison is very tenuous because of the differences in which the questions were stated in the two studies, it suggests that those pilots who attended the safety seminars might have had a more accurate estimate of their personal capabilities.

Table 3.6   **Opinions about flying (N = 2,548)**

	Strongly Agree	Agree	Neither Agree nor Disagree	Disagree	Strongly Disagree
I would duck below minimums to get home	1%	3%	8%	27%	61%
I am capable of instrument flight	23%	29%	15%	17%	17%
I am a very capable pilot	22%	56%	21%	2%	0%
I am so careful I will never have an accident	1%	8%	43%	33%	15%
Most accidents are beyond the pilot's control	1%	2%	13%	54%	31%
Weather forecasts are usually accurate	2%	48%	32%	16%	2%
I am a very cautious pilot	33%	57%	9%	1%	0%
Pilots should have more control over how they fly	7%	26%	54%	11%	1%
It is easy to understand weather information	8%	57%	18%	16%	2%
It is unlikely I would have an accident	1%	12%	39%	38%	11%
I fly enough to maintain proficiency	9%	44%	20%	20%	8%
There are few situations I couldn't get out of	3%	21%	45%	26%	6%
You should push yourself and aircraft to find limits	1%	11%	22%	43%	24%
Sometimes you have to depend on luck	1%	2%	8%	35%	54%

Source: Hunter, 1995

The latter data were obtained from an evaluation of the aviation safety seminars conducted by the FAA throughout the United States which further describes the general aviation pilot population (Hunter, 1997). Data were obtained from 5,615 participants at 226 seminars. Mean attendance was 57 and the median was 36. Four evaluation forms, each with a common set of questions on the first page and unique questions on the second page were distributed at the seminars. The evaluation forms assessed participant satisfaction with seminars, perceptions of the seminar content and the presenter, frequency of attendance at seminars, aviation qualifications and experience, seminar content, format and venue preferences, access to computer and video technology for training delivery, self-perceived knowledge and proficiency, training activities, and maintenance activities.

Of the participants, 99% reported that they were satisfied with the seminar and 96% either definitely or probably would attend another seminar in the next year. While these data were of interest (and comfort) to the FAA, more relevant for the present were the other results pertaining to participant activities. With the exception of the purely hardware-based solutions to improving human performance. (such as described in Part 5), almost all the other research described in this volume will be applied through some sort of training provided to general aviation pilots. Certainly this is true of the research dealing with decision-making described in Part 4. The chapters in Part 3 address instruction and training directly; however, for these interventions to be effective they must be keyed to the characteristics of individuals receiving the instruction. More than that, training, and even the hardware performance enhancers, must be marketed successfully in order to have any effect. Unattended seminars, unread manuals, and unsold heading bugs. do not improve performance. Pilots must be "sold" on the need to invest their time and money in attending seminars, reading manuals, and buying heading bugs through effective marketing of the advantages that these activities and devices offer. In order to do this, detailed information is needed on what pilots. perceive as important, what their preferences are for format and venue of training, and what alternatives might exist to the traditional training delivery systems.

An examination of the data on the primary reason for attending a safety seminar (Table 3.7) clearly shows that the predominant factors were the topic to be discussed and the participant's positive prior experiences with seminars. What is of particular interest here is the two-edged sword nature of the primary reason for attending—the topic to be discussed. Just as the topics perceived as interesting and relevant by the pilots. draw attendees, topics perceived as uninteresting and irrelevant repel them. Given the heterogeneity of the general aviation pilot population, developing a programme that balances the attraction-repulsion effect is clearly a difficult undertaking. Topics that might be relevant for half of the pilots. (for example, instrument landing techniques) will likely be perceived as irrelevant by the half of the pilots. who do not have an instrument rating.

Although air carrier pilots and many corporate pilots. participate in regular, formal training programs to ensure that they maintain skills and are kept abreast of the latest developments, most general aviation pilots. do not have access to this training. Neither, in contrast to the air carrier pilots. is training required by regulation for general aviation pilots. Lacking ready access to training and without a legal mandate to attend, general aviation pilots. are left to their own devices to obtain what, if any, training they think is needed. Fortunately, large numbers of pilots. avail themselves of the safety seminars conducted by the Federal Aviation Administration. In addition, many pilots attend seminars conducted by

Table 3.7   Primary reason for attending safety seminars (N = 4,999)

Curiosity	8%
Friend's recommendation	8%
Topic to be discussed	32%
Professional obligation	16%
Reputation of speaker	7%
Good previous seminar	29%

Source: Hunter, 1997

various private organisations, most notably the Aircraft Owners and Pilots Association and the Experimental Aircraft Association. About two-thirds of pilots. surveyed in a nationwide poll (Hunter, 1995) reported having attended at least one FAA safety. seminar over the preceding two years; while less than half had attended two or more seminars. This suggests that there is a sizeable portion of the general aviation population (perhaps a third) who simply do not attend seminars, while about a half of the population attends about one a year, with the choice being principally driven by the topic to be discussed.

The survey of pilots attending FAA safety seminars (Hunter, 1997) showed that the attendees were approximately evenly split into three groups: those who had not been to another seminar in the previous year, those who had been to one seminar, and those ("seminar junkies") who had attended several seminars (See Table 3.8). To some extent this is encouraging news to those who produce safety seminars, since they had previously been of the belief that they "preached to the choir" with the same pilots. attending all their seminars. While it is true that there is a substantial proportion of the pilots. who frequently attend the seminars, there is also a substantial influx of new participants at each seminar. Since the general goal of government sponsored programs like the FAA safety seminars is to produce the greatest good for the greatest number of people, as opposed to producing a few very well trained pilots. these data suggest that goal is being achieved. Still, the data are hard to interpret and it is difficult to decide whether one should be happy or sad over the fact that so many new participants show up at each seminar, because, if the total number of participants remains relatively constant (as it has) then it means that an equal number of former participants are staying away. Regardless, it seems clear that a fairly substantial number either do not attend seminars at all or only attend irregularly. If receiving the training that attending safety seminars provides confers some safety benefit in terms of reduced risk of accident involvement then some means must be developed to make that training available to all general aviation pilots. even those who cannot or will not attend safety seminars.

Table 3.8   **Number of seminars attended in previous 12 months (N = 5,493)**

None	31%
1	29%
2 to 4	31%
5 to 7	6%
8 to 12	2%
More than 12	2%

Source: Hunter, 1997

The data shown in Table 3.9 address the issue of instructional method. These data show, for example, that pilots. perceive the flight instructor to be the most effective method of learning about aerial manoeuvres. Conversely, human factors are best learned in a safety seminar setting, and regulations are best learned directly from books or manuals. If pilots. expectations regarding the effectiveness of various training methods for different topics influence their decisions to avail themselves of training (and there is every reason to believe that they would) then training designers must take these expectations into account when deciding upon which topic will be covered by a given method. On the face of it, a seminar that advertised a discussion of aerial manoeuvres seems much less likely to draw participants than one that discussed pilot decision making.

Another factor that will influence pilots.' decisions to attend seminars or make use of other training is their perceived need for it on a personal basis. That is, is the subject to be covered one which addresses an area in which he or she feels a need for training, or is it an area in which the pilot feels completely competent. The data in Table 3.10 indicate that the area in which pilots feel least competent is instrument flying—perhaps because less than half of the pilots. had an instrument rating. Other topics in which pilots felt less competent were air space regulations, air traffic control procedures, and weather. These are also among the topics which pilots also feel can be effectively addressed in a seminar setting. Therefore, they might be among the first choices of seminar designers, while basic VFR techniques should be avoided, since pilots do not feel they are lacking in these skills and furthermore do not see them as appropriate topics in a seminar setting.

These recommendations are reflected in the data given in Table 3.12. The topics of human factors, aircraft systems, and weather are three of the four most often presented topics at seminars and, as might be expected from the data in Tables 3.9 and 3.10, are generally considered useful by pilots. However, takeoffs and landings, which was the second most often presented topic, was next to last in perceived usefulness, perhaps reflecting the expectation pilots have regarding the effectiveness of discussing aerial manoeuvres in a seminar setting.

Table 3.9   **Perception of most effective method for learning about aviation topics (% by row) (N = 1,050)**

	Other pilots	Safety seminars	Flight Instructor	Books or Manuals	Videotape	Computer based training
Aerial Manoeuvres	6%	10%	69%	7%	6%	2%
Emergency Procedures	3%	20%	44%	27%	4%	2%
Pilot decision making	10%	46%	26%	10%	6%	1%
Flight hazards (e.g., weather)	7%	47%	13%	18%	14%	1%
Human factors	11%	52%	10%	19%	6%	1%
Federal Aviation Regulations	1%	32%	8%	52%	5%	2%

Source: Hunter, 1997

One alternative method of training delivery aimed at the large number of pilots who do not attend safety seminars is computer based training (CBT). Although Table 3.9 shows that very few pilots considered CBT an effective means of learning about aviation topics, it seems likely that this effect is partially due to the lack of experience with CBT by general aviation pilots. As shown in Table 3.11 about two-thirds of pilots have a computer system at home. Of those pilots who do not presently have a computer at home, approximately one-third indicated that they intend to purchase a home computer in the next year. This suggests that the basic mechanism required for CBT will be available to well over three-fourths of all pilots within a short time.

In addition, 75% (Rakovan, Wiggins, Jensen, & Hunter, in preparation) to 85% (Hunter, 1997) of the pilots indicated they would use computer-based safety training programs developed by the FAA. Further, most would be willing to pay for the programs, with 78% (Rakovan, et al.) willing to pay $10 or more for a programme. CBT has many unique capabilities that recommend it as a training medium for aviation—most notably the capacity for real-time simulation of flight that highly engages the pilot in the learning process. These capabilities, combined with the availability of the technology among the target group and the willingness of pilots to utilise the technology, make CBT an outstanding contender for future training development and dissemination.

Table 3.10   Self-rating of knowledge and proficiency (% by row) (N = 1,280)

	Very Good	Good	Adequate	Somewhat Rusty	Very Rusty
Weather	20%	40%	27%	11%	3%
ATC procedures	18%	35%	28%	13%	6%
Air space regulations	16%	35%	30%	16%	4%
Basic VFR flying techniques	33%	43%	19%	4%	2%
Takeoff and landing procedures	31%	46%	17%	4%	2%
Emergency procedures	18%	36%	31%	13%	3%
Instrument flying	12%	23%	24%	20%	20%
Preflight planning	29%	42%	21%	6%	2%
Ground handling	31%	44%	20%	3%	2%
Radio navigation	27%	37%	23%	9%	4%
Navigation by pilotage	26%	40%	23%	7%	3%
Aviator decision making	24%	43%	24%	6%	3%
Cross-wind landing	22%	34%	29%	10%	5%

Source: Hunter, 1997

Table 3.11   Computer usage (N = 2,054)

	Yes
Use a computer at home	67%
Have used computer flight simulation programme	54%
Would certainly or possibly use FAA computer safety programs	85%
Would pay $5 or more to purchase an FAA computer safety programme	67%

Source: Hunter, 1997

Table 3.12   **Frequency of presentations at seminars and perceived usefulness of topics (N = 1,822)**

Topic	Frequency of Presentation	Mean Usefulness (S.D.)
Human Factors	13%	3.76 (1.12)
Pilot Decision Making	11%	3.73 (1.03)
Aircraft Systems	22%	3.67 (1.08)
Weather	13%	3.64 (1.13)
ATC Procedures	12%	3.64 (1.02)
Air Space Classifications	8%	3.60 (1.12)
Operating Procedures (VFR or IFR)	11%	3.53 (1.10)
Takeoffs and Landings	19%	3.46 (1.24)
FAA Regulations	8%	3.38 (1.19)

Source: Rakkovan, et al., in preparation

Note: 5-point scale (5 = high, 1 = low).

Summary

In the foregoing pages I have placed a great deal of emphasis on the numbers that describe the general aviation pilot population because I believe that in order to develop methods to improve human performance. among general aviation pilots, it is of central importance to understand the sort of humans with whom we are dealing. As the military services have found (Booher, 1990), systems must be designed to fit the people who use and maintain them. The capacities and characteristics of the general aviation pilot population have implications for the design of future aircraft technologies and training systems. Aircraft systems and training systems that do not take these factors into account run the risk. of being at best unused, and at worst unsafe. To ensure that this does not happen, reliable data on GA pilots far in excess of that presented here are needed so that we may understand their capacities, needs, preferences, skills, and experiences and plan our research and development accordingly.

Operational Implications and Recommendations

General aviation represents a heterogeneous mix of aircraft and pilots. Training and technology developed to improve human performance for one sector of general aviation may not be suitable for other sectors. Therefore, researchers and systems developers must closely attend to the characteristics of the target population for whom their products are intended.

Further improvements in the safety of general aviation depend, to some extent, on ensuring that all pilots receive appropriate training in an effective medium. Because the current training efforts do not reach the total pilot population, alternative methods must be developed to disseminate training information to those pilots not currently being served by seminars.

Computer-based training, that utilises the unique capabilities of this medium and capitalises upon the availability of an installed base of delivery systems accessible to pilots has the potential to fill the training gap and significantly raise the skill levels of large numbers of pilots who would not otherwise avail themselves of training.

Just as all topics are not relevant for all pilots no single delivery medium is appropriate for all pilots. In addition to utilising computer-based training in new systems development, alternative delivery mediums must also be explored to ensure that every pilot has the opportunity to become safer and more competent.

Safety training must be approached from a marketing perspective. Safety training should be viewed as a product that must compete against other activities for a pilot's resources—time and money. To sell safety training we must understand the factors that mediate pilot reaction to training and must design our products and delivery systems accordingly.

Research Agenda

Research is now underway which is expected to culminate in a system which will be analogous, at the general aviation level, to the LOFT (line-oriented flight training) provided to airline flight crews. This Small Aircraft Flight Environment (SAFE) Training will utilise training scenarios conducted in computer based flight simulations of realistic general aviation flights. In this approach to training, instead of telling students about the effects of weather on flight, students will experience the effects of weather during their flights. The use of computer based training as an adjunct or substitute for training required for pilot certification or currency is also being evaluated. In addition to these efforts, more research is needed which will fully explore the unique capabilities of computer based training. Lintern (1995) has argued that optimal training benefit in aviation accrues in a situated learning environment—that is, one that duplicates as nearly as possible the situation in which the learning will later be demonstrated or applied. In addition, the benefits of training programs that effectively engage the learner's interest in the learning process are well established (Gagne, 1977). The dynamic capabilities of the computer show promise for the realisation of both of these recommendations through the use of simulation and realistic flight event training.

Simultaneously with the development of new uses for computer-based training, other innovative approaches to the marketing of safety training must be developed. Some of these approaches will depend upon the availability of new technology, such as the Internet, while others may simply be more intelligent uses of traditional training media. Research is certainly required to establish the best mix of delivery methods that optimise the delivery of training to the largest possible segment of the pilot population. Part of that research must include additional studies of the pilot population to further establish the characteristics of population segments with differential training preferences and needs.

Research should be conceived and executed with an explicit marketing focus. It must recognise that the potential of any human performance enhancement, however effective it might be, is only achieved when it is used. And, that use depends upon its being successfully marketed to the target population.

References

Barnett, B. (1989). *Modeling information processing components and structural knowledge representations in pilot judgment.* Unpublished doctoral dissertation, University of Illinois, Urbana-Champaign.

Booher, H. R. (1990). *People, machines, and organizations: The Manprint approach to systems integration.* New York: Van Nostrand Reinhold.

Buch, G. D., & Diehl, A. E. (1982). An investigation of effectiveness of pilot judgment training. *Human Factors, 26,* 557-564.

Diehl, A. E. (1991). The effectiveness of training programs for preventing aircrew error. In R. S. Jensen (Ed.), *Proceedings of the 6th International Symposium on Aviation Psychology.* Columbus, OH: Ohio State University.

Diehl, A. E., & Lester, L. F. (1987). *Private pilot judgment training in flight school settings* (DOT/FAA Report AM-86/6). Washington, DC: Federal Aviation Administration, Office of Aviation Medicine.

Ericsson, K. A., & Smith, J. (1991). *Toward a general theory of expertise: Prospects and limits.* New York: John Wiley.

Gagne, R. M. (1977). *The conditions of learning.* New York: Holt, Rinehart and Winston.

Giffin, W. C., & Rockwell, T. H. (1984). Computer-aided testing of pilot response to critical in-flight events. *Human Factors, 26,* 573-581.

Guilkey, J. E., Jensen, R. S., Tigner, R. B., Wollard, J., Fournier, D., & Hunter, D. R. (in press). *Intervention strategies for aeronautical decision making* (DOT/FAA/AM-pending). Washington, DC: Federal Aviation Administration, Office of Aviation Medicine.

Hunter, D. R. (1997). *An evaluation of safety seminars* (DOT/FAA/AM-97/16). Washington, DC: Federal Aviation Administration, Office of Aviation Medicine.

Hunter, D. R. (1995). *Airman research questionnaire: Methodology and overall results* (DOT/FAA/AM-95/27). Washington, DC: Federal Aviation Administration, Office of Aviation Medicine.

Jensen, R. S. (1997). A treatise on the boundaries of aviation psychology, human factors, ADM, and CRM. In L. Rakovan & R. S. Jensen (Eds.), *Proceedings of the 9th International Symposium on Aviation Psychology.* Columbus, OH: Ohio State University.

Jensen, R. S., & Benel, R. A. (1977). *Judgment evaluation and instruction in civil pilot training* (FAA-RD-78-24). Springfield, VA: National Technical Information Service.

Kirkbride, L. A., Jensen, R. S., Chubb, G. P., & Hunter, D. R. (1996). *Developing the personal minimums tool for managing risk during preflight go/no-go decisions* (DOT/FAA/AM-96/19). Washington, DC: Federal Aviation Administration, Office of Aviation Medicine.

Kochan, J. A., Jensen, R. S., Chubb, G. P., & Hunter, D. R. (1997). *A new approach to aeronautical decision-making: The expertise method* (DOT/FAA/AM-97/6). Washington, DC: Federal Aviation Administration, Office of Aviation Medicine.

Lintern, G. (1995). Formal instruction: The challenge from situated cognition. *International Journal of Aviation Psychology, 5,* 327-350.

McElhatton, J., & Drew, C. (1993). Time pressure as a causal factor in aviation safety incidents: The "hurry-up" syndrome. In R. S. Jensen & D. Neumeister (Eds.), *Proceedings of the 7th International Symposium on Aviation Psychology* (pp. 269-274). Columbus, OH: Ohio State University.

NTSB (1989). *Annual review of aircraft accident data: U.S. general aviation calendar year 1987* (NTSB/ARG-89/01). Washington, DC: National Transportation Safety Board.

Rakovan, L., Wiggins, M. W., Jensen, R. S., & Hunter, D. R. (in preparation). *A national pilot survey to enhance the development and dissemination of safety information* (DOT/FAA/AM-pending). Washington, DC: Federal Aviation Administration, Office of Aviation Medicine.

Stokes, A. F., Kemper, K. L., & Marsh, R. (1992). *Time-stressed flight decision making: A study of expert and novice aviators* (Technical Report ARL-93-1). Savoy, IL: University of Illinois, Aviation Research Laboratory.

Wickens, C. D., Stokes, A., Barnett, B., & Davis, T. (1987). *A componential analysis of pilot decision making* (Technical Report ARL-87-4). Savoy, IL: University of Illinois, Aviation Research Laboratory.

Wiggins, M., & O'Hare, D. (1995). Expertise in aeronautical weather-related decision making. A cross-sectional analysis of general aviation pilots. *Journal of Experimental Psychology: Applied, 1,* 305-320.

4 Grace under Fire: The Nature of Stress and Coping in General Aviation

Alan Stokes and Kirsten Kite

Introduction

The need to perform adequately under challenging circumstances is hardly confined to the high drama worlds of combat aviation, hijackings, and catastrophic failures. Stress, like pain, is an equal-opportunity condition. It comes to all, in chronic and acute forms, occasionally in dramatic, but mostly in mundane, even banal contexts. And, like pain, stress is aversive and unwelcome, but has a purpose. If the conditions that give rise to it cannot be avoided, it must be managed. With these points in mind, consider the following incident, selected precisely because it will never be the basis of an 'Airport 2000' type docudrama. The incident report was filed in January, 1992, by a pilot in Hawaii:

> *I pre-flighted the airplane and removed two wasps from the cabin. I was practising a soft field take-off when ... a wasp flew into my field of vision and distracted me. I tried to hold altitude at about 200 ft to prevent a stall (from pulling up unknowingly in panic), as I gathered my wits together. The wasp stung me on the hand as I tried to wave it away, but I was able to correct my flight-path (which had diverted right of the runway, close to a skydiving operation) and make a sharp climbing turn to the left away from a building and some trees ....*
>
> *Just because you fly out of an airport frequently does not mean that you can expect your cockpit to be free of insects (or worse). It only takes a minute for a small creature to move their 'home'—a wasp could build a good size nest overnight. We need to remember that pre-flighting the cabin does not mean just removing the gust lock. I assumed that since I saw only the two wasps, that that was all there was. I should have looked even harder to see if there were more.*
>
> *The next day I flew, I found a centipede under the step outside the door—I'm glad I am looking harder now as I could have found him hidden in my plush sheepskin cover as I sat down—you can never be too careful. We read stories of people who have had their airplanes sitting in fields or hangars for long periods of time and had passengers of this nature which they didn't*

expect but that is not the only place it happens. I was lucky there wasn't a big nest full of them circling me at once. I don't know if I could have handled that. A funny side note: being a female, I wear perfume every day—and I've found that wasps and bees are attracted to me when I wear flower scents.

[Aviation Safety Reporting System, Report Number 199301]

This account is, of course, one of those stories that tend to sound the most amusing to those at the safest distance from the episode to which it refers. In fact the report documents a rapid cascade of events which placed the pilot at real risk, and, it appears, close to the limits of her performance abilities. Beginning with a trivial and common experience—an encounter with a pesky insect—matters quickly deteriorated and became genuinely dangerous. Distraction, sharp manoeuvres and (as the incident report notes) partial loss of control at low level, amid buildings and other aerial activities, is life-threatening—and certainly stressful, by almost any everyday definition. Judging by the account, this particular pilot may have had a pre-existing unease about insects. Perhaps at some point in her life she had experienced a traumatic encounter with bees or suchlike. In any case, the reported experiences seem to have left her sensitised, apprehensive, and vigilant for "worse" stowaways.

Given all this, the incident has a number of features that raise intriguing questions about the nature of stress and performance. These include the potential role of pre-existing, or 'trait' anxieties; the effects of significant ('life stress') events upon human performance; and the importance of the interaction between trait anxiety and inflight incidents and the acute stress reactions that accompany these. Moreover, the incident raises questions about other potentially interacting variables, such as training, knowledge, confidence and preparation.

The incident provokes the consideration of the way in which differing threats (collision risks, terrain, wasp stings, etc.) would be appraised, prioritised and handled by differing individuals with various personalities and styles of thought. Most of all, perhaps, we find ourselves facing the challenge of pinning down the notion of 'stress' itself. After all, if we adhere to some older ways of speaking about stress, that is, in terms of 'stressors' (things external to humans that cause an internal 'stress response'), we are going to have to add 'wasps' to the list of putative stressors. If wasps, then why not add 'dogs' (for mail carriers), 'food' (for bulimics), 'children' (for W. C. Fields), and so on. Clearly, we first need to step back and take a hard look at the concept of stress, before going on to explore its effects and discussing what we, as pilots, can do about it.

What is Stress?

In considering this question, a helpful approach may be to think in more general terms for a moment and first ask, what is emotion, any emotion? There is a tradition in Western philosophy and literature of considering emotions as cognitive 'spoilers', unbridled passions from the baser part of our souls that undermine and confound rationality and the exercise of intelligent reason. Although placed on a pedestal by the nineteenth century Romantics, emotion has hardly been associated with good decision making. After all, Byron himself, in a rush of Hellenistic passion, impulsively decided to go to fight with the Greeks and was promptly killed at Missolonghi—not a model we want flight students to emulate. However, there is every reason to reconsider the role of emotion in intelligence. After all, humans have done fairly well so far from an evolutionary point of view. It is instructive to subject the traditional western view of emotion to the 'evolutionary plausibility' test. In a Darwinian world, traits that tend toward the propagation of genes into the future are selected for, while traits that do not are, of course, not passed on to future generations. They die out. Death, as one flight instructor tried to impress on his youthful and all-too-fearless students, is bad for your sex life (Hyman, 1985). Thus it cannot be plausible that emotion, in the Romantics' sense of irrational passion, noble impulsiveness, etc., has been selected for over the millennia. After all, Romeo and Juliet didn't leave any offspring!

This seems to suggest that emotion should be looked at from a different perspective—as a trait that, while troublesome at times, has on average been a positive contribution to human intellect and functioning, at least as far as these relate to reproductive success. Emotion can be viewed as a powerful signal system which provides (roughly as pain does for the body) an early warning to the cognitive system as to whether things are going well or badly and indicates which things need attention. In fact, evidence is accumulating that suggests that the brain routes certain types of important information directly through the emotion processing system even before it is passed on to the higher cortex for interpretation (Damasio, 1994). The advantage of this is speed. Those who took fear and started running before they interpreted the threat as a sabre-toothed tiger presumably had a slight, split second edge over those who interpreted first and ran later. Over evolutionary time periods, even tiny advantages tend to be preserved, and even minor disadvantages can die away. The fact that the run-first-and-ask-(cognitive)-questions-later strategy must occasionally lead to false positives (and the needless expenditure of metabolic energy) appears to have been voted on by nature, and found to be a worthwhile trade-off (c.f., Fodor, 1983, on reflexes). Note, however, that evolution doesn't, so to speak, care about anyone's peace of mind, happiness or performance (other than that which, in the past, led to breeding success). Evolution hasn't had time to change us to better fit the engineered environment we have so recently created. Paul Griffiths (1997) puts it this way: "Evolutionary psychology claims that the human mind is a bundle of cognitive adaptations. It does not claim that these adaptations are currently adaptive" (p. 107).

So while our emotional system may, for example, provide that 'sixth sense' that something is not right about a planned flight (before we can put a finger on exactly what is wrong), it may also have a pilot 'tunnel' her attention onto just one instrument (e.g., the altimeter) when she is under stress. Cognitive and perceptual tunnelling might be perfectly appropriate when the threat is a tiger, but not nearly as adaptive when responding to a systems failure requiring the diagnostic integration of data from multiple instruments and other sources. With these considerations in mind, we can better tackle the meaning and applicability of stress as a term.

A look in any dictionary confirms that this word has a bewildering array of meanings. A musician stresses a note by playing it with greater intensity. A writer stresses a point by devoting particular attention to it. An environmentalist might speak of the stresses placed upon the ecosystem by industrial development. And when an engineer refers to stress, she is referring to an external physical phenomenon that exerts some kind of mechanical force on a part. This chapter, of course, is concerned with psychological stress, defined variously as "distress" (i.e., "pain or suffering"), or as "a physical, chemical, or emotional factor that causes bodily or mental tension," or as "a state resulting from stress, especially ... from factors that tend to alter an existent equilibrium" (Merriam-Webster, 1993). The psychological and non-psychological senses of the word actually overlap to a considerable extent, and many notions about psychological stress show the effect of this linguistic influence. For example, the engineering usage of the term echoes a common conception (some would say, misconception) of stress as being some kind of external force that produces 'strain' within an individual. (Indeed, one book on pilot performance makes explicit use of this analogy by stating that "the simplest way to understand human stress is to relate it to the material strength of an aircraft"; Campbell & Bagshaw, 1991, p. 105.) One definition of stress, therefore, can be phrased as follows:

Definition 1: "Stress is something that happens to a person."

This stimulus-based view of stress was neatly summarised by Sir Charles Symonds (1947), who stated that "stress is that which happens to the man, not that which happens in him; it is a set of causes, not a set of symptoms." In other words, stress is seen (in this view) as an external event or circumstance that affects "the man" in a negative manner. We might ask, which man (or woman)? For particular pilots external stressors might well include bad weather, heavy air traffic, or equipment failure. For others, it will include wasps, inadequate heaters, missing mealtimes, etc. In the fields of aviation psychology and aviation medicine, a good deal of the existing research has focused on the effects of various kinds of adverse conditions on aircrew well-being and performance. In typical studies of this sort, pilots are asked to complete a flight (or simulated flight) in the presence of, say, high levels of background noise, cabin heat, or high workload. and the noise or heat or workload is then described as the 'stressor'. It is true that such studies have yielded useable information about the kinds of factors that can affect aircrew performance (that is, noise, heat, and workload, not to mention sleep deprivation, circadian rhythm disruption, medications of various kinds, etc.). However, these studies have much less to tell us about stress itself. After all, increased workload that you can cope with may have one set of effects on performance, while workload you cannot cope with (and which perhaps engenders a real sense of anxiety) may have quite different effects.

Here are some of the research problems. First, in the stimulus approach, 'stress' can be just about any factor or condition that the investigator thinks might be aversive. This may lead to a kind of circular reasoning in which the condition becomes labelled a stressor a priori, without a systematic attempt being made to determine whether the pilots concerned are actually experiencing any sense of stress or not. Almost any factor can conceivably be interpreted as stressful under some kind of circumstance: hot weather or cold weather (or perhaps even average weather); too much cockpit workload or not enough workload; flying solo or flying with a socially unpleasant crew member; or, perhaps, the presence of a wasp in the cabin. However, if any factor under the sun *can* be stressful, then no special claims are really being made about any one in particular. The focus on 'stressors' is in some ways beside the point—which is (as we shall see) more to do with the pilot's interpretation of events or circumstances as threatening.

Second, the engineering analogy approach to stress has little if anything to say about the role of individual differences (Glass & Singer, 1972; Grinker & Spiegel, 1945). Conditions that might be interpreted as stressful by pilot Smith might not be so interpreted by pilot Yeager or pilot Glenn. Many individuals reading this chapter may feel that they, personally, would have been able to cope with the presence of a wasp in the cabin without nearly flying the aircraft into the ground. And so they might—but what is equally clear is that the feeling of alarm experienced by this particular pilot was quite genuine, and endangered not only her own safety but that of others in the surrounding airspace. As such it deserves our serious consideration.

A related point is that not only do individuals vary in their responses to the same conditions, but that the same individual may react differently at different times. In other words, what stresses Smith this week might not stress her next week. (Those who advocate meditation or other relaxation techniques as a way of 'reducing stress' recognise this; if their assumptions were wrong, there could be no logical value in trying to moderate one's internal responses.) There is something to be learned from this 'external stressors' approach, however. Pilots can try to identify events and circumstances that they might find stressful at their particular state of training (perhaps crosswind landings, or night time arrivals), and make plans to cope with, or to avoid them. The idea of 'personal minimums', limits, or restrictions is a useful one.

Having said this, it still remains the case that to view stress only in terms of external forces or conditions is to ignore the actual *experience* of stress. Let us then consider the following proposition:

Definition 2: "Stress is something that happens within *a person."*

This, of course, represents the opposite way of looking at stress: what Symonds would call the 'set of symptoms'. It would seem, on the face of it, to make intuitive sense. Who, after all, would argue with the idea that stress is a personal reaction? However, this definition does not, by itself, tell us much. If stress is something that happens within a person, then it is reasonable to ask, "What is the precise nature of this 'something,' how can it be observed and measured, and how can a pilot avoid or change it?"

Approaches to this question could, in principle, consider a wide range of phenomena, including behavioural, emotional, and cognitive processes. Overwhelmingly, however, the focus has been on physiological and biochemical changes associated with various external conditions. Such changes include increases in, for example, heart rate, respiration rate, adrenaline output, and a variety of other metabolic and endocrine functions. Collectively, they constitute a generalised, nonspecific physiological reaction that is commonly labelled as 'arousal'.

Increases in arousal level have been extensively documented in aviation settings. These studies are far too numerous to review in their entirety, but the following findings are typical: increased heart rate in helicopter pilots undergoing checkrides (Melton & Wicks, 1967), elevated testosterone levels in F-16 fighter pilots during simulated emergencies (Vaernes, Warncke, Myhre, & Aakvaag, 1988), heightened urinary catecholamine excretion in student pilots practising spins (Krahenbuhl, Marett, & Reid, 1978), and increased production of plasma phospholipids in Boeing 737 pilots during a simulated birdstrike emergency (Sive & Hattingh, 1991). While all this biochemical terminology is most impressive, it is not clear what such findings 'buy' us. As pilots we might be forgiven for muttering something akin to a "so what" response. In fact, it's a good question, and it's worth looking at how we come to be scrutinising phospholipids and urinary catecholamines when our interest is in stick-and-rudder performance, judgement, airmanship, training, and safety. The answer lies in a history of research quite outside the preoccupations of aviation.

Studies of arousal actually date back almost to the turn of the century (Cannon, 1915), but it was in the 1950s, with the work of famed stress researcher Hans Selye (1956), that arousal came to be identified with stress. Selye noted that these physiological responses were sometimes adaptive in the sense of enabling the individual to respond more energetically to an external threat. (Most readers are probably familiar with this as the 'fight or flight' reaction.) However, they also imposed a metabolic cost. This was especially true over the long term, Selye argued: continued activation of arousal responses might lead to physiological depletion, exhaustion, and an inability to respond to any additional threats.

Studies of arousal do represent a useful contribution to our understanding of human reactions to extreme or threatening conditions. However, there has also been a certain amount of conceptual confusion. Many researchers, both in aviation and other fields, have come to assume that signs of physiological arousal are also evidence of the condition called 'stress,' which most people understand as having some kind of important psychological dimension. Indeed, every one of the physiological studies cited above is explicitly couched as a 'stress' study. Yet it cannot be that physiological arousal necessarily equals psychological distress. Many 'arousing' situations are not psychologically unpleasant in the least: obvious examples include sexual activity, thrill rides, sports, and, to pilots, flying! Indeed, people often deliberately seek out experiences such as bungee jumping specifically for the 'adrenaline rush' they provide. On the other side of the coin, an absence of physiological arousal cannot be dismissed as an absence of stress. Underarousal can be very threatening in certain situations: there is a considerable literature on boredom and underload, although it is probably only necessary here to recall Charles Lindberg's vivid account of his struggle to stay alert during his famous transatlantic flight.

While the physiological approach to stress is no longer considered viable (see, for example, Stokes & Kite, 1994), there is something to be learned from it, just as the notion of 'personal minimums' is reinforced by the stimulus-based approach to stress. First, modern medicine confirms Selye's notion that prolonged activation of the fight-or-flight response can be damaging. The immune system is weakened, and individuals become prone to opportunistic diseases. We also know that the body secretes endorphins—home-grown opioids that help buoy up feelings of well-being and 'disguise' depletion of resources. This may counter the effects of adversity until complete exhaustion 'catches up with you' and even a little more demand may, like the proverbial straw, break the camel's back. This may account for the common research findings that performance can continue largely unchanged, despite increased challenges to the individual—even when these are electric shocks, for example (Neiss, 1988).

The danger to bear in mind, then, is that of 'pushing oneself' too hard—adopting the attitude that a 'real' man/woman could hack it. This outlook is often encouraged and reinforced by peer pressure and by employers ("Real pilots don't bring their domestic troubles to work ...") Moreover, the normal pilot can-do attitude may become more like machismo when taken too far. Indeed, machismo is one of the famous 'five deadly attitudes' often taught in aviation (see, for example, Lester & Bombaci, 1984). However (as with the other four attitudes), it is more often thought of as a personality type and thus dismissed by the majority of pilots ("I'm in the Rotary Club—I don't even own a chest medallion ..."). But the temptation to push on regardless is perhaps better viewed as a transient attitude that can affect any of us when the circumstances are right—that is, when they are supported by rationalisation, peer expectations, and so on. (Indeed, in wartime we reward such persistence with medals.)

To summarise, we may not see much of direct operational significance to GA in the scrutiny of urinary biochemicals and the like, but there is a valuable legacy from Selye's work. It lies in our better appreciation of the negative effects that accumulate when the body stays at its biochemical battle-stations for too long. We might put it this way: a psychophysiological debt can be built up over months of overwork and worry. Note that everyday hassles and frustrations can be quite sufficient—major life events such as divorce or job loss need not be invoked (Stokes & Kite, 1994). Masked for awhile by our endorphin system, the accumulated debt may finally be 'called in', so to speak, on that bumpy night-time ILS approach down to minimums at an unfamiliar airport.

External Cause or Internal Effect?

As we have seen, stress has been defined both in terms of environmental stimuli and internal (mainly physical) responses. (These two definitions have also been described as stimulus-based and response-based models, respectively.) Both approaches do tell us something about stress, but they also suffer from similar shortcomings. Both tend to consider stress in terms of a simple stimulus-response or direct cause-and-effect relationship (see upper part of Figure 4.1): a 'black box' view of stress that essentially bypasses the role of the individual as a thinking, feeling participant in the process (Lazarus, DeLongis, Folkman, & Gruen, 1985; see lower part of Figure 4.1). As such they are both strangely non-psychological, especially given that stress is a phenomenon normally defined in psychological terms. (People who feel themselves to be suffering from 'excessive stress' may on occasion consult a psychologist about it, but it would be considered odd behaviour indeed to seek out a cardiologist in order to complain about one's increased heart rate, or an endocrinologist about one's urinary catecholamine output. In any case, the bewildered physician would be quite likely to promptly refer the patient back to a psychologist!) A theoretical approach to stress which retains its psychological core is clearly desirable, from the standpoint of throwing light upon human decision making and other aspects of cognitive performance.

A more current and more 'psychological' way of looking at stress, then, is the so-called transactional approach, which considers both the person and the environment, as well as the relationship between them (Welford, 1973; McGrath, 1976; Cox, 1978; Fisher, 1983). In this view, the individual brings to the interchange (the 'transaction') his own set of capacities, beliefs, motivations, worries, etc., and the environment likewise offers threats, opportunities, and so forth. Stress, in this view, can be defined as follows:

Definition 3: Stress results when there is a mismatch between the demands imposed by the environment and the individual's ability to cope with these demands.

This is a stronger definition in that it incorporates both environmental and personal dimensions of stress. However, there is a further refinement to made, and it concerns the dimension of individual thought, perception, and appraisal. For it is not necessary for there to actually be a mismatch for stress to occur—it is only necessary that the individual (rightly or wrongly) perceives that such a mismatch exists.

This important point can be illustrated by the real-life example of a pilot who experienced an alternator failure while flying at night. Viewed objectively, this was not a serious emergency, because the pilot had enough battery power to get to his (large and well-lit) destination airport without difficulty. Unfortunately, however, he overestimated the seriousness of the situation and, in the resulting stress, elected to land immediately at a small, unlit airport—with fatal results as the aircraft overshot the runway (Simmel, Cerkovnik, & McCarthy, 1987). The urgency, in this case, existed only in the pilot's mind, but his reaction to it was anything but imaginary.

The 1985 crash of Galaxy Airlines Flight 203 in Reno, Nevada, provides another case in point. Shortly after takeoff the crew heard a strange vibration that reverberated throughout the aircraft. Unable to determine its source, and suspecting an engine mount failure, the crew became highly alarmed and attempted to turn back. However, they failed to control the aircraft's flight path and airspeed, and crashed before reaching the runway with fatal consequences. Upon investigation the vibration proved to have been caused by an improperly closed air start access door that was fluttering against the wing. This posed no threat whatsoever to air safety; thus, in the objective sense, there was no actual mismatch between the demands of the flight and the crew's capacity to cope. It was entirely a matter of perception.

(a)

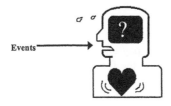

TRANSACTIONAL (COGNITIVE APPRAISAL) MODEL

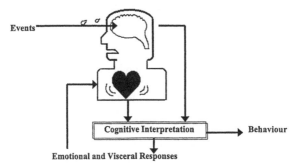

Figure 4.1   **Stress as stimulus, response or transaction**

This insight is reflected in a more sophisticated formulation of the transactional model (Cox, 1978), which runs as follows:

Definition 4: Stress results when there is a mismatch between how an individual perceives *the demands of a situation, and how he* perceives *his ability to meet these demands.*

This is the view of stress that we will be making use of later in the chapter, when we discuss stress countermeasures. It is often described as a cognitive appraisal model. Here we should recall the earlier discussion about direct evaluation of stimuli by the emotional system prior to cognitive interpretation. This, it has been suggested, is a quite independent "automatic appraisal mechanism", or AAM, which operates without our awareness, although we are aware of the emotional states, the outputs of the system (Ekman, cited in Griffiths, 1997). However, the definition above still doesn't cover all the cases it should cover; there are a couple of additional dimensions that need to be appended to the basic formulation. Consider, for example, two different pilots, each of whom is flying in a flight simulator. The first pilot, Jane, performs fairly well, but the second pilot, Tony, who is much less experienced, flies so poorly that he actually 'crashes' the simulator. Which pilot feels more stress in this situation?

The definition given above would, in its unmodified form, seem to predict that Tony will experience more stress. After all, the flight is far more difficult for him than for Jane. In fact, however, Tony is a casual flight student who has stepped into the simulator for fun, has no particular expectation of performing well, and walks away from the 'crash' unconcerned and joking with his friends. Jane, however, is performing the flight as part of an employment interview with a major airline. She is competing for the position with ninety-seven other, equally skilled aviators, and she has a mortgage to pay and a family to support. She might well experience considerable stress during the flight—not because of any lack of flying skill, but because her success is so critical to other life goals. In other words, the importance of success (or, more precisely, the perceived importance of success) is an important mediating variable, and needs to be included in the definition.

Now consider a different scenario. Three university flight students are about to take their first end-of-semester ground school examination. The first student is extremely well prepared, has passed all the tests and practice exams with flying colours, and anticipates no real difficulty. She feels 'in control'. The second student is moderately prepared, has had some marginal scores, and is not sure whether he will pass or not. The third student has studied hard for the exam but finds the material too difficult to comprehend, has no realistic hope of passing, and knows it. (In fact, he has already decided that he would rather go to library school.) Which of these student pilots is likely to experience the most stress? The third student may be experiencing the greatest objective mismatch between situational demands and personal capacities, but it is of course the second, the one who is not sure of himself, who may well be worrying the most about this exam. Uncertainty can add considerably to the experience of stress, other factors aside.

Summary

Bringing together the various elements discussed above, the following view of stress emerges. "Stress" is a term with multiple, often related meanings that we often use to describe certain mental, physical and emotional states and behaviours. These result from a combination of factors, including:

- The pilot's perception of demand (i.e., what the situation demands of him or her).
- The pilot's perceived ability to meet that demand (i.e., what resources in time, skill, knowledge, etc. does the pilot feel he or she can bring to meet the perceived level of demand).
- The perceived importance of success (how critical is success, judged in terms of overall goals and objectives—including retaining your job, life, limbs, self-respect, etc.), and
- Uncertainty (which makes it difficult to compute the first three factors, and tends to leave open the potential threat to overall goals). However, we can readily recognise the role of personality and prior experiences in the interpretation of uncertainty—some will assume the worst, others will be optimists, as it were.

With this framework in hand we can go on to examine the actual effects of stress upon the kinds of things that pilots must do to succeed as pilots. Fortunately there is no shortage of research results to inform us. Unfortunately, the studies represent a minefield of problems, and many may be misleading or simply irrelevant to real flying. We will try to sift through the evidence, categorising and evaluating it by reference to 'real world' operational issues.

What Stress does to Information Processing

After nearly a century of research, not to say millennia of experience, the question of what stress does to human performance. remains a difficult one. This is not because there are few research results to cite, but because there are so many, springing from a bewildering variety of studies built on a range of assumptions. Some studies are laboratory based, experimental, and well-controlled; however, these tend to be narrow and difficult to generalise to the complex, multivariate world. Other studies are observational and set in the 'real world'; these may perhaps be ecologically more valid, but also tend to be poorly controlled because of the many variables influencing results. Still other studies represent a compromise between these two types, often using simulation for ecological validity combined with experimental techniques for better control over the variables influencing results.

To complicate matters further, the studies collectively labelled as 'stress research' are by no means designed around a consistent, unified model of stress.

On the contrary, they run the full gamut of assumptions and approaches discussed in the first part of this chapter. Indeed, many studies do not refer explicitly to any particular model of stress whatsoever. Given these circumstances, to ask for the 'results' of this corpus of research is a bit like asking for the results of a soccer championship in which the various games have been played according to several different sets of rules, and in some games, no clear rules at all!

A second issue, when considering the effects of stress on performance, is to define what is meant by 'performance' (or 'flying performance'), and to determine how this is to be measured. One method of evaluating pilot performance is to observe pilots during flight. This, of course, is what flight instructors and checkride pilots do. This approach offers the highest degree of ecological validity; that is, it is the most directly relevant to actual flight (because it is actual flight). From a practical point of view, however, naturalistic observation is not always the most feasible or appropriate method of research. First, the inevitable variations in wind, weather, and traffic limit precise comparisons between flights. Second, many kinds of stress manipulations are difficult or dangerous to implement in an actual flight. Third, the kinds of observations possible are necessarily somewhat subjective.

One solution to these problems is to have pilots fly simulators rather than actual aircraft. Simulator-based studies offer greater uniformity in experimental conditions and a higher level of control over the kinds of manipulations that can be implemented. Simulators can also measure the physical parameters of flying performance with far greater precision (e.g., the ability to hold to a particular course), and store these measurements electronically for subsequent analysis.

However, there are some things that a simulator cannot do. It can measure performance, and it can impose workload. What it does not do so easily is create a sense of actual threat: even the most high-fidelity device ever built does not place pilots at genuine risk of life and limb. Within the context of stress research this is a significant consideration, particularly in light of the cognitive-appraisal model discussed above. Earlier we cited the example of Galaxy Airlines Flight 203, in which the crew became highly alarmed by a mysterious noise and vibration occurring shortly after takeoff. Unable to determine the cause, they became so distressed that they failed to control flight path and airspeed, and crashed fatally not far from the runway. Ironically, as it turned out, the vibration was trivial in origin and posed no danger to the flyability of the aircraft. More to the point, for the purposes of this discussion, the stress experienced by these pilots would hardly have occurred in a simulator: although the noise and vibration could have been reproduced, the mortal threat that these signified could not have been (Stokes & Kite, 1994).

There is another consideration connected with simulator- and aircraft-based studies, and that is that both look directly at complex operational performance. Sometimes this is desirable (as, for example, with a checkride), but sometimes it is not. In the case of Galaxy Flight 203, for example, it might be interesting to know the extent to which the crew's loss of control was occasioned by fear, and to what extent the physical effects of noise and vibration might have been responsible. Equally, did they suffer a loss of psychomotor coordination, or of decision making ability, or, perhaps, of the ability to perform more than one task at a time?

Stated another way, complex skills such as operating an aircraft actually require a number of other, more basic cognitive abilities. Consider, for example, the kinds of basic abilities needed to drive a race car, or to play chess, or to cater a six-course dinner for two hundred guests. The driver needs good eyesight, fast reflexes and excellent psychomotor coordination. The chess player must be able to identify configurations of pieces, visualise other configurations, and compute their effects. The caterer should be good at planning and prioritising, at carrying out multiple tasks simultaneously, and at coping quickly with unexpected crises.

What kinds of basic skills does a pilot need? Like the race car driver, he needs good visual perception (to read the instruments, monitor air traffic, navigate VFR, and land on the runway), and good psychomotor coordination (for 'stick and rudder' operations). Like the chess player he must be able to visualise the position and movements of surrounding objects (i.e., to anticipate the flight paths of other aircraft). Like the caterer he must be good at prioritising, at coping with emergencies, and at overseeing multiple operations (e.g., instrument scanning, managing high workload phases of flight). Moreover, a pilot needs to be able to:

- Hear radio communications and extract relevant information,
- Retain and process multiple bits of information in short term memory,
- Navigate and orient himself with relatively few visual cues, and
- Calculate headings, glideslope, etc.

Many laboratory-based studies of stress and pilot performance have adopted a componential approach centred on basic skills such as these. For example, psychomotor performance might be measured not by how well a pilot executes a crosswind landing, but by how well she can track a moving target on a computer screen. By the same token, dual-task performance might be assessed by the ability to track a moving target AND memorise a list of words at the same time. To a pilot, laboratory tests such as these may seem too abstract, too remote from the actual business of flying, to reveal much that is useful about pilot performance in the real world. Indeed, such concerns may sometimes be justified: not all psychologists are as conservative as they perhaps should be about generalising their findings from the laboratory to the flight deck. Nevertheless, the componential approach does allow us to see rather clearly which of the relevant mental processes are affected by stress, and how.

Perceptual-Motor Coordination

Some studies of perceptual-motor coordination are simulator based, while others (as suggested above) measure computer based tracking performance, where error (time off target or distance off target) is analogous to the time the needle is 'out of the doughnut' in VOR tracking or on an ILS approach. Performance on tasks such as these has been adversely affected by external physical 'stressors' such as high G forces (Frazier, Repperger, Thoth, & Skowronski, 1982), heat (Hancock, 1982), and carbon monoxide exposure (Gliner, Horvath, & Mihevic, 1983). However, there is also evidence that perceptual-motor impairment can be induced by psychological threat—for example, in paratroopers prior to jumping (Hammerton & Tickner, 1969). Distraction, speed, workload, and secondary tasks have also been shown to have negative effects on tracking (Garvey, 1957; Stokes, Belger, & Zhang, 1990). Perceived social threats may also play a role: Cohn (1994) describes an accident in which a relatively inexperienced pilot apparently became flustered and intimidated by the voice of an air traffic controller and subsequently lost control of the aircraft.

There is much that remains unclear about the effects of stress on tracking and other psychomotor performance. For example, some pilots have a tendency to overcontrol (i.e., to 'fight' the aircraft) when under stress, whereas others undercontrol or even 'freeze'. The reason for these differences is currently unknown. Stokes and Kite (1994) have suggested that a model able to predict which way a given pilot was likely to respond, and under what circumstances, could have considerable value for both selection and training.

Memory

Working memory For any given flight there is a good deal of specific information that must be acquired and retained over the short term. Examples include step down fixes, decision heights, and the call signs of other aircraft. This information must be integrated with instrument readings of navigational position, attitude, altitude, speed, direction, etc.—not to mention system variables such as fuel state—to form an overall mental model of the flight. In short, pilots are required to keep in mind a rather daunting number of variables (which, of course, are constantly changing). This can impose a significant load on working memory functions.

Unfortunately for the pilot, working memory is markedly degraded by anxiety, stress, and fatigue (Kleitman, 1963; Wachtel, 1968; Davies & Parasuraman, 1982). This degradation takes place in several ways. Perhaps most obviously, there may be a reduction in memory capacity, or the amount of new information that can be retained (see, for example, Stokes et al., 1990). One possible reason for this, it has been suggested, may be that stressed individuals divert part of their available capacity into worrying (Eysenck, 1979, 1982).

Separate from the issue of capacity is that of strategy: it appears that persons under stress have more difficulty manipulating information in working memory (Daneman & Carpenter, 1980; Darke, 1988). In connection with this comes a tendency to simplify information, to overgeneralise, and to ignore details (Keinan, Friedland, & Arad, 1991); presumably this is done in order to make information easier to assimilate. This kind of mental 'chunking' presumably has an evolutionary origin, and if it helps the pilot to disregard irrelevant or unnecessary information it may actually aid performance. However, sometimes critical information may be ignored, particularly if it does not comport with the pilot's prior expectations. This phenomenon is known as confirmation bias, and is what leads stressed aircrew to land at the wrong airport (because its details match those of the correct airport). It may also account for 'friendly fire' accidents in wartime.

Long term memory Less is known about the effects of stress on long term memory. In general, long term memory functions appear to be far less affected by stress than are working memory functions, a fact with important implications which will be discussed below. There is some evidence to suggest that stress impairs the transfer of newly acquired information from working memory to long term memory: that is, it may impede learning (Keinan & Friedland, 1984). If this is the case, then 'realistic' military flight training programs, which seek to duplicate the stresses of actual operations, may not represent the optimal training approach. In terms of recall from long term memory, one effect that has occasionally been observed is the phenomenon of stress related regression. This refers to the tendency to forget recently learned behaviours, strategies, and schemata, and to revert to older, often cruder patterns (Barthol & Ku, 1959). The authors have previously described an incident involving a flight student who, in the face of a door open emergency, responded by diving the aircraft to low level; subsequent debriefing revealed that he was acting on an earlier, nonpilot's schema, i.e., 'low equals safe' (Stokes & Kite, 1994).

Attention

People pay more attention to those things that they perceive to be most important. This much is stating the obvious; we now consider the effects of stress on this process. Stress appears to have two main effects on attention allocation: it concentrates attention on 'important' elements at the expense of more peripheral information, and it influences which elements are perceived to be important. The former effect is known as perceptual tunnelling, and may involve an actual narrowing of the visual field (Hockey, 1970; Broadbent, 1978). This is an example of a feature that may have been more adaptive in our ancestral environment than it is in an aircraft cockpit, where it may be vital to maintain a full scan of instruments and airspace and not to become overly fixated on particular elements (for example, an unwinding altimeter). In so-called 'cognitive tunnelling', the focus is on a particular thought, task, or perceived source of danger. Tunnelling may be an important source of performance impairment in dangerous environments (Baddeley, 1972), and appears to have been a contributing factor in the Galaxy Airlines crash cited earlier.

A related phenomenon is task shedding, which involves not only the ignoring of information but the actual abandonment of certain tasks or subtasks (Sperandio, 1978). It is not unusual, for example, for beginning pilots to become so preoccupied with monitoring and responding properly to radio calls that they neglect the primary task of flying the aircraft. Like tunnelling, task shedding may sometimes be adaptive, if less important tasks are shed in favour of more central ones. However, this is not always the case, as the above example suggests.

Communication

The same example also highlights the general importance of communication skills in aviation. For many pilots, comprehending and responding to radio communications represents one of the more intimidating aspects of learning to fly. Equally significantly, for the purposes of the present discussion, both speech production and the reception of information are affected by stress in a number of important ways. Acoustical changes in speech have been well documented, for example, and can include increases in pitch, vibration, amplitude, frequency, and speech rate (Williams & Stevens, 1981). The latter, in particular, may well have an effect on communication efficacy, although specific operational implications are not clear at this point. Stress-related phonetic changes include a blurring of vowel sounds (Tolkmitt & Scherer, 1986), which may well make it more difficult for listeners to distinguish one word from another—not least on the flight deck, where radio communications may be none too intelligible in any case (Stokes & Kite, 1994).

The effects of stress on communication do not end here, however. Stress affects not only how people speak, but what they say. This subject has received much less investigation (Ruiz, Legros, & Ruell, 1990), but such findings as do exist suggest that stress-related changes in lexis and syntax may be of considerable importance to communication as a whole. For example, Wanta and Leggett (1988) have reported an increase in the use of clichés. There may also be a tendency to regress to childhood dialects and other modes of speech used earlier in life (Weick, 1991). Stokes, Pharmer, and Kite (1997) report a decrease in the use of sentence connectives and an increase in the use of pronouns. This is particularly true of pronouns whose meaning is clear to the speaker but not necessarily to anyone else, as in the following exchange:

"Look out! Over there!"
"Where?"
"There!....No, not there, *there!*"

Taken together, these kinds of changes suggest that under stress, humans revert to ways of speaking that are "easy" for them—i.e., that require little in the way of intellectual effort. Unfortunately, what is easier for the speaker may not be for the listener: effective communication (oral, and indeed, written) involves understanding one's audience, monitoring reactions for signs of incomprehension, articulating what may not be clear, and so forth. The Stokes et al. findings in particular suggest that people may, under stress, be more likely to abandon the "mental models" they carry of other individuals' knowledge states, attitudes, etc., and to operate as though others think and see as they do. In some kinds of situations (for example, team sports), this may actually represent an efficient conservation of time and effort: where individuals share the same goals and operate within the same environmental context, simplified forms of speech may be quite adequate. However, this is unlikely to be true in general aviation, where the pilot is communicating with ground personnel who may be miles away and in a very different kind of environmental context.

What Stress does to Decision Making

How Pilots Really Make Decisions

Just as the operationally relevant effects of stress depend in part upon how one views or models stress, so the effects of stress upon decision making depend somewhat upon how one views decision making. The most familiar approach (known as the classical model) assumes that decision making is primarily a matter of option selection. That is, an individual is seen as being faced with a number of possibilities or potential courses of action, with one of them being the best for any particular set of circumstances. Historically, the classical model didn't set out to be a *de*scription of human decision making. Rather it was developed as a *pre*scription for it (an 'ought' rather than an 'is'). In this view the decision maker ought to weigh or evaluate each option against the others, perform a sort of cost-benefit analysis, and then home in on the optimal one. Which choice is to be regarded as optimal is a function of how useful it is perceived to be by the decision maker for the purpose at hand. In the language of analytical decision making theory, the decision maker is said to be trying to 'maximise subjective utility.'

At first sight the analytical model is a very intuitively plausible and universally applicable approach. It feels familiar, and to pilots it has a satisfyingly procedural, almost checklist quality to it. The trouble is found in the conceptual small print, so to speak, as it insists that the process only works properly if all the viable alternatives are considered, each being exhaustively analysed and compared (using Bayesian statistics for the real sticklers).

In exploring the likely effects of stress on this process, we have to consider the heavy demands on working memory that such a thoroughgoing analysis of options would impose. Stokes & Kite (1994) suggest, as a comparable thought experiment, trying to analyse the costs and benefits of several competing life insurance policies *in your head.* But as we saw earlier, stress tends to decrease working memory capacity and to cause a certain cognitive and perceptual 'tunneling' or narrowing of attention. Both effects would have ominous implications for any process dependent for its success upon broad, inclusive, and exhaustive analyses. This is especially so when these mental gymnastics have to be carried out 'in real time', under workload, fielding ATC calls, and all the while covering terrain at two to three miles a minute or more.

Nevertheless, studies of stress using the classical decision model have certainly been done. For example, Keinan has reported several studies of the effects of stress on the selection. of alternatives, and suggests three specific mechanisms by which decision quality might degrade under stress (Keinan, 1986; 1987; Keinan, Friedland, & Ben-Porath, 1986). The first of these, *premature closure.* refers to the forming of a decision before all possible options have been evaluated. The second is *nonsystematic scanning..* This refers to a frantic and disorganised, rather than a logical and orderly review of decision options. It may even involve shuttling repeatedly between the same options without settling on one—a symptom of 'hypervigilance', a stress condition described by Janis (1982). Finally, in what Keinan calls *temporal narrowing,* the decision maker does consider all options, but devotes insufficient time to each. The studies conducted by Keinan and colleagues required subjects to solve a series of logical analogy tests under threat of electric shock (a stimulus based approach to stress). The results demonstrated the presence both of premature closure and of nonsystematic scanning, but they failed to demonstrate the expected temporal narrowing.

We began this section by pointing out that the kind of effects one might anticipate stress having upon decision making is to some extent a function of how one views decision making. Most pilots are probably familiar with the classical view of decision making, and many may have even been taught to attempt to use some algorithmic version of it in practice. These training attempts are often recognisable by their preference for checklists in the form of acronyms like DECIDE, PASS or SAFE. Each letter represents a step in the recommended process, such as Select a set of alternatives, List the advantages and disadvantages of each, Opt for the one that maximises expected subjective utility, and Whisk it into action if there's any time left. The limitations of this approach were neatly summarised by Lofaro, Adams, and Adams (1992):

> *Such models are time-consuming, cumbersome and inadequate to the dynamic, often time-compressed, decisional situations that aviators face when flying. In point of fact, it seems fairly accurate to say if a linear model was taught on the ground, it was rarely used aloft (p. 5).*

The research evidence agrees with Lofaro et al. Analytical strategies have been shown experimentally to be ineffective in situations where time limits constrain the amount of cognitive processing that can be done, and so prevent a fully comprehensive analysis of the situation (Rouse, 1978; Howell, 1984; Zackay & Wooler, 1984). Analytical decision making strategies are surely used from time to time by pilots (e.g., in pre-flight route comparisons, purchasing an aeroplane, selecting hangarage, etc.), but there is good reason to doubt whether most pilots, most of the time, really do go about *inflight* decision making in the 'traditional', prescriptive way.

To the extent that this is true, stress effects of the type described by Keinan et al. may be somewhat beside the point, and devising coping skills and counter-measures to them of very limited utility. Indeed, the well meaning proponents of decision checklists may be exacerbating the memory overload problem—a case of the medicine worsening the disease. Fortunately there is an alternative approach. This approach suggests that practical decision making prowess may stem less from exhaustive analysis of options (with its locus in short-term memory), than from applying the lessons of experience (a more long-term memory oriented approach). What's more, if we can abandon the search for the holy grail of classical decision making theory, that is, the best possible decision option, we may be able to cast the question of coping strategies in a rather different light.

Consider the following proposition: A workable course of action now, is better than the best one too late. Shanteau (1988) has suggested that "although expert decision makers may make small errors, they generally avoid large mistakes. They seem to have discovered that for many decisions coming close is often good enough: the key is not to worry about being exactly right, but to avoid making really bad decisions" (p. 208). For example, verbal protocols obtained from fireground commanders have shown that these expert firefighters do not spend precious time weighing alternative options or evaluating probabilities (Klein, Calderwood, & Clinton-Cirocco, 1986). Rather, they report seeking 'workable', 'timely', and 'cost effective' courses of action. Similarly, expert aviators may, instead of investing substantial mental resources in identifying an optimal solution, invest the minimum resources consistent with arriving at a strategy that is workable and adequate for the extant problem, even if it is not necessarily the most elegant one. This decision strategy is sometimes referred to using the ugly but apt neologism 'satisficing' (Simon, 1955). An alternative model of decision making that is more consistent with this view, therefore, focusses on the experienced pilot's ability to use long term memory to invoke a decision response without major recourse to deliberations in (highly stress vulnerable) working memory.

The Aerial Chessboard: Recognising Patterns

The highbrow world of chess may not seem like the obvious place to look for inspiration about pilots decision making. It's a good place to start, however, in part because chess is not a physical skill, but is all about situational awareness and decision making. In addition, chess clubs carefully classify players' skill levels from beginners to Grandmasters—so the experts are known and their expertise rated. Finally, in the research community chess expertise has been well studied and a considerable literature on its nature has been developed.

One particularly interesting study, for example, showed that chess masters (an advanced skill classification) were far more capable than novices of accurately replacing chess pieces removed from a board, *but only if those pieces had formerly been arranged in a coherent game position*. The 'only if' is key here: when the pieces were randomly placed, masters' memories were no better than those of the novices. In addition, chess masters tend not to be affected by time pressure. This would be odd if they are carrying out complicated analyses of alternative moves, visualising the game many moves ahead and so on. In contrast novices are badly affected by time-restrictions (Chase & Simon, 1973; Calderwood, Klein, & Crandall, 1988)

These and related findings provide little support for what might be termed the 'egghead' theory of chess expertise (prodigious memories, superior logical reasoning, etc.). Rather, the research suggests a more mundane source of chess expertise—experience. Ace chess players have played so many games over the years that they can recall and recognise board configurations very well, plus the moves that tend to work for each board configuration. A 'been there, done that' theory of expertise seems to fit the evidence better than the 'egghead' theory— good news for those of us for whom 'eggheadedness' is not an option!

This is how chess expertise is thought to work: masters match the actual board layouts that they see with 'mental layouts'—patterns or templates based on games played in the past and stored and classified in long term memory. These are quickly recognised when they occur. Sometimes they even have a name: "Oh he's trying the Sicilian Defence". In a study of communication, Stokes, Pharmer, and Kite (1997) were able to observe the efficiency with which teamed chess experts used technical terminology to label entire game states and strategies. It is an interesting exercise to reconsider professional jargons in this light—not least our own 'aerospeak'.

Another outcome of the expert's efficient pattern matching strategy is its invulnerability to time pressure. This is, in part because slow computational processes of option comparison in working memory are largely bypassed. Novices cannot use this overview approach as they typically do not have the huge mental repertoire of 'game states' that experts have. Thus, novices are stuck with 'egghead' strategies—trying to compare all the permutations of possible moves and countermoves, analysing from first principles the great heap of information that the chessboard in play represents.

To encapsulate the research findings, then, we might say that the decision performance of experts is characterised by a fast, recognitional and 'top down' knowledge representational model, rather than by the painstaking, inferential and 'bottom up' information processing model that beginners in many fields appear to be stuck with. (Stuck with, that is, until experience and/or training comes to the rescue. Some recommendations for flight training are explored toward the end of this chapter.)

In considering what all this means to aviation, just think of pilots as players on the 3-D gaming board of the airspace, and events or situations as aerial 'game states' or board positions. Now we can ask, what makes someone an aviation Grandmaster? On the one hand, it could be that expert pilots—highly selected personnel, after all—have superior cognitive capacities which help them to exhaustively analyse situations and review all alternatives before selecting the best course of action (the analytical model). Alternatively, in parallel with chess masters and novices, proficient high time flyers may differ little from low time pilots in basic information processing skills. Rather, the seasoned fliers could be seen as accessing a large repertoire of well organised situational schemata—the aerial game states, as it were, which, once recognised and categorised, confine the consideration of action alternatives to a very few high value 'moves'. (By high value we mean, of course, options in which the probability of success is high.)

This 'Recognition Primed Decision making', or RPD model as it has become known (Klein 1989), is concerned less with finding the 'best' solution to a given decision problem than merely a workable and timely one. Indeed, there is evidence to indicate that both skilled pilots and chess players generate an adequate 'option' as the very first course of action they consider. Often, this, the first solution that presents itself, is immediately adopted as the most efficient course of action, in a "no alternative options need apply" kind of way (Klein, Wolf, Militello, & Zsambok, 1995; Stokes, Kemper & Marsh, 1992).

An Integrated Model of Pilot Decision Making Under Stress

Stokes has argued that experienced pilots routinely function by using the pattern recognition strategy in flight management decisions (Stokes, 1991; Stokes & Kite, 1994). Figure 4.2 presents a diagrammatic model of pilot decision making under stress, in which the large upper box represents the pilot's long term store of domain specific information: flight situations experienced, procedural 'scripts', and mental models (of aerodynamics, systems, the airspace, and so forth). The large lower box represents processes within working memory: non-domain specific, analytical processes such as logical inference, mental arithmetic, and the like. Other processes include the 'metacognitive' processes of appraisal (such as the evaluation of situational demands and personal coping resources) that, as we discussed earlier, are associated with the level of stress actually felt or experienced.

High stress levels are assumed to significantly reduce working memory capacity, not least by using capacity for 'worry work', that is, for non-task related self referential thoughts. (For more on this, see the discussion of stress induced inaction in Stokes & Kite, 1994). The decision making process is represented by the bold outline boxes. Going from left to right, the pilot 'samples' the world, registering instrument readings, ATC calls, the external scene, and so on. The 'picture' of the situation (equivalent to the chess master's board layout) is then matched to a situational 'schema' or template in long term memory. Most of the time an adequate match is found. With it comes at least one workable option (or, at least, one that has worked before), as well as the procedures to implement it. Part of the decision making process involves setting priorities; thus, the pilot slots the chosen procedures into an appropriate place in his flight management sequence and executes the decision.

In the model represented by the diagram, the long term memory strategy (RPD) is the 'default' strategy—the routine way of doing things. Pilots resort to the analytical approach (with its heavy call on working or short-term memory) only when the 'pattern matching' strategy fails or cannot be used. This, in turn, occurs only when an appropriate schema is not evoked by the environmental cues.

It is easy to see how failure to 'pattern match' may be due to the inexperience of the pilot. Novice aviators, like novice chess players, have had fewer hours 'on the job', and simply have a smaller stock of situational schemata to play with. They are, therefore, more often forced to drop down into slow, inefficient, and stress-prone analytical processing. This is why we try to help out beginners with checklists, catch-phrases, mnemonics, etc. The model predicts that, other things being equal, novice flight decision making should be particularly badly affected by stress. A flight simulation study reported by Stokes, Belger, and Zhang (1990) demonstrated that this does in fact occur, even though the novices and experts were evenly matched by cognitive testing under stress, and by trait anxiety measures.

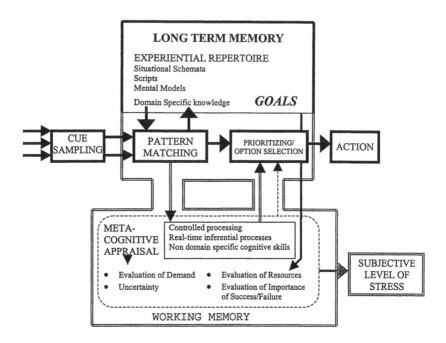

Figure 4.2   A model of decision making under stress (from Stokes and Kite, 1994)

Even where the pilot is very experienced, the environmental cues may not 'pop up' or trigger an appropriate schema from long-term memory because of the unique features of the circumstances. For example, in 1979 the crew of an American Airlines DC-10 was faced with the complete loss of power on the left engine at take-off. What had really happened was the physical loss of the whole engine and nacelle, which had rolled up and over the left wing. This had, in turn, ripped out the hydraulics to a leading edge high lift device, causing it to retract. Nothing in the crew's experience—certainly, no simulated engine failure emergencies—had prepared the crew for such an eventuality. They followed normal 'engine out' procedures. When the aircraft began to roll left uncontrollably on climbout, the crew were taken by surprise. With sufficient time any crew could probably have 'figured out' what had to have happened. Unhappily, the length of time that it takes to run analytical diagnostics in working memory was more than the time it took for the aircraft to sustain the power loss, stall the left wing, and hit the ground.

The Catch-22 of Stress and Decision Making

In evaluating the circumstances in which pilots might drop into a slow analytic decision making process, one element is striking, whether the discussion is of novices 'out of their depth' in ordinary circumstances, or experienced fliers faced with extraordinary circumstances. That is, the conditions are often precisely those calculated to maximise the mismatch between perceived demand and perceived resources to cope, where failure really matters and uncertainty is high: in short, conditions of stress as defined in the transactional model described in the first sections of this chapter. It is an ironic 'Catch 22' that the very circumstances that tend to favour a switch to intensive analytical processes in working memory are also those that reduce the capacity of working memory and the efficiency of processes in it. In stress, then, an inefficient decision mechanism (time hungry option comparison) is rendered yet more inefficient.

Stress and Decision Performance

Taken together, these considerations strongly suggest that the apparent 'stress resistance' often ascribed to experienced pilots may come from their ability to utilise recognition primed decision making strategies, and to bypass the computational strategies stipulated in (stress-prone) classical decision making models. This hypothesis has, in fact, been investigated in a series of studies beginning in 1988. In the first study, novice instrument pilots flew a simulated flight either in a nonstressed condition or with several (putative) stress manipulations, including time pressure, a simultaneous secondary task, the risk of penalty, and irritating noise. These manipulations did seem to provoke significantly poorer flight decision making, but only for decisions involving visuo-spatial operations (very memory hogging processes, as those who try to store and manipulate computer graphics images—especially 3-D images—will appreciate). Decisions depending heavily on the simple retrieval of textbook knowledge from long term memory, although not (in these particular novices) very good to begin with, were no worse in the stress condition (Wickens, Stokes, Barnett, & Hyman, 1988). So the results of the study were certainly consistent with the 'bypass' hypothesis.

However, this study did suffer from a number of methodological limitations. First, stress effects were inferred from decision performance on problems rated subjectively (albeit by pooled expert ratings). While it is relatively easy to rate a problem as being high or low in demand for spatial skills or declarative knowledge, it is much harder to rate the working memory demand of a problem scenario. Thus, these ratings may simply have been inaccurate. Alternatively, the decrement in decision performance observed for spatial problems may have been partially due to the visual/verbal (screen displayed) nature of the workload increasing secondary task. Since most of the spatial problems involved instrument scanning, it is possible that these were disrupted more by the need to visually time share with the secondary task than by a reduction in working memory capacity. On the other hand, very similar stress manipulations have been shown to give rise to significant decrements in performance on working memory and spatial tasks as tested in an aviation relevant test battery (Stokes & Raby, 1989). Moreover, in this latter study, declarative knowledge scores were not significantly different in the stress and nonstress conditions, consistent with the findings of the simulation experiment.

A second stress study was accordingly designed which, in addition to replicating the first, included a number of modifications intended to permit a clearer test of the hypothesis that the source of stress resistance in expert aviators is indeed 'recognition primed decision making' (Stokes, Belger, & Zhang, 1990). For example, the earlier work had only inferred stress effects upon putative components of decision making (such as working memory and spatial ability) from decision making performance on specific problem types (memory problems, spatial problems, etc.). This placed a great burden upon the accurate content analysis of the problems. In the second experiment, however, stress effects were measured directly and outside the criterion task. This was done by utilising the same stress manipulations used in the flight simulation during administration of the cognitive test battery. In fact, stress effects on the performance of knowledge representation tasks were also measured separately. Also, the study used a within-subjects design and two different simulated flights (stressed and nonstressed, in a counterbalanced design). This eliminated the difficult task of trying to match pairs of pilots +by skill and flight hours in order to run stressed and unstressed groups on a single criterion flight. Third, the stress manipulations used in this study are believed to adhere more closely to a cognitive appraisal model of stress, and involved a closer replication of noise stress conditions that have been found to simulate anxiety in their effects upon performance (Hockey, 1986). Last, the study contrasted novice and expert IFR pilots, as in the initial experiment in the series, rather than observing a relatively homogeneous set of low time pilots. It was anticipated that these changes in the research design would permit the experimenters to observe the extent to which stress gave rise to (domain independent) cognitive performance decrements in experts and novices, as well as the extent to which these effects were either reproduced or resisted by experts and novices in the simulated operational setting.

In the experiment, thirty instrument pilots flew two simulated IFR cross-country flights (one stressed and one unstressed). The battery of automated tests was also administered under stress and control conditions, providing data on the effects of the stress manipulation upon the putative cognitive components of decision making, independent of the criterion task. The results were dramatic. They clearly showed significant decrements in performance under stress for both experienced and inexperienced pilots in non-aviation specific information processing tasks. These decrements were of the same magnitude for experts and novices, and involved the same tasks. However, as hypothesised, this was *not* associated with any performance decrement in simulated flight decision making by experienced pilots. Although decision quality was about equal for the two groups in the control condition, only the novice pilots made significantly poorer decisions under stress, and only on "dynamic" scenarios (problems involving attention to moving display indicators). This not only comports with earlier results showing the vulnerability of spatial processes to stress, but is consistent with the view that novice pilot performance is more dependent than expert performance upon an analytic decision making strategy that is particularly susceptible to disruption by stress.

Perhaps more significant than this, it was found that high anxiety was associated with markedly worsened decision performance only in novice aviators exposed to acute stress. No impairment of flight decision performance was observed in experts exposed to acute stress. This is all the more interesting insofar as the trait anxious group within the experts actually exhibited higher trait anxiety scores than the trait anxious group within the novices. Domain specific knowledge, the researchers concluded, may not only provide a 'stress inoculation', but may be a more important variable in resistance to acute stress than personality—at least insofar as trait anxiousness is concerned. Put another way, the hypothesis that trait anxiety predisposes individuals to respond to stressful circumstances with more marked state anxiety was found to be true for the low time aviators in the study, but untrue for the high time aviators.

The third study in this series investigated the relationship of cue recognition and hypothesis generation to proficient aeronautical decision making in the presence of one particular stressor, time pressure (Stokes, Kemper, & Marsh, 1992). As in the preceding studies, low and high time pilots were administered tests both of general (domain independent) information processing ability and of experience related (domain specific) cognitive schemata. The flight simulation portion of the study differed, however, in that the problem scenarios permitted open ended (as opposed to multiple choice) responses. In addition, subjects were required to identify the cues underpinning their decisions and to list all viable decision alternatives. Time limits were also imposed.

Results showed that under time pressure, the experienced pilots detected significantly more relevant cues than did their junior colleagues. They also chose to execute their first alternatives significantly more often, suggesting that their decision making was also more efficient. Pilot's certification level was easily the most powerful predictor of decision performance. (Total flight hours, in contrast, had no predictive value.) Among the psychometric variables, the best predictors of decision making proficiency included spatial memory skills and the knowledge representation test measures.

Further support for this model of analytical versus experience based decision making comes from Cohen (1993), who conducted a study of decision strategies adopted by a group of airline pilots. The experiment was a paper and pencil exercise involving a series of inflight problem scenarios. In each scenario a flight was defined, followed by the announcement of unexpected weather problems requiring a diversion/no diversion decision to be made. The weather problem consisted of a fog bank that might obscure both destination and alternative, one but not the other, or neither. As an additional variable, dispatch recommended continuation in half the cases and diversion in the other half.

Examination of the strategies adopted by the pilots revealed three groups: risk takers, less experienced non-risk takers, and experienced non-risk takers ('experience' being defined as having twenty years of experience or more). Only the latter pilots took the dispatch advice seriously. While not following it slavishly, they did tend to begin their decision making from the dispatch recommendation—utilising a 'provisional acceptance' and attempted rebuttal approach.

Cohen (1993) notes that the decision process observed in these pilots was at variance with the analytical model in several respects. First, the options were evaluated sequentially, rather than being compared concurrently. Second, pilots analysed potential outcomes selectively, not exhaustively, concentrating on the worst case and the expected case. Third, potential courses of action were evaluated using an 'accept and critique' strategy, instead of by reference to quantitative weightings, as attached to decision alternatives in the analytical model.

Cohen makes the important observation that while the decision making strategies of experienced and inexperienced pilots may indeed be qualitatively different, this is not merely the result of the repertoire of situational experiences ("specialised recognitional templates"—p. 247) that experience brings. Rather, experience in a particular domain (for example, aviation) also brings with it characteristic strategies for using that information. These could be described as metacognitive skills which only evolve over extended periods of time. This insight may have a number of important implications for pilot training, a subject to which we return later in this chapter.

Operational Implications

Training Pilots to Make Good Decisions

In an earlier section of this chapter we considered the nature of human decision making (the real vs. the ideal), and reviewed several studies of decision making in pilots. Clearly, in the light of these and other studies not reviewed here, there remains a lot of research work to be done, especially on the domain specific skills that evolve over time, and the specialised reasoning strategies appropriate to differing contexts. From the operational and training point of view, however, there is already sufficient evidence to reconsider many of the traditional ways of proceeding. The value of procedural decision aids and acronymic decision checklists needs to be reconsidered. They are prescriptive, not descriptive, and the prescription may worsen the disease, making decision making yet more wooden, slow and non-intuitive, and even more dependent upon fragile memory processes which may fail only too readily under stress. Recall the (doubtless apocryphal) tale of the paratrooper briefed to cry 'Geronimo' after exiting the aircraft, in order to permit the correct time to elapse before pulling his ripcord. The soldier is found crumpled on the ground, parachute still packed, and down to his last breath still mumbling "what was the name of that --------- Indian".

Rather than relying on catch-phrase mnemonics, training might benefit far more from attempts to build the repertoire of 'game states' which characterises expert's knowledge of a domain. This might be done by packaging vicarious experience, as it were, into Event-Based Simulations (Stokes, Kemper, and Kite, 1997). These could take the form of desktop (or laptop) PC flight simulators in which the input is via decision rather than stick, rudder and switches. The role of the PC is to present events in context, and let the trainee decide her way through the flight, as in the experimental MIDIS system used in many of the experiments discussed above.

It is important that the events are presented in such a way that the correct classification and categorisation of them in the trainee's mind is facilitated, together with the high value (i.e., appropriate) response. Feedback can be immediate and comprehensive, or delayed and compact. What works best is an empirical matter, open to study. Stress need not be imposed in order to train stress-resistance. As discussed above, knowledge, organised as a readily accessible recognitional repertoire, complete with ready-to-hand responses, *is* the stress resistance factor that we wish to maximise.

Note too, that the simple expedient of presenting a flight in its phase by phase sequence, as in normal simulation, is not necessarily the best way to implement EBS. The point is to build the expert's knowledge structure efficiently, not by merely replicating the hit-or-miss, uncategorised sequence of events in the real world. The expert's knowledge representation is a product of costly, time-consuming iterative re-evaluation and re-classification over the years.

Coping with Stress

Many methods exist for coping with stress and anxiety, from discussing one's problems with a psychologist, to taking up meditation, drinking alcohol or changing one's career. Some approaches focus on external, environmental 'stressors'—that is, on altering conditions that are felt to cause difficulty. (The career change would be an example of this.) Other approaches focus more on altering internal, physiological responses to stressful conditions. Meditation and other relaxation techniques come under this category; so does the use of alcohol, nicotine, and other drugs. Yet another way of coping with stress is to change one's perspective—i.e., to find new ways of viewing one's situation. This is what people are often attempting to do when they consult a psychotherapist.

Although all of these varied approaches may be useful at times, what has often been lacking is a shared conceptual underpinning, a coherent basis common to all. However, Cox (1978) has suggested that the transactional model of stress can be used to form the basis of a systematic approach to stress reduction. If, as we have discussed, stress arises when an individual perceives a mismatch between the demands of a situation and his ability to cope, then the possible avenues for stress reduction are several. One approach is to address the transaction itself, either by reducing situational demands or by increasing coping ability. It is also sometimes possible to focus not so much on the objective circumstances as on one's appraisal of them, i.e., by reducing perceived demand or increasing perceived coping ability. Other aspects of cognitive appraisal may also be amenable to intervention, including perceived uncertainty and perceived importance of success. Yet another type of approach (which is probably the one many people think of first) is to intervene directly in physiological 'stress' responses, either with drugs or with self-help relaxation techniques. Which of these approaches is the most feasible or appropriate for a given situation depends, of course, on many factors; below, we discuss some issues relevant to general aviation.

Stress antidote no. 1: reducing demand What, for a pilot, constitutes a stressful situation? Most would probably answer this question in terms of inflight emergencies of one sort or another: equipment failure, perhaps, or deteriorating weather conditions, or navigational difficulties (being 'temporarily unsure of one's position', as the preferred wording goes). The ASRS database of incident reports certainly contains many accounts of stress associated with factors such as these. However, pilots may also experience stress from domestic problems, health worries, financial difficulties, job-related anxieties, and clashes with coworkers—in short, from the same 'slings and arrows' that bedevil the rest of the population. Nor are pilots immune from the effects of these difficulties when they step into an aircraft. In some kinds of circumstances, the most obvious and effective antidote is prevention, that is, recognising situations likely to be stressful and taking steps to prevent or avoid them. For example, inflight equipment failures may not be completely preventable, but the likelihood of one occurring can be reduced by completing the proper preflight checks and paying attention to the aircraft's maintenance log. By the same token, it is usually a relatively simple matter to choose not to fly in marginal weather conditions.

Stress antidote no. 2: increasing coping ability How can a pilot (or any person) increase his or her ability to cope with stress? The average member of the public would probably maintain that some people are naturally 'stress resistant', while others are not. As we shall see, this idea does have merit: certain personality traits do appear to be associated with more positive responses to adversity. However, any pilot, regardless of personality makeup, can increase his coping abilities by obtaining additional flight training and experience. Even on a day-to-day basis, however, there are a number of measures which any pilot can adopt to maximise performance. These include reviewing unfamiliar procedures before flying, and mentally rehearsing difficult manoeuvres. Adequate rest and regular meals also come under this category.

Stress antidote no. 3: reducing perceived demand There are many circumstances in which it is more feasible, effective, or appropriate to address the pilot's cognitive appraisals than to change the situation itself. First, some stressful situations cannot easily be avoided or predicted ahead of time. Second, cognitive appraisals can sometimes be inaccurate. Yet a third limitation to the avoid-the-stressor approach is that in some instances, removing the apparent 'stressor' may fail to eliminate stress. Consider the pilot quoted at the beginning of the chapter, who removed a few wasps from her aircraft but apparently lived in fear of encountering more in the future. Indeed, many individuals find it difficult to overcome the effects of frightening past experiences. The condition known as post-traumatic stress disorder was first identified in connection with combat veterans, but has been generalised to many other kinds of situations in recent years. When a source of threat has been removed but anxiety nevertheless persists, a more effective approach to stress management may be to focus on one's appraisal of the situation.

If, as is sometimes the case, the individual perceives a situation as being more threatening or demanding than it actually is, she may be able to reduce her stress or anxiety by adopting a more realistic view of the demands or risks involved. We have seen how this approach might work with a parachute jumper; it is also a common coping mechanism for student pilots. For many students, for example, stalls and spins represent one of the more nerve-wracking aspects of flight training, involving as they do manoeuvres quite alien to our everyday experience, and which our instincts would normally have us avoid. One way of reducing this visceral anxiety is through education: a student who understands the aerodynamics involved and the capabilities of the aircraft is more likely to perceive stalls and spins as 'do-able'. Yet another type of individual who may benefit from this perspective is the more experienced pilot who suddenly develops a disabling fear of flying. The problem is not that aviation has become more hazardous, but that his appraisal of it has changed. Judging from the aviation medicine literature, this phenomenon is not altogether uncommon. It may be precipitated by an aviation-related trauma (for example, an accident occurring to a friend), or by a personal difficulty of some kind. Professional pilots may be referred to a flight surgeon or psychiatrist; recreational GA pilots are more likely to be left to their own devices.

Stress antidote no. 4: increasing perceived coping ability As regards perceived coping ability, a more familiar synonym is 'self-confidence'. Often, of course, this overlaps with 'actual' coping ability: an effective way to feel competent is to be competent. However, the two are not identical. Consider a student pilot about to make her first solo flight. She knows all of the necessary procedures and has successfully carried them out in her instructor's presence. Even so, she is feeling somewhat intimidated by the prospect of flying alone. She might, in this instance, remind herself that since her instructor has signed her off for the flight, it is reasonable to suppose that she is capable of performing it.

There are some indications that an individual's perception of his ability to cope with external demands may actually relate to certain stable personality characteristics. One such characteristic has been labelled 'locus of control' (Rotter, 1966). Individuals with an internal locus of control have a general belief in their ability to influence events; they see themselves as 'making their own luck'. People with an external locus of control are more fatalistic, and tend to attribute events to chance or to the actions of others. They are also more likely to show feelings of resignation when confronted with difficult circumstances.

Closely related to locus of control is another personality variable known as 'self efficacy' (Bandura, 1977). It, too, refers to a belief in one's ability to influence events, but differs from locus of control in being more situation specific. Thus, a pilot might feel highly self efficacious when operating an aircraft, but less efficacious when, for example, relating to the opposite sex. The relevance of this concept to a cognitive appraisal view of stress should be obvious:

It seems sensible to assume that persons who believe that they can master most demands and threats by doing what is needed or by discovering what to do and how to do it are less likely to be threatened or to feel helpless or hopeless in stressful transactions. The obverse of this is the chronically anxious person who can be thought of as someone who maintains a general belief both that the environment is hostile, and that he or she is incapable of mastery. Such persons are anxious even in situations in which the ordinary person does not experience threat because the very act of engaging the environment carries with it the implication of danger (Folkman, Schaefer, & Lazarus, 1979, p. 286).

An internal locus of control and a strong sense of self efficacy are clearly advantageous qualities for a pilot to have. An important question, then, concerns the stability of these characteristics. Are they fixed and unchangeable personality 'traits', or are they merely 'attitudes' which can be modified? It would certainly be useful to know whether a pilot with a highly externalised locus of control can be taught to modify his world view, and if so, how. An alternative possibility might be to make pilots aware of maladaptive cognitive appraisal patterns, and, perhaps, to suggest compensatory strategies.

This having been said, it must be pointed out that in the case of perceived coping ability, more is not always better! (Some teenaged flight students, in particular, need to be introduced to the idea that they are not, in fact, immortal.) However, the problem of over-confidence is hardly confined to adolescents. Cohn (1994) relates a remarkable tale in which an alcoholic, unlicensed student pilot and three of his friends fraudulently rented a Cessna Skywagon, loaded it with beer and shotguns and, amazingly, proceeded to shoot at birds from the aircraft during flight. The Cessna crashed when a shotgun pellet hit one of the wings and tore it open. Clearly, it is possible to be over-confident, and a positive appraisal of one's coping skills is only advantageous if it bears some relationship to reality.

Stress antidote no. 5: Decreasing the importance of success This could also be termed 'decreasing the cost of failure'. Although the most extreme type of failure within an aviation context is of course the loss of the flight, this is not the only type of threat that pilots may face. This is certainly true for pilots involved in emergency medical evacuation (medevac) services, whose job is to airlift critically ill or injured persons to hospitals. Although television portrayals of such services tend to stress the heroism of the pilot who goes to extraordinary lengths to deliver the patient in time for life-saving surgery, real-life medevac pilots are often given no medical or personal details about their passengers, precisely to reduce their sense of urgency and to prevent inappropriate risk-taking. What is being manipulated, in other words, is the pilot's perception of the importance of success. For a more everyday example consider that well-known pilot's syndrome, 'get-home-itis'. This is the overriding desire to get to one's planned destination, come what may and whatever the risk. In such circumstances it may be helpful for the pilot to adopt a more relaxed perspective about the importance of being 'on time', and to remember the old motto, "better to be late in this world than early in the next". It can also be helpful to form contingency plans or otherwise arrange matters so that unplanned delays do not cause serious inconvenience.

Social and other 'life related' factors may also play a role in 'upping the ante'. Flight checkrides again provide a useful case in point. The authors have elsewhere related the story of a certain university flight student who stood to lose, if she failed one more checkride, not only her place in the flight training programme, but also her standing in the university, and, by extension, the lease on her apartment—not to mention her chosen career (Stokes & Kite, 1994). One way of coping with this kind of threat is to try to 'mentally downplay' the importance of passing (at least during the flight)—a sort of positive self deception.

This leads to the subject of defence mechanisms, which can be defined as mental compromises and other processes, often unconscious, that allow individuals to make positive cognitive appraisals in situations where they otherwise would not. The concept has its origins in the psychoanalytic literature, which describes a wide variety of mental manoeuvres that people utilise in order to avoid psychological distress. Three in particular are relevant to the present discussion: repression, denial, and rationalisation. Repression refers to the banishing of threatening emotions and memories from conscious awareness; thus, a flight student might "forget" an angry reprimand from an instructor. Denial is similar but involves the exclusion of external rather than internal stimuli, so that the threatening situation apparently does not exist; in this instance the student might not hear the instructor's words at all. Rationalisation is a less extreme defence in which the individual creates reassuring but false explanations for events that would otherwise be emotionally threatening. ("He's not really angry at me; he's just having a bad day.")

Defence mechanisms can certainly be maladaptive and even dangerous, particularly if they prevent the pilot from taking constructive action. David Beaty (1991) gives a horrific example of this in his account of the 1980 crash of a Lockheed TriStar in Saudi Arabia, in which the cabin caught fire shortly after takeoff (apparently one of the passengers had lit a gas stove on board). The captain turned back but, disregarding both the smoke alarm and repeated entreaties from the crew, gave orders *not* to evacuate. Eventually the ground crew managed to open the doors, but not before all 301 people on board had perished. From a medical perspective, too, defence mechanisms are often considered to be maladaptive, because they do not address underlying causes of anxiety (see, e.g., Spielberger, 1979). However, we would point out that if a source of threat cannot be avoided in any case, a useful coping strategy may be to close one's mind to it. In some highly stressful situations, such as military combat, this may be the only way that many individuals can cope at all. Even in less extreme conditions, however, people who can disregard certain aspects of their reality may be able to function more effectively. Thus the student described earlier might pretend to herself that this flight is not a checkride at all but an ordinary lesson; alternatively, she could persuade herself (temporarily) that even if she fails, all will be well nevertheless.

Actually, in many respects, a thorough and realistic accounting of all the real risks life offers is actually associated more with psychological depression than with healthy functioning. Indeed, it has been suggested that clinical patients suffering from psychological depression may actually have a more accurate (if pessimistic) worldview than nondepressed individuals. This phenomenon is known as depressive realism (Alloy & Abramson, 1988), and seems to indicate that a 'healthy' degree of self delusion is an essential component of normal psychological functioning. From a Darwinian perspective, too, it might be noted that a blanket dismissal of defence mechanisms reflects an older, prescientific conception of the origin and function of the human mind. Although the poets may proclaim us noble in reason, infinite in faculties, and godlike in apprehension, humans were not designed by nature to be abstract thinking machines dedicated to the search for objective truth. Rather, human cognitive processes, like those of other species, have been shaped by natural selection to aid and promote adaptive behaviours—i.e., behaviours which, over evolutionary millennia, have been associated with successful reproduction. A predisposition to look on the sunny side of the street appears to have been sanctioned by evolution. Good stress coping seems to tap into this resource.

Stress antidote no. 6: decreasing uncertainty Uncertainty, in the context of transactional models of stress, is usually understood to refer to uncertainty about the outcome of one's efforts, that is, whether one will succeed or fail. We observed earlier that a student who *might* pass an examination may actually experience more stress than a student who definitely will not. However, there is also another type of uncertainty, which concerns the consequences of failure. This may be easier to manipulate than the first type of uncertainty. For example, the student we cited earlier, who faced dire consequences if she failed her checkride yet again, might try forming some contingency plans in the event things go badly once more. She may not be able to predict whether she will pass, but she can reduce her uncertainty about what will happen if she does not.

A more useful distinction, perhaps, might be between short-term and long-term appraisals, 'defensive' or otherwise. The captain of the ill-fated Lockheed TriStar, for example, may well have benefited *in the short term* from his appraisal of the cabin fire as relatively non-threatening. That is, as far as can be determined he was able to turn back and land the aircraft without being unduly distracted by anxiety. Once on the ground, however, this mellow attitude did nothing to improve his performance, and he, his crew, and passengers would all have been better off if he had, at that point, reappraised the situation as more threatening than previously. In general aviation, too, shutting one's mind to risk or threat is clearly more useful at some times than at others. We have already suggested that a student pilot worried about her checkride might, during the ride itself, pretend that it was an ordinary flight; however, during the days and weeks leading up to the test, a certain degree of anxiety would more likely motivate her to be well prepared. Indeed, a recent study of students in a university flight school showed that trait-anxious individuals were more likely to succeed than their more blasé counterparts (Stokes & Bohan, 1995).

Cognitive vs. Noncognitive Appraisal: Should we Listen to our Emotions?

At the beginning of this chapter we made the point that human emotions should not be dismissed as awkward, embarrassing, and unhelpful substitutes for 'sweet reason'. They are not additions to be circumvented, controlled, suppressed, or ignored. Far from being the untamed passions so beloved of the Romantic poets and despised by Mssrs. Spock and Data of Star Trek fame, emotions are an important signal system without which our decision making would be seriously damaged. Indeed, experiments with 'Spock-like' frontal-lobe-damaged patients have demonstrated this very point (Bechara, Damasio, Tranel, & Damasio, 1997). What is more, counter to popular conceptions, our emotional system recognises threats and prompts us to respond to them *before* the higher cortex of the brain has even perceived the nature of the threat.

A Civil Air Patrol pilot once told a group of emotion researchers that sometimes his commanding officer would call him up to conduct a search mission. The trouble was, he explained, that he sometimes had a 'bad feeling' about the proposed flight, but couldn't always set out his reasons to his superior's satisfaction. The pilot wanted to know how to make his feelings explicit, and whether he should fly in these situations. As one, the emotion researchers (who were pilots too) cried "No! Go with your gut!" They pointed out that the limbic system—a brain region associated with emotional processing—may well be responding to information that the cortex and its speech centres do not have access to. Where a life-or-death, go/no-go decision is concerned, the limbic system should *sometimes* be given the benefit of the doubt.

We say sometimes, of course, because not all emotional reactions are, in every case, reliable indicators of the actual state of affairs. For example, for most beginning parachute jumpers, the act of stepping out of an aircraft into empty space requires a concerted mental effort not to listen to one's emotions. The pilot cited at the beginning of this chapter, who was worried about wasps, might also benefit from learning to react more calmly to that particular stimulus.

An additional factor to consider, in connection with this point, is that humans did not evolve in an environment that included cockpit instrument panels and air traffic controllers. Some features of the stress response, e.g., attentional tunneling, may have had a highly adaptive function in the hunter-gatherer society that moulded our bodies and minds over the course of hundreds of thousands of years. However, what may have been good for keeping a Flintstonian eye on the ball (or sabre-toothed tiger) may not be adaptive on the flight deck, where the correct response to threat may require attention to a broad range of cues.

Research Recommendations

Naturalistic vs. classical decision strategies Research in this area has thus far been primarily descriptive rather than prescriptive. A number of researchers, including Wickens et al. (1992) have identified situations in which one or the other type of decision making is likely to occur, but much less is known about when particular decision strategies should, optimally, occur.

Generating and evaluating decision options There is some evidence to indicate that more experienced decision makers generate higher-quality options at the outset of the decision process than do inexperienced individuals. Less is known; however, about the later options that experts generate (if any): are these less optimal, or should they be given equal consideration? Answers to this question would help to indicate whether satisficing strategies are, in fact, associated with optimal decision making. Also of interest is the question of how expert and novice decision makers decide which, among generated options, are the most desirable, whether differences exist between these groups, and what implications exist for training.

Identifying errors and judging outcomes One potential pitfall associated with pattern-matching decision strategies is the possibility that previous experiences could, conceivably, unduly bias current expectations. That is, if a situation appears to match others recalled from long term memory, to what extent might the pilot be less receptive to information that contradicts this expectation? To our knowledge, this question has yet to be addressed within the context of aeronautical decision making.

Stressors Although some interesting findings exist concerning the effects of 'stress' (variously defined) on classical vs. naturalistic decision strategies, little is currently known about the specific effects of different sources of stress. How, for example, might recognition primed decision making be affected by time pressure, and does the presence of factors such as time pressure influence the choice of decision strategies?

Stress effects on LTM A great deal of research has evaluated the effects of stress on short term memory. Relatively little has examined the effect of stress upon long-term memory related processes such as the retrieval of pattern-matched information, analogue situations, etc. How robust is RPD under real threat for the individual?

Training Many of the findings described here, particularly those concerning expert vs. novice decision making strategies, have powerful implications for pilot training. If, as these findings suggest, the observed 'stress resistance' of experienced pilots derives in part from recognition-based decision strategies, it becomes important to consider how these pilots' experiential repertoires can most effectively be conveyed to their junior colleagues. Traditionally, expertise has been 'measured' primarily in terms of flight hours, and the accumulation of varied flying experiences within these hours has been somewhat of a random matter. However, as we have suggested elsewhere (Stokes, Kemper, & Kite, 1997), a case can be made for considering expertise in terms of the quality (as opposed to the quantity) of experience. Flight instruction curricula could be adapted in a number of ways to reflect this, in ground school but especially in simulator training. Event based simulators (EBS) can be structured in a manner that explicitly adds to students' experiential repertoires. Unlike conventional simulators, with their emphasis on stick-and-rudder skills and use of cockpit instrumentation, these decision trainers would not necessarily need either high fidelity or conventional controls, and thus could be implemented at relatively little expense.

References

Alloy, J. R., & Abramson, L. (1988). Depressive realism: Four theoretical perspectives. In L. Alloy (Ed.), *Cognitive processes in depression* (pp. 223-265). New York: Guilford Press.

Baddeley, A. D. (1972). Selective attention and performance in dangerous environments. *British Journal of Psychology, 63,* 537-546.

Bandura, A. (1977). Self-efficacy: Toward a unifying theory of behavior change. *Psychological Review, 84,* 191-215.

Barthol, R. P., & Ku, N. D. (1959). Regression under stress to first learned behavior. *Journal of Abnormal and Social Psychology, 59,* 134-136.

Beaty, D. (1991). The worst six accidents over the last thirty years from a human factors point of view. *Flight Deck, 2,* 19-25.

Bechara, A., Damasio, H., Tranel, D., & Damasio, A. R. (1997). Deciding advantageously before knowing the advantageous strategy. *Science, 275*(5304), 1293-1294.

Broadbent, D. E. (1978). *Decision and stress,* New York: Academic Press.

Campbell, R. D., & Bagshaw, M. (1991). *Human performance limitations in aviation.* Oxford: BSP Professional Books.

Calderwood, R., Klein, G. A., & Crandall, B. W. (1988). Time pressure, skill and move quality in chess. *American Journal of Psychology, 101,* 481-493.

Cannon, W. B. (1915). *Bodily changes in pain, hunger, fear, and rage.* New York: Appleton.

Chase, W., & Simon, H. (1973). Perception in chess. *Cognitive Psychology, 4,* 55-81.

Cohen, M. S. (1993). Taking risks and taking advice: The role of experience in airline pilot diversion decisions. In R. Jensen (Ed.), *Proceedings of the 7th International Symposium on Aviation Psychology* (pp. 244-247). Columbus OH: The Ohio State University.

Cox, T. (1978). *Stress.* London: Macmillan.

Cohn, R. L. (1994). *They called it pilot error.* New York: McGraw-Hill.

Damasio, A. (1994). *Decartes' error: Emotion, reason, and the human brain.* New York: G. P. Putnam.

Daneman, M., & Carpenter, P. (1980). Individual differences in working memory and reading. *Journal of Verbal Learning and Verbal Behaviour, 19,* 450-456.

Darke, S. (1988). Anxiety and working memory capacity. *Cognition and Emotion, 2,* 145-154.

Davies, D. R., & Parasuraman, R. (1982). *The psychology of vigilance.* London: Academic Press.

Eysenck, M. W. (1979). Anxiety, learning, and memory: A reconceptualization. *Journal of Research in Personality, 13,* 363-385.

Eysenck, M. W. (1982). *Attention and arousal cognition and performance.* Berlin: Springer.

Fisher, S. (1983). Memory and search in loud noise. *Canadian Journal of Psychology, 37,* 439-449.

Folkman, S., Schaefer, C., & Lazarus, R. S. (1979). Cognitive processes as mediators of stress and coping. In V. Hamilton & D. M. Warburton (Eds.), *Human stress and cognition: An information processing approach* (pp. 265-300). Chichester: John Wiley and Sons.

Fodor, J. A. (1983). *The modularity of mind: An essay in faculty psychology,* Cambridge, Massachusetts: MIT Press.

Frazier, J. W., Repperger, D. W., Thoth, D. N., & Skowronski, V. D. (1982). Human tracking performance changes during combined -Gz and +Gy stress. *Aviation, Space, and Environmental Medicine, 53,* 435-439.

Garvey, W. D. (1957), *The effects of task induced stress on man-machine system performance* (NRL report no. 5015). Washington, DC.

Glass, D. C., & Singer, J. E. (1972). *Urban stress experiments on noise and social stressors.* New York: Academic Press.

Gliner, J. A., Horvath, S. M., & Mihevic, P. M. (1983). Carbon monoxide and human performance in a single and dual task methodology. *Aviation, Space, and Environmental Medicine, 54,* 714-717.

Griffiths, P. E. (1997). *What emotions really are: The problem of psychological categories.* Chicago: University of Chicago Press.

Grinker, R. R., & Spiegel, J. E. (1945). *Men under stress.* McGraw-Hill, New York.

Hammerton, M., & Tickner, A. H. (1969). An investigation of the effect of skill upon stressed performance. *Ergonomics, 12,* 851-855.

Hancock, P.A. (1982). Task categorization and the limits of human performance in extreme heat. *Aviation, Space, and Environmental Medicine, 53,* 778-784.

Hancock, P. (1986). Stress, information-flow, and adaptability in individuals and collective organizational systems. In O. Brown, Jr. & H. Hendrick (Eds.), *Human factors in organizational design and management, II* (2nd International Symposium on Organizational Design and Management). Amsterdam: North-Holland.

Hockey, G. R. J. (1970). Effects of loud noise on attentional selectivity. *Quarterly Journal of Experimental Psychology, 22,* 28-36.

Hockey, G. R. J. (1986). Changes in operator efficiency as a function of environmental stress, fatigue, and circadian rhythms. In K. R. Boff., L. Kaufman & J. P. Thomas (Eds.), *Handbook of perception and human performance,* Vol. 2. New York: Wiley.

Howell, G. E. (1984). *Task Influence in the Analytic Intuitive Approach to Decision Making.* Final Report. (Office of Naval Research contract N00014-C-001 Work Unit [NR197-074].) Houston, Texas: Rice University.

Hyman, F. (1985). Personal communication.

Janis, I. (1982). Decision-making under stress. In L. Goldberger and S. Breznitz (Eds.), *Handbook of stress: Theoretical and clinical aspects,* (pp. 69-80). New York: Free Press.

Keinan, G. (1986). Confidence expectation as a predictor of military performance under stress. In N. A. Milgram (Ed.), *Stress and coping in time of war: Generalizations from the Israeli experience.* New York: Brunner/Mazel.

Keinan, G. (1987). Decision making under stress scanning of alternatives under controllable and uncontrollable threats. *Journal of Personality and Social Psychology, 52,* 639-644.

Keinan, G., & Friedland, N. (1984). Dilemmas concerning the training of individuals for task performance under stress. *Journal of Human Stress, 10,* 185-190.

Keinan, G. Friedland, N., & Arad, L. (1991). Chunking and integration: Effects of stress on the structuring of information. *Cognition and Emotion, 5,* 133-145.

Keinan, G., Friedland, N., & Ben-Porath, Y. (1986). Decision making under stress: Scanning of alternatives under physical threat. *Acta Psychologica, 64,* 219-228.

Klein, G. A. (1989). Recognition-Primed Decisions. In W. Rouse (Ed.), *Advances in man-machine systems research, 5* (pp. 47-92). Greenwich, Connecticut: JAI Press, Inc.

Klein, G. A., Calderwood, R., & Clinton-Cirocco, A. (1986). Rapid decision making on the fire ground. *Proceedings of the Human Factors Society 30th Annual Meeting* (pp. 576-580). Santa Monica, CA: Human Factors Society.

Klein, G. A., Wolf, S., Militello, L., & Zsambok, C. (1995). Characteristics of skilled option generation in chess. *Organizational Behavior and Human Decision Processes, 62,* 63-69.

Kleitman, N. (1963). *Sleep and wakefulness,* University of Chicago Press, Chicago.

Krahenbuhl, G. S., Marett, J. R., & Reid, G. B. (1978). Task-specific simulator pretraining and in-flight stress of student pilots. *Aviation, Space, and Environmental Medicine, 49,* 107-110.

Johnston, N. (1991). Organizational factors in human factors accident investigation. In R. Jensen (Ed.), *Proceedings of the 6th International Symposium on Aviation Psychology* (pp. 668-673). Columbus, OH: The Ohio State University.

Lazarus, R. S., DeLongis, A., Folkman, S., & Gruen, R. (1985). Stress and adaptational outcomes: The problem of confounded measures. *American Psychologist, 40,* 770-779.

Lester, L. F., & Bombaci, D. H. (1984). The relationship between personality and irrational judgement in civil pilots. *Human Factors, 26,* 565-572.

Lofaro, R. J., Adams, R. J., & Adams, C. A. (1992). *Workshop on aeronautical decision making* (ADM), Federal Aviation Administration Research and Development Service (Report DOT/FAA/RD-92/14,1).

McGrath, J. E. (1976). Stress and behaviour in organizations. In M. D. Dunnette (Ed.), *Handbook of industrial and organizational psychology* (pp. 1351-1396). Chicago: Rand-McNally.

Melton, C. E., & Wicks, S. M. (1967). *In-flight physiological monitoring of student pilots* (FAA Office of Aviation Medicine Report AM-67-15).

Merriam-Webster's collegiate dictionary (10th ed.). (1993). Springfield, MA: Merriam-Webster.

Neiss, R. (1988). Reconceptualizing arousal: Psychological states in motor performance. *Psychological Bulletin, 103,* 345-366.

Rotter, J. B. (1966). Generalized expectancies for internal versus external control of reinforcement. *Psychological Monographs, 80* (Whole No. 609).

Rouse, W. B. (1978). Human problem solving performance in a fault diagnosis task. *IEEE Transactions on Systems, Man and Cybernetics, 4,* SMC-8, 258-71.

Ruiz, R., Legros, C., & Guell, A. (1990). Voice analysis to predict the psychological or physical state of the speaker. *Aviation, Space, and Environmental Medicine, 61,* 266-271.

Selye, H. (1956). *The stress of life.* New York: McGraw-Hill.

Shanteau, J. (1988). Psychological characteristics and strategies of expert decision makers. *Acta Psychologica, 68,* 203-215.

Simon, H. A. (1955). A behavioral model of rational choice. *Quarterly Journal of Economics, 69,* 99-118.

Sive, W. J., & Hattingh, J. (1991). The measurement of psychophysiological reactions of pilots to a stressor in a flight simulator. *Aviation, Space, and Environmental Medicine, 62,* 31-36.

Sperandio, J.-C. (1978). The regulation of working methods as a function of work-load among air traffic controllers *Ergonomics, 21,* 193-202.

Spielberger, C. (1979) *Understanding stress and anxiety.* London: Harper & Row.

Stokes, A. F. (1991). Flight management training and research using a microcomputer flight decision simulator. In R. Sadlowe (Ed.), *PC based instrument flight simulation a first collection of papers* (pp. 25-32). New York, American Society of Mechanical Engineers.

Stokes, A. F., Belger, A., & Zhang, K. (1990). *Investigation of factors comprising a model of pilot decision making: Part II anxiety and cognitive strategies in expert and novice aviators.* Savoy, IL: University of Illinois Aviation Research Laboratory.

Stokes, A. F., & Bohan, M. (1995). Academic proficiency, anxiety, and information-processing variables as predictors of success in university flight training. In R. Jensen (Ed.), *Proceedings of the 8th International Symposium on Aviation Psychology* (pp. 1107-1112). Columbus, OH: The Ohio State University.

Stokes, A. F., Kemper, K., & Kite, K. (1997). Aeronautical decision making, cue recognition, and expertise under time pressure. In C. E. Zsambok & G. A. Klein (Eds.), *Naturalistic decision making* (pp. 183-196). Mahwah, New Jersey: Lawrence Erlbaum Associates.

Stokes, A. F., Kemper, K., & Marsh, R. (1992), *Time-stressed flight decision making: A study of expert and novice aviators.* Savoy, IL: University of Illinois Aviation Research Laboratory.

Stokes, A. F., & Kite, K. (1994). *Flight stress: Stress, fatigue, and performance in aviation,* Aldershot: Ashgate.

Stokes, A. F., Pharmer, J. A., & Kite, K. (1997). Stress effects upon communication in distributed teams. *Proceedings of the IEEE Transactions on Systems, Man, and Cybernetics* (pp. 4171-4176). New York: IEEE.

Stokes, A. F., & Raby, M. (1989). Stress and cognitive performance in trainee pilots. *Proceedings of the Human Factors Society 33rd Annual Meeting* (pp. 883-887). Santa Monica, CA: Human Factors Society.

Symonds, Sir C. P. (1947). Use and abuse of the term flying stress. In *Air Ministry, psychological disorders in flying personnel of the Royal Air Force, investigated during the war, 1939-1945*. H.M.S.O., London. (Cited in Cox, 1978).

Tolkmitt, F. J., & Scherer, K. R. (1986). Effect of experimentally induced stress on vocal parameters. *Journal of Experimental Psychology, 12,* 302-313.

Vaernes, R. J., Warncke, M., Myhre, G., & Aakvaag, A. (1988). Stress and performance during a simulated flight in an F-16 simulator. In *AGARD Conference Proceedings No. 458, Human Behaviour in High Stress Situations in Aerospace Operations*. Neuilly-sur-Seine, France: NATO.

Wachtel, P. L. (1968). Anxiety, attention, and coping with threat. *Journal of Abnormal Psychology, 73,* 137-143.

Wanta, W., & Leggett, D. (1988). Hitting pay dirt: Capacity theory and sports announcers' use of clichés. *Journal of Communications, 38,* 82.

Weick, K. E. (1991). The vulnerable system: An analysis of the Tenerife air disaster. In P. J. Frost, L. F. Moore, M. R Louis, C. C. Lundberg, and J. Martin (Eds.), *Reframing organizational culture* (pp. 117-130). London: Sage Publications.

Welford, A. T. (1973). Stress and performance. *Ergonomics, 15,* 567-580.

Wickens, C. D, Stokes, A. F., Barnett, B., & Hyman, F. (1993). The effects of stress on pilot judgment in a MIDIS simulator. In O. Svenson & J. Maule (Eds.), *Time pressure and stress in human judgement and decision making* (pp. 271-292). New York: Plenum Press.

Williams, C. E., & Stevens, K. N. (1981). Vocal correlates of emotional states. In J. Darby (Ed.), *Speech evaluation in psychiatry* (pp. 221-240). New York: Grune & Stratton.

Zackay, D. and Wooler, S. (1984). Time pressure, training and decision effectiveness. *Ergonomics, 27,* 273-284.

Part 3
Instruction and Training

5 Integrating Human Factors Education in General Aviation: Issues and Teaching Strategies

Irene Henley, Prue Anderson and Mark Wiggins

Introduction

While crew resource management (CRM) and human factors training has flourished in the airline environment, the deliberate teaching of human factors in general aviation (where the foundation for a professional career in aviation is laid) has been sporadic or almost non-existent (Henley, 1995; Jensen et al., 1995). With additional research and the reassessment of the effectiveness of human factors training has come the realisation that the development of human factors knowledge, attitudes and skills and the teaching of the principles of CRM should not be left to the stage at which a pilot joins an airline. At that stage, it may be too late for some to change the habits and thinking patterns that have been learned in general aviation (Diehl, 1991a, 1991b; Jensen, 1995; Jensen et al., 1995; Johnston, 1993; Pinet & Enders, 1994). Consistent with this perception, Helmreich, Wiener and Kanki (1993) noted that "it seems counterproductive to delay the introduction of human factors training until crew members are experienced in line operations. A more useful approach should be to include basic instruction in human factors theory and applications as part of the curriculum for *ab initio* training" (p. 481). Ideally, these concepts and skills should be introduced early in flight training, reinforced during upgrade training, and reviewed during recurrent training (Diehl, 1991b; Pinet & Enders, 1994). The belief is that, as a result of earlier training, human factors skills, like the learning of any skill, will become fully developed after continued practice (P. M. Anderson, 1993).

At the Third ICAO Global Flight Safety and Human Factors Symposium held in Auckland in 1996, both researchers and practitioners once again stressed that human factors education must be integrated into all levels of flight training, including training in general aviation. More recently, civil aviation authorities in several countries have legislated that human factors must be included as part of the training syllabi for most aviation licences. However, if flight instructors and, more importantly, the trainers of flight instructors, are not adequately prepared for their role as professional educators, then the success of the implementation of human factors training in general aviation could be severely jeopardised.

The first part of this chapter raises some of the issues that need to be addressed in order to successfully implement human factors training in general aviation. Some of these issues include: the preparation of flight instructors; the need to approach human factors training as the development of skills and attitudes as well as knowledge; the importance of integrating human factors into the flight training sequences; and the need to explore different teaching approaches in order to help students develop human factors knowledge, attitudes and skills. The second part of the chapter outlines some teaching approaches and instructional strategies that have been used to successfully integrate human factors into basic flight instruction as well as into flight instructor education.

Issues Relating to the Implementation of Human Factors Training in General Aviation

Despite the requirement to include human factors in most flight training syllabi, little concerted action has been taken to effectively implement human factors training programmes in general aviation. In the past twenty years, for example, articles and books on pilot judgement training, aeronautical decision making, human factors and CRM have proliferated. However, very little of that material has been systematically incorporated into civilian *ab initio* flight training programmes. One of the major stumbling blocks to this process has undoubtedly been the lack of preparation of flight instructors.

At least two decades ago, Jensen and Benel (1977) stressed that because the primary responsibility for pilot judgement training rests with the flight instructor, a major part of the implementation of pilot judgement or human factors training and evaluation is the training of flight instructors. Jensen and Benel (1977) also claimed that unfortunately, "most instructors are not equipped to communicate this type of material effectively" (p. iv). The authors further recommended that future research programmes be designed to examine strategies to train instructors and evaluators in terms of pilot judgement and human factors.

The researchers who conducted the pilot judgement training validation projects in the United States, Canada and Australia in the early 1980s heeded Jensen and Benel's (1977) advice and provided training for the flight instructors who were involved in those research projects (Berlin et al., 1982; Buch, 1983; Buch & Diehl, 1984; Telfer & Ashman, 1986). Unfortunately, the training of all the general aviation flight instructors in human factors has not followed. A few manuals for flight instructors have been produced, however, strategies for training flight instructors to teach and evaluate the acquisition of pilot judgement and human factors knowledge and skills have not been forthcoming. It seems expected that flight instructors in general aviation will incorporate new material into their courses without themselves receiving additional training.

ICAO Circular 227—*Training of Operational Personnel in Human Factors* (1991) recommends the inclusion of crew resource management concepts into the flight instructor training syllabus, stressing that the training for the instructor rating should explicitly address issues relating to crew communication, coordination and management. However, according to current research into the quality of flight instructor training, these topics are usually not covered at all, or are poorly covered, in flight instructor courses (Henley, 1995). Furthermore, the majority of flight instructors have indicated that flight instructor training has, up to now, mostly failed to prepare them for their role as educators, and more specifically as educators of adult learners (Harris & Conrad, 1997; Henley, 1995). As a result, most flight instructors remain unprepared to teach human factors within the aviation environment. Nevertheless, the successful integration of human factors into flight training in general aviation is fundamentally dependent upon the knowledge and skills of the flight instructor within the human factors domain. Moreover, one of the main impediments is that the trainers of flight instructors have not been adequately prepared to teach human factors and CRM principles nor to teach trainee instructors how to teach (Henley, 1995).

The importance of flight instructors with regard to the implementation of human factors education and their influence on flight safety cannot be overstated. If flight instructors are not adequately trained, there are repercussions throughout the flight training system. However, there persists the viewpoint that if the requirements are changed periodically to include additional materials or new knowledge and skills, instructors will automatically incorporate these principles into their training syllabus and will intuitively teach any new concepts and skills effectively. However, as one senior instructor correctly pointed out: *"instructors cannot teach what they don't know"* (Henley, 1995).

The introduction of human factors throughout the various training syllabi would probably not pose a particularly significant problem, if flight instructors were taught how to research and prepare lesson plans effectively, and if they had received human factors training and had been exposed to a variety of teaching strategies during their flight instructor course. For the most part, however, flight instructors have been taught only to replicate another instructor's lesson; they have not been given the tools with which to effectively incorporate human factors into their teaching (Harris & Conrad, 1997; Henley, 1995).

Moreover, most flight instructors have never received pilot judgement training or decision making training as an integral part of their own training. They have been taught rules, procedures, systems, manoeuvres and techniques—the disciplines of flight. While they become familiar with the functioning of the aircraft and with the environment in which they fly, they rarely develop insight about themselves, the pilot (Henley, 1995). Flight instructors are not usually asked to consider why pilots make faulty decisions or flaunt rules and regulations; nor are they taught how to recognise and cope with stress or how to operate effectively as a team member. Therefore, unless flight instructors are provided with additional training in the process of teaching human factors and pilot judgement or aeronautical decision making to adult learners, the implementation of human factors in private and commercial pilot training is unlikely to be particularly successful.

If the quality of flight training is to improve and if the implementation of new programmes such as human factors and CRM is to be successful, flight instructors need to be thoroughly schooled in human factors skills and concepts.

Moreover, flight instructors need to be taught "how to teach" and need to be familiar with different teaching approaches more appropriate for adult learners. This will assist student pilots to develop practical human factors skills, such as stress coping and decision-making strategies, self-awareness and effective team skills in addition to piloting skills. While there has been a plethora of books and articles about human factors, very little has been produced on "how to teach" human factors within the aviation environment. Consequently, the first step in the implementation of human factors training will need to be the more thorough training of flight instructors in instructional principles and methods in addition to human factors.

A related issue is the need for human factors training to be context-bound instead of being treated as a theory subject only. Human factors training needs to stem from a practical application. Pilot training is typically divided into two distinct components: ground and flight training. The former involves the development of factual knowledge, or a declarative understanding of the principles of aviation operations, while the latter involves the development of applied, or procedural knowledge and skills. From a skill acquisition perspective, the aim is to facilitate the development of both the knowledge and the skills necessary to operate efficiently and effectively within a dynamic and uncertain operational environment.

Until now, human factors education has usually been a concept which is firmly based within ground training and has yet to be integrated within the flight training programme (a notable exception being the recent material produced by Transport Canada, 1996a, 1996b, 1996c). Therefore, while students may possess a sound theoretical understanding of the concepts associated with human factors in aviation, they may lack the skills necessary to implement human factors within the operational environment. Until this has been achieved, it is unlikely that human factors education will impact significantly upon the proportion of accidents and incidents for which human factors are a significant factor. Consequently, human factors should not be seen as a separate subject—it is an approach to flying (Transport Canada, 1996c). Moreover, instructors need to approach the teaching of human factors not only as the acquisition of knowledge but as the development of skills and the fostering of safety conscious attitudes.

In order to successfully introduce human factors skills, attitudes and knowledge into *ab initio* flight training, instructors will need to adopt a variety of teaching strategies. Human factors and CRM principles such as self-awareness, decision making strategies and team skills cannot be developed effectively using only the lecture-based approach or the "show and tell" approach that most flight instructors have been trained to use (Henley, 1995). Instructors need to become proficient in using a variety of teaching strategies, such as problem-based learning (PBL), cooperative/interactive learning, case studies and scenarios, simulation, small group dynamics, reflective journals, and computer-assisted learning, if they are to fulfil the new expectations of *ab initio* flight training (P.M. Anderson, 1993; Anderson & Henley, 1994, 1997; Henley & Wiggins, 1997; Jensen, 1995; Smith, 1993, 1995; Trollip, 1995; Wiggins, 1996; Wiggins & O'Hare, 1993; Wiggins, O'Hare, Jensen & Guilkey, 1997).

Three teaching strategies that have proven successful in the development of human factors. knowledge, attitudes and skills such as self-awareness, decision making strategies and team skills include the use of reflective journals, computer-assisted learning programmes, and problem-based learning (PBL. The following section describes the use of both reflective journals and PBL to develop team skills, self-awareness and self-directed learning skills. (For an overview of the application of computer-assisted learning programmes in aviation education, see chapters 7 and 8 in the present volume).

Teaching Strategies to Integrate Human Factors Training into *Ab Initio* Flight Training

Flight instructing is a demanding profession not only because instructors must impart the necessary theoretical knowledge and practical flying skills to their students in a dynamic teaching environment, but also because instructors must help students develop the knowledge, attitudes and skills which will form the foundation for their future training in human factors. and in resource management.

A few studies have recently investigated the extent to which team skills could be developed in *ab initio* students using a PBL. approach based mainly on group work (P.M. Anderson, 1993; Anderson & Henley, 1994). Other studies have examined whether undergraduate students could learn CRM principles by using self-analysis of Line-Oriented Flight Training (LOFT.) as an active-learning strategy (Smith, 1993, 1995) or by using computer-assisted learning to develop decision making skills (Wiggins, Taylor & Nendick, 1996) or by using reflective journals to foster the development of self-awareness and self-appraisal skills (Anderson & Henley, 1997). These research projects have shown that innovative teaching approaches can be effective in developing human factors. concepts and skills in undergraduate pilot training (also see Jensen et al., 1995; Hunter, 1995). However, as mentioned previously, the success of the implementation of any of these programmes is heavily dependent on the competence of the instructor (P.M. Anderson, 1993; Henley, 1995).

Developing Self-awareness through the Use of Reflective Journals

One of the aims of aviation education is to help student pilots develop the skills necessary to make reasoned decisions in unfamiliar situations, deal with problems, adapt to, and participate in change, make self-evaluations, and work effectively as team members. To facilitate the development of these skills, however, pilot training should first aim to develop self-awareness or metacognition, including self-evaluation skills, and self-directed learning skills. An effective way of developing some of these skills can be through the use of reflective journals (Calderhead & Gates, 1993; Clift, Houston & Pugach, 1990; Sparks-Langer et al, 1990). The main purpose of using reflective journals in aviation education is for students to be more aware of what they do, how they do it, why they do it, and to be able to identify their own strengths and weaknesses with regard to their understanding of content knowledge, procedures and practical skill development and application. Moreover, the aim is to develop an orientation toward open-mindedness and a willingness to accept responsibility for self-directed learning as well as foster a keen sense of observation, critical thinking and reasoned analysis (Anderson & Henley, 1997; Brookfield, 1995).

To operate effectively in an ever-changing environment, professional pilots need to become reflective decision makers who make decisions consciously and rationally, even with limited information. However, sound judgement and effective decision making strategies are usually the product of extensive practical experiences. Reflective practice can provide a means to help accelerate the development of the experiential base amongst student pilots.

The aim of reflection is to prepare, or make us ready for new experiences, to bring to light different ways of doing things, to open our minds to a new approach to a task, to clarify an issue, or to resolve a problem (Boud, Keogh & Walker, 1988; Montgomery, 1992). Hence, reflection is a key component of learning. Through reflection, new knowledge, skills or perspectives may be developed or new ideas may emerge. One of the aims of reflection is self-awareness (metacognition), which in effect, is the first step toward situation awareness (Beaumont, 1998). Traditionally, students are taught about the aircraft and the environment in which they operate but very little about the human element. However, the first step in becoming a reflective practitioner is the development of metacognition, which includes planning, reflecting, monitoring, evaluating, self-regulation and independent thinking (Cole & Chan, 1994; Hong, 1995; Schön, 1987). Students should be aware of how they learn and they also need to know about themselves and become fully aware of their personal limitations as well as their strengths. Furthermore, if we want professional pilots who are reflective, self-directed learners, it is important that they be encouraged to develop these skills during their initial training.

The importance of self-awareness is that students become actively involved in the evaluation of their understanding and of their performance, and come to realise that they are responsible for their own professional growth. By identifying their strengths and weaknesses, students come to realise that they need to improve themselves, thus becoming more intrinsically motivated to act on their lack of understanding and/or inadequate performance. The result of being actively involved in evaluation is that students are forced to decide what they need to do to improve their situation, which in turn, promotes self-improvement (Schön, 1987).

According to Boud, Keogh and Walker (1988), "one of the most important ways to enhance learning is to strengthen the link between the learning experience and the reflective activity which follows it" (p. 26). In flight training, this link is normally planned to be achieved during the debriefing period at the end of each flight. However, because of time constraints, the debriefing is often shortened or curtailed. Yet, this is a crucial element of flight training. If time is a factor, students should be encouraged to write a reflective journal which can then be discussed soon after the event or during the pre-flight briefing for the next lesson.

Journal writing is an effective way to structure self-evaluation so that students will learn from their experiences (Zeichner & Liston, 1987). Learning can be augmented and more profound when students are encouraged to reflect on the learning event and exercise their judgement about the content and the processes of learning. The benefits of this "consciousness raising" are that students learn to chart their development, identify patterns, challenge their own beliefs and avoid repeating the same mistakes. Reflection can lead toward greater confidence and assertiveness or to a change in attitude, perspective or priorities. A reflective journal should, therefore, answer questions such as:

- What happened?—what did I do?
- How did I do it?—why did I do it?
- How did I feel?—what did I learn?—what does it mean?
- What should I have done differently and what do I need to do next time?
- Which aspects were successful?
- What specific areas do I need to improve or extend?
- What do I want to investigate further?
- How does this relate to previous knowledge and experience and to future learning activities?

According to Boud, Keogh and Walker (1988), there are three stages in the reflective process:

- The returning to experience,
- Attending to feelings, and
- Re-evaluating experience.

The first stage in the reflective process, *the returning to experience* (*what*), is the recollection of salient aspects of the episode or of the experience, the recounting of what happened. This phase includes a narration or a description of an event or a learning situation, but it usually remains strictly factual and is devoid of any reference to personal concerns, attitudes or emotions and does not reveal the student's understanding of the situation or what was learned from the experience. For example, Anderson and Henley (1997) cited one student who wrote: *"The flight planning stage began in the early hours of the morning. I obtained the weather forecast and started [calculating] my headings and groundspeeds. . . . Departure was about half an hour late . . . My taxi and take-off was good but it was close to the top of climb before I realised that my flaps were still extended"*.

This stage is useful since it provides a detailed description of what occurred during the flight. However, the student does not expand further. He does not explore why he did not retract the flaps, what factors made him forget to retract the flaps, and what should he do to improve this for the next flight? Nevertheless, the narrative element is important because it serves to "contextualise" what happened, and it provides students with a better understanding of what took place. However, reflection is more than simply "thinking about" a host of events. It is the logical and analytical thinking that takes place when faced with a problem we need to resolve in a rational manner. Consequently, in the feedback provided by the instructor, the student should be encouraged to address the points mentioned above in order to develop his or her reflective skills to a higher level.

The journal entry included above is representative of what many students tend to do when first asked to complete a reflective journal; they usually treat the reflective journal as a diary, only including descriptions, details and facts regarding their activities (*what*) instead of going beyond that stage to consider the reasons for their actions. Therefore, in the feedback provided, the instructor needs to prompt the student to go beyond the description (*what*) to consider *how, why, so what,* and *what if?* Asking these questions is crucial in the development of reflection in students. By asking questions that provoke reflection, students begin to realise the importance of context. That is, students come to the realisation that theory is not isolated from practice.

Attending to feelings and developing an awareness of feelings is the second stage in becoming reflective. Students need to consider the affective dimension of their experiences. Reflection includes the need to emphasise the positive feelings and remove impediments or obstructing feelings in order to examine the experience more objectively and more thoroughly without blinders. In relating the event, it is possible to reconsider and re-examine the experience afresh. Students can realise how they were feeling and examine what prompted their action. This stage helps to clarify the student's personal perception of the experience. The student is allowed to stand back from the experience and review it and can view it from different perspectives or in a wider context (Boud, Keogh, & Walker, 1988). Students also need to be encouraged to integrate elements of description, observation, analysis, interpretation and introspection. That is, the student's experiences, concerns and feelings should all contribute to the reflection.

The following journal entry illustrates how a student is starting to incorporate material covered in class in her reflection as well as trying to integrate this with the practical application. The student also describes her emotions, the outcome of the situation and the difficulty involved in the situation. She has also grasped the fact that human factors are not limited to factual material covered in class; human factors include skills and attitudes as well:

Two bolts of lightening struck in front of us the instant [X] looked up for a second and we were concerned we had flown into one of the monsters. [X] stated on the radio 'can someone please help us, we've just flown into a thunderstorm'. [Instructor Y] instructed us very firmly to do a 180 turn that instant and I was told by [my instructor] later in the day that that was the second time that morning he had been truly horrified . . . By this time the knot in my stomach was worsening, not with fear I don't think, but worry and I thought I was going to throw up, so [Z] handed out the sick bags. Somehow we managed to get out of the weather, I don't know how long we were in cloud but it felt like forever. Having all three of us there was a major contributing factor to the fact that we are all still alive. Even in such a high pressure situation we were still considering so many vital things. . . . [Now out of the cloud] We joked about our multi-crew environment and how it took three people to fly a single engine aircraft and the fact that's how many it takes to fly something as large as a 747 . . . We were all so glad that we had each others company, something between us clicked and we worked so well together, neither one of us put any pressure on the other and any suggestion put forward was not taken in a manner as if we were telling each other what to do. We wondered if the same thing would have happened had we been with a different crew of students .... The importance of understanding human factors was highlighted to me. The good relationship the three of us had in the aircraft the entire time and the knowledge we collated was the reason for our success. Throughout this journal entry I have been mainly using the word 'we' as opposed to singular names. This is because nothing on the flight was without consultation, assessment and support. I personally find it astounding what we achieved and the manner in which we achieved it. Nothing in class, while it may bring to awareness the importance of relationships and communication, can totally prepare anyone for a real life situation (cited in Anderson & Henley, 1997).

Self awareness through journal writing is one way to incorporate human factors into all facets of flight training. As Kolb (1984) noted, "Learning is the process whereby knowledge is created through the transformation of experience. Knowledge results from the combination of grasping experience and transforming it" into building blocks for future learning (p. 41). One student, who had to land in a paddock because of deteriorating weather, wrote: *"Probably the best things to be learned from this is not the appreciation of the weather but the human factors, like the 'push-on-itis' and 'anticipation' which clouded our views. Each one of us knows we shouldn't have left that day and it is important to recognise that thought processes can be very dangerous when pressure is applied to get away"* (cited in Anderson & Henley, 1997).

The third stage, *re-evaluating experience*, is that of consolidation and application. This stage is not always achieved if the preceding two stages are not completed. However, instructors can help students achieve this stage through supportive questioning; they can act as a sounding board, and can help students clarify their understanding, their interpretation and their intentions. Re-examining the experience is essential to appropriate new knowledge and skills to previous conceptual framework and to guide future action (Boud, Keogh, & Walker, 1988). In short, students should be encouraged to describe the event and then reflect ("*look back*") on what they have done and then "*look ahead*" to what they should do the next time (see Figure 5.1). In addition, they should be encouraged to acknowledge the reasons for their actions and consider how they will approach similar tasks in the future. Reflection, in essence, is proactive; students are encouraged to reflect not so much to revisit past experience or dwell on the past, but to guide future action. Retrospectively, after the student has completed a flight or a learning activity, he or she must think about what took place in a critical and analytical way. In other words, pilots must engage in reflection about their practice.

In the final stage of reflection, students should be able to describe and analyse a learning situation, issue or problem, gather and evaluate information regarding the issue or problem under consideration, generate several alternative solutions, look at their potential implications and finally, integrate all of the information into a tempered conclusion or solution for the problem identified or for future application or actions. The following student discusses an article associated with her new knowledge (*"the article says"*) and relates this information to previous experience (*"thinking through I can begin to see why"*). In addition, she views her situation (past or present) and relates her learning to it. The student tries to explain past experiences in the light of this new knowledge. Furthermore, she acknowledges that she may experience some difficulties in the future as a flight instructor but she is willing to look for strategies to overcome future problems:

> *I wanted to record my interest in the Learning Style Inventory that we completed. When we started to discuss the different types of learners I could see pretty quickly which category I fell into and I was very interested to see if the Inventory would indicate this—and it certainly did!! ... Kolb's article says of Type Two learners that they are often not concerned with 'why' or 'how' and this certainly applies to me ... Thinking this through I can begin to see why I find it easier to study for tests than some of my friends, and why it doesn't take me as long ... I can now understand why my younger sister could never just accept what I told her when I was helping her with her work. Looking back now I can remember saying over and over—'I don't know why—why isn't important—just accept it!!'. I gave her the Inventory to complete and she's a Type 1 learner!! ... Although I consider myself lucky in some ways to be a Type Two learner ... I can see the difficulties this may pose for me when I go to teach someone else. As with my sister, I can see myself becoming impatient with a student's inability to understand just the bare facts or need to explore the how or why of a situation. It will be important for me to remember that not all—in fact 70%—of my students will not be gaining full advantage of my teaching if I simply present them with the facts and expect them to fill in the gaps through their own thinking. Although at this stage of the course I can't really form any strategies to cope with these differences I suppose I have taken the first step to overcoming them by gaining an awareness of the problems posed. Perhaps by the end of the semester I will have a few more ideas on ways to adapt my teaching to suit a larger proportion of my students (cited in Anderson & Henley, 1997).*

Since writing a reflective journal is a skill, it is imperative that instructors provide timely feedback to students to help them develop their reflective thinking. The feedback needs to be non-threatening, constructive and supportive. It is also important to include questions that will allude to, or prompt students to make connections between theory and practice. With appropriate feedback and prompting questions, students can move from simple description and unsubstantiated judgements to a greater frankness, open-mindedness and thereby acquire the ability to link theory to practice and even demonstrate a willingness to search for alternatives. As students develop their ability to reflect, they can be challenged to question their understanding and their perceptions, thus becoming more critical.

Reflective journals can also successfully promote self-directed growth of students and thereby contribute to their professional development. As evidenced in the following comments, reflective journals also helped to raise students' awareness of the causes and consequences of their actions:

> *I had a real sense of achievement. We had completed the task in time, beat other groups and, most of all, I had fitted in. We all agreed that next time, we would do some aspects differently; take more time to set a firm plan and stick to it. We learnt that when doing a group task, you allocate tasks based on the skills each person has ... I need to have more faith in myself (cited in Anderson & Henley, 1997). Another student noted: From this experience I learned that in order to work as a team, one must plan ahead as well as construct some sort of strategy in order to effectively complete the set task. I have also been made aware that a career in aviation will involve working within a group and becoming a team player (Ibid).*

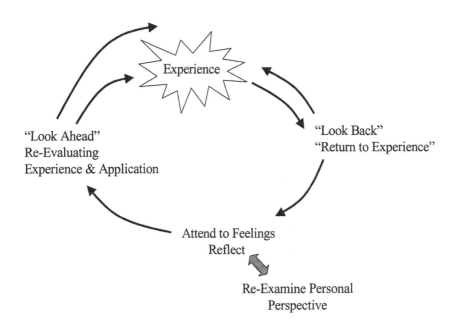

Figure 5.1   The reflective process
Source: Adapted from Boud, Keogh, & Walker, 1988

In summary, reflective journals can assist students in developing metacognition; it gives them a way of organising their ideas and of relating knowledge and experience; and it enables students to analyse and evaluate their own practice, adopting an analytical approach towards their learning. When students reflect, they become researchers in the practical context. They do not separate thinking from doing, and thus, reflection becomes the core of practice. Furthermore, through reflection, students can also make new sense of situations of uncertainty or uniqueness. However, in order to assist students in becoming more reflective, it must be stressed that feedback is of paramount importance. Through insightful feedback, instructors/facilitators can respond to students' accounts, providing alternative interpretations and responses to students' writing that challenges their views and beliefs and encourages them to question and analyse their perspectives or preconceptions.

Some additional benefits in using reflective journals are that it permits instructors to gather progressive feedback, as well as obtain an overall evaluation of the quality of the ground theory or flight instruction. In addition, journal entries provide a means of obtaining course feedback and provide insights into the thoughts and feelings of students as they progress through the course (Stockhausen & Creedy, 1994).

Using Problem-Based Learning to Develop Human Factors Knowledge, Attitudes and Skills

Many practitioners have claimed that graduates from traditional flight training programmes are inadequately prepared when it comes to the practical application of the knowledge gained during their training (Ross, 1989). Indeed, traditional flight training courses have been criticised for their lack of attention to human factors, for placing little emphasis on encouraging teamwork, for poor attempts at developing skills of enquiry and communication in students, and for the inadequate portrayal of the context of major safety issues and problems (Henley, 1995; Ross, 1989).

Another teaching approach that has proven successful in developing CRM concepts and human factors knowledge and skills has been the use of problem-based learning (PBL). Although problem-based learning is not new, it resurfaced three decades ago because of the growing realisation that traditional teaching often falls short of the fulfilment of many educational objectives. In 1965, in an attempt to overcome discrepancies between tertiary education and the requirements of the professions, the Faculty of Health Sciences at McMaster University in Canada implemented what became known as a Problem-Based approach to learning.

Curriculum changes have occurred at a number of universities because of the outcomes suggested by PBL research (Anderson, 1993; Barrows & Tamblyn, 1980; Boud, 1985, 1988) and the identification of the needs of adult learners and of graduates in the workplace. According to Engel (1991), some of the objectives of a PBL course are to develop high professional competency, foster reasoned decisions in unfamiliar situations, deal with problems, adapt to, and participate in change, develop self-evaluation skills, appreciate another person's point of view and work productively as a team member, all of which are of paramount importance in the education and development of professional pilots.

PBL is defined as an activity in which the trigger for learning is an issue, a query or a problem the learner wishes to solve (Barrows, 1985; Boud, 1985). Learners develop an understanding of content (expand their knowledge base) through direct involvement with problems of professional realism and of relevant context. Moreover, learners are actively involved and are responsible for solving problems which require the use of analytical, problem-solving, and problem definition skills (Anderson & Henley, 1994). Therefore, PBL is not simply the addition of problem-solving activities to otherwise discipline-centred curricula, but is a method of integrating and coordinating subject disciplines, where the curriculum is centred around key problems of professional life (Boud & Feletti, 1991).

Objectives of PBL

Practical application Barrows (1985) identified a number of key educational objectives considered pertinent to medical education and which hold similar value for other professions, including aviation. These objectives, he believed, could best be achieved using a PBL approach. The first objective involves the structuring of knowledge for use in practical applications. Tulving and Thomson (1973) found that the more closely the learning situation resembles its practical application, the better the performance.

Supporting this perspective, research by R. C. Anderson (1977) and Willems (1981), for example, found that students familiar with content knowledge were often unable to apply their knowledge given a practical situation. These students wrote highly knowledgeable reports and solved complex problems, but appeared unable to take the next step and use the acquired knowledge in a practical context. These two studies led Schmidt (1983) to conclude that "people can possess knowledge which they seem unable to apply. They know information but can not use it" (p.11). Furthermore, adult educators have repeatedly stressed the importance of both practical application and relevancy for adult learners (Brookfield, 1986; Cross, 1982; Houle, 1980; Kidd, 1973; Knowles, 1980; Rogers, 1977). Hence, there needs to be an integration of practical skills and theoretical knowledge in teaching the professions.

Presently, most of the subjects in aviation studies are taught as separate entities and often without immediate practical application. Professional pilots, however, do not operate in a neatly compartmentalised environment. A normal charter flight, for example, would require the integration of knowledge and skills gained in flight training, principles of flight, engine, systems and instrumentation, aircraft performance and operations, navigation, meteorology, flight rules and procedures, and human factors. A PBL approach would seem more apt to facilitate the integration of these disciplines than would a traditional lecture-based approach because PBL is triggered by real life situations where practical skills and theoretical knowledge are closely intertwined.

Reasoning process Glaser (1982, as cited in Barrows, 1986) stresses that "as the student acquires knowledge, he or she should also be empowered to think and reason" (p. 74). According to Schmidt (1983), PBL is "an instructional method that is said to provide students with knowledge suitable for problem-solving" (p. 11). Barrows (1985) recognised that factual knowledge was not the most valuable contribution to be made to a doctor's education. On the other hand, problem-solving skills required in the reasoning process were acknowledged as being of paramount importance to the student doctor. Similarly, aviation presents a situation which consistently requires the use of problem-solving skills. Every contingency cannot be planned for, or foreseen in aviation's complex environment (Foley, Anderson, & Henley, 1995).

For example, there was no specific training for the two pilots of an Aloha Airlines B737 who found themselves faced with a unique situation when the upper section of the fuselage tore off mid-flight causing a flight attendant to be sucked out and multiple systems to fail. Nor was there specific prior training for the crew of United Airlines Flight 232 that crashed at Sioux City in 1989 where the pilots managed to crash land the jet airliner using differential power when all the hydraulic systems had failed (National Transportation Safety Board, 1990). These predicaments were unprecedented and, as such, the pilots had no formal training in how to deal with an aircraft falling apart in-flight, or with the total loss of hydraulic systems. However, the pilots used refined problem-solving skills and an effective reasoning process to adapt all their knowledge and experience to a unique and unpredictable situation. These skills enabled the crew to handle the multiple systems failures and unusual aerodynamic characteristics of a crippled aircraft so that they could land the aircraft with a minimum number of casualties (National Transportation Safety Board, 1990; O'Lone, et al., 1989). The skills used by the pilots included the generation of hypotheses, inquiry, data analysis, problem synthesis and finally, decision-making. Through the presentation of problem simulations and real-life scenarios, PBL has the potential to refine the problem-solving skills and the team skills of students so that the knowledge and concepts gained in the classroom can be applied in a practical context.

Self-directed learning skills Aviation is a dynamic environment. Technological advances are forcing increasingly rapid changes with more far-reaching implications than ever before. Information is quickly rendered redundant, and therefore, professionals must continually update their knowledge. Thus, the professional pilot is a perpetual student. Accordingly, some educators have recognised that their primary goal is not only to teach specific amounts of knowledge, but also to foster the development of self-directed learning skills and professional competence. PBL students are required to independently locate and use resources in order to understand new concepts and solve problems, just as the professional pilot is required to do when working in the industry. PBL is, therefore, a mechanism that can enhance the development of self-directed learning skills by making students responsible for their own learning in preparation for that time when they join the workforce and become truly independent learners.

Motivation for learning Integrating the theoretical knowledge required to fly an aircraft with the practical aspects of flying can provide an extremely motivating force in aviation studies. The level of interest generated by lecturing to students about such subjects as aerodynamics, flight rules and procedures, meteorology, aircraft fatigue management, and human performance and limitations is probably limited when compared with providing students with a real-life problem through a PBL approach. For example, the following statement could be used as a trigger for learning: "Following is a passenger's eyewitness account of what occurred on an Aloha Airlines B737 which lost part of its fuselage in flight. Find the probable causes of the accident". As the scenario is developed, the student will need to research areas as varied as crew resource management, flutter and material fatigue (Foley, Anderson & Henley, 1995). They may require more information such as fatigue testing records prior to the incident and the Company Policy for fatigue testing. Because students are actively involved, and the problems are realistic, the PBL approach usually generates more interest than the traditional lecture-based approach which, according to research on learning styles, appeals to approximately 20% of learners (Burns, 1989; Kolb, 1984). The increased intrinsic motivation to learn about the subject is likely to drive the student and subsequently enhance the learning as well as develop skills of inquiry and critical thinking.

Problem-Based Learning versus Traditional Teaching

The respective advantages and disadvantages of PBL and traditional teaching have been collated by P.M. Anderson (1993) from various sources (see Table 5.1). Proponents of PBL argue that it allows for a broader, more relevant and integrated knowledge base than the traditional, lecture-based approach (Boud, 1988; Coles, 1985; Engel, 1991). While traditional teaching tends to burden the student with large amounts of facts to be assimilated (memorised) and subsequently regurgitated for the purpose of passing examinations, PBL emphasises knowledge which is pertinent to the profession by placing it in a practical context. Traditional teaching approaches also usually focus more on the teachers and what they teach, whereas in a student-centred approach, students have more responsibility for their learning (Harden, Sowden & Dunn, 1984). Furthermore, PBL provides an excellent medium for developing team skills because it is very suitable to group work.

Aviation is an industry where effective, co-ordinated teamwork is essential. Commercial pilots commonly fly with different crews and need to develop team skills to ensure that they can effectively adapt to new situations and work with new crew members. Even single pilot operations involve the combined resources of pilots (including instructors), controllers, and ground staff (refuellers, maintenance engineers, flight service officers, etc.). It is important, therefore, that team skills (effective communication, cooperation, conflict resolution, resource management, etc.) be learned early in a pilot's career so that ineffective behaviours will not interfere with the safe operation of the aircraft. Indeed, both researchers and practitioners point out that training in team skills should be developed during the *ab initio* stage of flight training (Diehl, 1991; Sams, 1989), since team skills, like any other skill, will only fully develop with continued practice (P.M. Anderson, 1993). In this respect, PBL can be seen as a precursor to CRM, and more specifically to Line Oriented Flight Training (LOFT), which is widely implemented in the airline industry.

A study by Coles (1985) compared the learning styles of PBL medical students with those of traditionally taught medical students both prior to, and subsequent to, their first year of study. Coles concluded that, although both groups of students showed similar approaches to studying prior to entering medicine (low reproduction/memorisation, high meaning and high versatility), by the end of their first year in medicine, the traditional students appeared to be detrimentally affected. Their studying approaches declined towards greater reproduction, lower meaning and lower versatility. However, PBL students maintained their desirable approaches to learning, even leaning towards less reproduction by the conclusion of first year. Coles concluded that traditional teaching appeared to have detrimental effects on learning approaches while PBL seemed to be "creating an educational climate [enabling] ... students to learn ... in a desirable manner" (p. 309). Similarly, Albanese and Mitchell (1993) found that PBL students were "more likely to study for understanding or to analyse what they need to know and study accordingly" (p. 52), which implies a deep rather than surface approach to learning (Moore & Telfer, 1993). According to Moore (1991), deep approaches to learning in aviation are more beneficial than surface learning.

Designing a Problem-Based Learning Course

Before looking at the applications of PBL in aviation, it is important to understand that PBL is not limited to one specific learning and teaching approach. PBL is an umbrella term widely used to describe a variety of problem-centred programmes. Barrows (1986) developed a taxonomy of PBL approaches to identify and create an awareness of the spectrum of PBL methods. PBL encompasses an entire range of methods which vary greatly in their degree of teacher or student direction. PBL, for example, can be situated anywhere from a fairly high teacher-directed approach using lecture-based cases to *closed loop* PBL, where students learn completely independently, such as professional pilots would do once they are in the industry (see Table 5.2).

In designing a PBL course, the first step is to develop clear instructional objectives to ensure that all the course content is covered and that the desired skills are developed through the use of appropriate triggers (see Figure 5.2). Having established educational objectives, the degree of teacher direction should be considered. For example, implementing a *closed loop* PBL approach with first year students might be unreasonable and ineffective, since students may not have the necessary skills, experience, or resources to be fully independent learners. However, creating PBL so that it becomes increasingly student-directed as the student nears graduation is one way in which the knowledge and skills gained in the educational environment and those needed in the workplace can be effectively integrated.

Table 5.1   Advantages and disadvantages of PBL and traditional teaching

PROBLEM-BASED LEARNING		TRADITIONAL TEACHING	
Advantages	**Disadvantages**	**Advantages**	**Disadvantages**
Facilitator:	*Facilitator:*	*Teacher:*	*Teacher:*
• Students prepared for continuing education	• Staff must adapt to role change	• Teacher experience limited to teacher-centred and feel comfortable with approach	• Overload students with need to know information
• Emphasis on student, more responsibility given to student	• May need to prepare teachers for PBL approach	• Fewer demands on teachers	• Majority of knowledge is irrelevant, useful to pass exams bit not in practice
• Greater relevance of content	• Considerable advanced preparation needed		• Place little emphasis on teamwork
• Greater provision of feedback to students	• Teacher expertise lacking in development & planning PBL curriculum		• Poor attempts at developing skills of inquiry in students
• Problems of same context as real-life			• Lack of staff/student contact
		Student:	*Student:*
Student:	*Student:*	• Students' experience limited & feel secure with how they learn	• Student boredom
• Student satisfaction	• Conflict student's own learning pattern & may feel threatened	• Understand fundamentals & vocabulary of discipline before tackling problems	• Emphasis on memorisation & regurgitation of large amounts of knowledge
• Increased student motivation	• Some students find PBL demanding	• Development of a logical progression of concepts in a discipline	• Lack of feedback on progress
• Development of an integrated body of knowledge			
• Develop problem-solving & self directed learning skills	*Department:*	*Department:*	
• Active participation by students	• Difficult to implement in faculties with strong departmental structures or external subjects	• Resource availability	
• Expanding knowledge base	• More resources required		

Source: Boud & Feletti, 1991; Harden et al., 1984; Schwartz, 1991; Todd, 1991

Table 5.2   Problem Based Learning spectrum

Lecture-Based Cases	Case-Based Lectures	Case Method	Modified Case-Based	Problem-Based	Closed Loop Problem Based
• Teacher-directed	• Teacher-directed	• Combines student & teacher directed	• Combines student & teacher directed	• Student directed & teacher facilitation	• Student directed & teacher facilitation
• Case Summary	• Case Summary	• Case Summary	• Partial Problem	• Full Problem	• Full Problem
• Limited hypothesis generation, data analysis and decision-making	• Limited hypothesis generation, data analysis & decision making	• Stronger hypothesis generation, data analysis & decision making	• Stronger hypothesis generation, data analysis & decision making	• Increased hypothesis generation, data analysis & decision making	• Increased hypothesis generation, data analysis & decision making
• Limited motivation	• Higher motivation	• Increased motivation	• Highly motivating	• Highly motivating	• Highly motivating
• Limited reasoning and reasoning and problem-solving skills	• Limited reasoning & problem solving skills	• Stronger reasoning & problem solving skills	• Stronger reasoning & problem solving skills	• Reasoning & problem solving skills developed	• Reasoning & problem solving skills further developed
	• Analyse case before new learning	• Analyse case before new learning	• Analyse case before new learning	• Analyse case before new learning	• Analyse case before new learning
		• Self-directed learning	• Self-directed learning	• Increased self-directed learning	• Increased self-directed learning
			• Limited Student inquiry	• Free inquiry	• Free inquiry
				• Activates prior knowledge for review & association	• Activates prior knowledge for review & association
				• Aids retention	• Aids retention
					• Closed loop since return to the problem to re-evaluate
					• Self evaluation

Source: Barrows, 1986, pp. 481–485

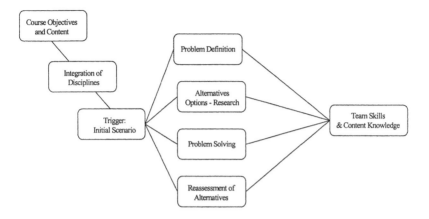

Figure 5.2   PBL scenario development

The degree of integration of disciplines also will need to be considered. For example, will the course be inter-disciplinary or intra-disciplinary? In most aviation programmes, subjects such as flight rules and procedures, meteorology, navigation, aircraft performance and operations, engine systems and instrumentation, and human factors are segregated from one another as individual subjects during the learning process. In practice, however, the professional pilot does not work in a compartmentalised environment, but integrates knowledge acquired from all the aviation disciplines (P. M. Anderson, 1993). Pilots conducting night freight operations, for example, are a long way from segmented courses in which they learned meteorology, human factors, systems and performance as separate units. Something must be done to link what is learned in the classroom to what is needed in the working environment.

PBL is based on problems taken from the working context in an attempt to bridge that gap. It also facilitates the integration of knowledge and practical skills and eliminates course segmentation to create a more practical and integrated approach to the study of aviation. In PBL, scenarios act as the catalyst for free inquiry and learning in any subject area. Scenarios also act as a stimulus for learning the basics of any discipline. In aviation studies, the appropriate choice of simulated situations can encourage students to learn all the relevant meteorological, navigation aerodynamic, performance and operational issues related to the scenario. Thus, in the process of solving a problem, students would review prior knowledge and learn a range of skills, in addition to gaining new concepts and knowledge. In this approach, the emphasis is on the students' gathering and sharing task-relevant information and making decisions to address a realistic situation or solve a problem that could be encountered in the work environment.

The next step in designing a PBL course is to develop scenarios, situations or problems that will serve as catalysts for learning. The problem or situation used as a trigger for learning can take any form, being limited only by the creator's imagination. For example, it might be an audio visual presentation, such as the "Time for Decision" (Lynch, 1988) exercise where a tape recording of a cockpit situation presents the problem to be solved. Similarly, a video covering any major accident can be the trigger for developing student inquiry, knowledge and problem-solving skills. Other mediums include card games, accident investigations, tasks such as a charter flight, an article to be written, a job interview or a debate.

The Night VFR (NVFR) syllabus, for example, could be well covered by using an accident investigation or a night charter operation which would prompt students to research the legal requirements, procedures and illusions associated with NVFR operations. A planned flight scenario could provide a study of navigation, meteorology, forced landings, aviation medicine, aerodynamics, systems and performance. The Aloha Airlines incident discussed earlier could also serve as the catalyst for an investigation into crew resource management and aircraft structures and materials and fatigue management. A scenario might begin with the students' gathering information about ageing aircraft and about Aloha's fatigue management policies, and researching the causes of material fatigue and structural failures (Foley, Anderson, & Henley, 1995). Similarly, a video-taped presentation or role-playing scenario could be used in part, or as the basis for a study of multi-crew operations in particular, with respect to personality traits, diplomacy, conflict resolution and crew coordination.

Above all, the aim of a problem or scenario is to arouse specific questions in the minds of the students, the answering of which will cause learning to take place and new skills to be developed. Barrows and Tamblyn (1980) have compiled a number of items to consider when designing problems or scenarios. They stress the need to:

- Consider the graduate, not the student. Think of the "end product" who will graduate from the course, not what the student needs to know now;
- Ensure the problem is one frequently experienced or poorly handled by pilots, air traffic controllers, aircraft maintenance engineers, flight attendants and/or managers;
- Ensure the problem is of vital importance to aviation;
- Ensure the problem is one in which intervention (preventative or corrective) can significantly alter the outcome;
- Decide on achievable specific learning objectives for each session;
- Decide how much information to give in the initial trigger (scenario) and ensure it will naturally elicit the learning issues/questions planned. Ensure the trigger is on-going and can be continued in another session; and
- Arrange appropriate resource material (maps, videos, articles and manuals) to be made available at the correct time.

(Adapted from Barrows & Tamblyn, 1980, p. 159)

The final step to be considered is the evaluation of PBL units. Albanese and Mitchell (1993) identified one of the major difficulties in assessing the success of PBL as being the unit evaluation. To be effective, the evaluation of PBL must assess the specific objectives of the unit. Since these objectives emphasise a deep, rather than a surface approach, this should be reflected in the evaluation. For example, the evaluation could encompass a personal interview, a group assignment, an essay or an examination which measures the student's critical analysis, synthesis, evaluation and application of the PBL unit rather than simply measure knowledge acquisition.

As mentioned earlier, PBL is more than the teaching of specific content. It develops critical thinking, problem solving, and team and communication skills. The development of these skills should also be measured for the assessment to truly reflect the objectives of the course. This aspect of the evaluation could encompass a group project or a more formative evaluation such as the ongoing review of team and communication skills as witnessed by the evaluator throughout the learning sessions. Another form of evaluation could be the group performance in a time-limited activity where team skills could be assessed (Anderson & Henley, 1995).

In summary, PBL has the potential to provide self-directed learning skills and contextual knowledge which combine to develop a graduate who is an independent learner, and a problem-solver who is better prepared for the industry. Furthermore, PBL can enhance team skills and build intrinsic motivation. For these reasons, PBL has the potential, if effectively implemented in aviation education, to yield definite benefits for future graduates and for the industry. In fact, from *ab initio* training right through to the Airline Transport Pilot Licence (ATPL) level and tertiary education, students can benefit from a PBL approach to the study of aviation.

Operational Implications and Future Research

The teaching strategies described here can easily be implemented without greatly increasing the cost to flight training institutions. However, to effectively implement human factors training into general aviation, there is a need to enhance flight instructor training. Instructors need to receive additional training in human factors and in teaching principles and methods. They also need to be provided with appropriate resource material to assist them in using more effective teaching strategies to integrate human factors into general aviation. Furthermore, the focus of aviation education and flight training will need to shift from a "teaching focus" to a "learning focus". In other words, aviation education must become more learner-centred.

The catalyst to improve the effectiveness of aviation education and flight training can be provided through research. However, researchers in the field of aviation education need to become more knowledgeable about the problems confronting practitioners in general aviation as well as help practitioners see the value of what is learned about aviation education in research. The relevance of research can be increased by facilitating communication between researchers and the possible end users. To address some of the needs and concerns mentioned here, future research should, therefore, aim to:

- Identify teaching competencies appropriate to flight instruction and to flight instructor training and evaluation;
- Implement and assess the effectiveness of innovative teaching strategies to develop human factors knowledge, attitudes and skills and take into account the way learners naturally learn (multiple learning styles, multiple learning orientations such as visual, aural, kinaesthetic, multiple preferences for problem-solving, etc.);
- Develop an educational package which integrates human factors education into the flight training syllabus using a variety of teaching strategies; and
- Conduct a validation of the impact of the educational package within the operational environment.

Presently, there is no definitive list of criteria of effective flight instruction. Therefore, there is a need to adapt teaching competencies, such as those identified by the Australian Teaching Council (1996) to flight instructor training and testing, and to flight instruction. Providing the trainers and the examiners of flight instructors with a list of teaching competencies would be a first step in improving the effectiveness of flight instructor training and testing so that teaching competence will come to be on a par with flying competence in the education and evaluation of flight instructors.

The aim of an educational package should be to base the principles of human factors education within the existing flight training sequences which comprise the private and commercial flight training syllabi. Instructors need to be provided with sufficient resources to facilitate the integration of human factors within the operational environment. In particular, there should be a flight instructor manual which incorporates concise pre-flight, in-flight and post-flight teaching objectives and strategies, as well as key questions and examples that might facilitate the process of skill development amongst student pilots. The manual, for example, also should be designed to emphasise the active participation of the student pilot and incorporate additional learning and teaching strategies germane to adult education. In addition, a number of broad principles should be applied so that human factors and adult education principles and strategies are fully integrated into the flight training syllabus.

There have been a number of limited attempts to integrate particular training initiatives into the training environment and where the initiative was integrated into the operational environment (such as that reported in Jensen, 1988; Smith, 1993; Wiggins, Taylor & Nendick, 1996), the outcome tended to be more successful than when the information was presented separately. This suggests that there is considerable value in following an integrated approach in the implementation of human factors training in general aviation.

A video component could complement the educational package. The aim of the video component would be to provide a visual example of the process through which human factors might be integrated into the operational aviation environment. This could be an important aspect of the educational package, since it would provide a series of practical examples which flight instructors might use to model their own behaviour.

However, a series of training seminars for flight instructors should be conducted before the educational package is released since feedback from the potential users is crucial and the appropriate training of flight instructors is fundamental to the success of such a programme. The seminars should also be designed to assist flight instructors to implement the educational package by providing them with background knowledge in human factors and CRM principles and by exposing them to a variety of teaching approaches useful in teaching human factors knowledge, attitudes and skills.

The importance of human factors and the development of teams skills should be particularly emphasised because flight instructors provide student pilots with their first experience in performing as a flight crew member. It is at this stage that instructors can instil in their students the fundamental knowledge and skills in aviation resource management. Indeed, the interaction between the flight instructor and the student pilot establishes the initial groundwork for future crew resource management (CRM) training (Petrin & Young, 1991). Consequently, flight instructors must be able to work effectively as a team with their students. This means that flight instructors need to encourage their students to voice their views and their concerns, make decisions and take advantage of all the available resources, including the instructor and air traffic controllers in order to participate effectively as a crew member. Initially, students are often intimidated by their instructors. However, instructors need to continually strive to overcome this impediment to effective teamwork. Flight instructors are, therefore, instrumental in helping students develop effective communication skills, decision making strategies, self-awareness and team skills. From the very beginning of training, students need to be made aware that "they are an important part of the educational team" (Zonnefeld, 1993, p. 51).

There is also a need for a student manual which would incorporate learning objectives for each flight training sequence, in addition to strategies by which these objectives might be achieved. These should be in the form of cognitive, affective and psychomotor objectives which could then form the basis for the validation process. Additional reference information should also be included such that the student has an opportunity to obtain additional insight concerning a particular interest area or issue. Emphasis should also be placed on the recognition of personal limitations and the coping strategies needed when operating in a dynamic and, at times, stressful environment.

Clearly, the success of any educational initiative will depend upon the extent to which it develops the skills, the attitudes and the knowledge intended. Moreover, there is a requirement to determine unequivocally, the extent of the transfer which occurs in terms of operational behaviour. This provides both feedback to improve the system and increases the probability that the industry will accept and implement the educational package.

However, in testing the validity of human factors education, it is important, if not essential, to examine the performance outcomes beyond the simple tests of declarative knowledge. According to a number of skill acquisition theorists including J. R. Anderson (1993) and Dreyfus and Dreyfus (1996), declarative knowledge is characteristic of novice performance, and it is unlikely that this type of information will facilitate the rapid development of accurate and efficient skills within the operational environment.

From a training perspective, far greater importance should be placed upon the evaluation of the skills associated with human factors education. These are the skills necessary for **implementing** declarative knowledge. Through task-related experience, these skills become more refined and efficient, such that eventually, having reached the expert stage, one possesses both the capacity for a rapid response and the capacity to generalise the knowledge and skills to relatively novel situations. Therefore, the validation process should comprise a cross-sectional analysis of pilot performance and a longitudinal analysis of incidents within selected flying training environments from general aviation. The validation of the educational package should also be based on sound methodological principles.

Conclusion

It should be stressed that traditional learning approaches also have their place in flight training and that teaching strategies are a means to an end, not an end in themselves. Our commitment to aviation education must therefore be a *critical* one. The whole purpose of educational innovation must be to improve learning, that is, to help the learner achieve objectives that would otherwise be difficult or impossible. Templer (1981) likened training to a mountain separating the students from the practical application of knowledge and skills (the ability to perform). The role of the educator or instructor is to find the shortest way across this obstacle—to drive a tunnel through the mountain so that, effectively, the obstacle disappears.

References

Albanese, M. A. & Mitchell, S. (1993). Problem-based learning: a review of literature on its outcomes and implementation issues. *Academic Medicine, 68*(1), 52-81.

Anderson, J. R. (1993). *Rules of the mind*. Hillsdale, NJ: Lawrence Erlbaum.

Anderson, P. M. (1993). *Problem-based learning and the development of team skills in Aviation Studies*. Unpublished honours thesis. University of Newcastle, Newcastle, Australia.

Anderson, P., & Henley, I. (1994). Problem-based learning and the development of team skills in Aviation Studies. In S. E. Chan, R. M. Cowdroy, A. J. Kingsland

& M. J. Oswald (Eds.), *Reflections on problem based learning* (pp. 319-345). Sydney: Australian Problem Based Learning Network.

Anderson, P. M., & Henley, I. (1995). Assessing the development of team skills using a time-limited exercise. In R. S. Jensen (Ed.), *Proceedings of the 8th International Symposium on Aviation Psychology* (pp. 625-630). Columbus, OH: The Ohio State University, Department of Aviation.

Anderson, P. M. , & Henley, I. (1997). Developing self-awareness through the use of reflective journals. In R. S. Jensen & L. A. Rakovan (Eds.), *Proceedings of the 9th International Symposium on Aviation Psychology* (pp. 1166-1171). Columbus, OH: The Ohio State University, Department of Aviation.

Anderson, R. C. (1977). The notion of schemata and the educational enterprise: General discussion of the conference. In R. C. Anderson, R. J. Spiro & W. E. Montague (Eds.), *Schooling and the acquisition of knowledge*. Hillsdale: Erlbaum.

Australian Teaching Council. (1996). *National competency framework for beginning teaching*. Leichhardt, Australia Author.

Barrows, H. S. (1985). *How to design a problem-based curriculum for the preclinical years*. New York: Springer Publishing Company.

Barrows, H. S. (1986). A taxonomy of problem-based learning. *Medical Education, 20*, 481-486.

Barrows, H. S., & Tamblyn, R. M. (1980). *Problem-based learning: An approach to medical education*. New York: Springer Publishing Company.

Beaumont, G. (1998). *Situation awareness or metacognition?* Paper presented at the 4th Australian Aviation Psychology Symposium, 16-20 March. Manly, Australia.

Berlin, J. I., Gruber, E. V., Holmes, C. W., Jensen, P. K., Lau, J. R., Mills, J. W., & O'Kane, J. M. (1982). *Pilot judgment training and evaluation* (Vols. 1-3). Daytona Beach, FL: Aviation Research Center, Embry-Riddle Aeronautical University.

Boud, D. J. (Ed.). (1985). *Problem-based learning in education for the professions*. Sydney: Higher Education Research and Development Society of Australasia.

Boud, D. J. (Ed.). (1988). *Developing Student Autonomy in Learning* (2nd ed.). London: Kogan Page Limited.

Boud, D. J. & Feletti, G. (Eds.). (1991). *The challenge of problem-based learning*. London: Kogan Page Limited.

Boud, D., Keogh, R., & Walker, D. (1988). *Reflection: Turning experience into learning*. London: Kogan Page.

Brookfield, S. (1986). *The skillful teacher*. San Francisco: Jossey-Bass.

Brookfield, S. (1995). *Becoming a critically reflective teacher*. San Francisco: Jossey-Bass.

Buch, G. (1983, April). Pilot judgment training validation experiment. In R. S. Jensen (Ed.), *Proceedings of the 2nd Symposium on Aviation Psychology* (pp. 307-316). Columbus, OH: The Ohio State University, Department of Aviation.

Buch, G., & Diehl, A. (1984). An investigation of the effectiveness of pilot judgment training. *Human Factors, 26*(5), 557-564.

Burns, S. (1989). There's more than one way to learn. *Australian Wellbeing, 33,* 42-44.

Calderhead, J., & Gates, P. (Eds.). (1993). *Conceptualizing reflection in teacher development.* London: The Falmer Press.

Clift, R. T., Houston, W. R., & Pugach, M. C. (Eds.). (1990). *Encouraging reflective practice in education: An analysis of issues and programs.* New York: Teachers College Press.

Cole, P. G., & Chan, L. (1994). *Teaching principles and practice.* Sydney: Prentice Hall.

Coles, C. R. (1985) Differences between conventional and problem-based curricula in their students' approaches to studying. *Medical Education, 19,* 308-309.

Cross, P. (1982). *Adult learners.* San Francisco: Jossey-Bass.

Diehl, A. (1991a, Fall). Cockpit decision making. *FAA Aviation Safety Journal, 1*(4), 14-16. Washington, DC: US Department of Transport.

Diehl, A. (1991b, April). The effectiveness of training programs for preventing aircrew "error". In R. S. Jensen (Ed.), *Proceedings of the 6th International Symposium on Aviation Psychology* (pp. 640-655). Columbus, OH: The Ohio State University, Department of Aviation.

Dreyfus, H. L., & Dreyfus, S. E. (1996). Why skills cannot be represented by rules. In N. E. Sharley (Ed.), *Advances in cognitive science* (pp. 315-335). Chichester: Ellis Harwood.

Engel, C. E. (1991). Not just a method but a way of learning. In D. J. Boud & G. Feletti (Eds.), *The challenge of problem-based learning* (pp. 23-33). London: Kogan Page Limited.

Foley, S., Anderson, P., & Henley, I. (1995). Preparing student pilots for their professions by implementing problem-based learning in aviation studies. In R. S. Jensen & L. Rakovan (Eds.), *Proceedings of the 9th International Symposium on Aviation Psychology.* Columbus, OH: The Ohio State University, Department of Aviation.

Harden, R. M., Sowden, S. & Dunn, W. R. (1984). Educational strategies in curriculum development: the SPICES model, *Medical Education,* Vol. 18, 284-297.

Harris, J., & Conrad, D. (1997). Adult learning in the aviation training environment. In R. S. Jensen & L. Rakovan (Eds.), *Proceedings of the 9th International Symposium on Aviation Psychology.* Columbus, OH: The Ohio State University, Department of Aviation.

Helmreich, R. L., Wiener, E. L., & Kanki, B. G. (1993). The future of crew resource management in the cockpit and elsewhere. In E. Wiener, B. Kanki & R. Helmreich (Eds.), *Cockpit resource management* (pp. 479-501). San Diego, CA: Academic Press.

Henley, I. (1995). *The quality of the development and evaluation of flight instructors in Canada and Australia.* Unpublished doctoral dissertation, University of Newcastle, Newcastle, Australia.

Henley, I., & Wiggins, M. (1997). Integrating human factors into ab initio flight training: Are flight instructors prepared for the task? In R. S. Jensen & L. Rakovan (Eds.), *Proceedings of the 9th International Symposium on*

Aviation Psychology (pp. 1213-1218). Columbus, OH: The Ohio State University, Department of Aviation.

Hong, E. (1995). A structural comparison between state and trait self-regulation models. *Applied Cognitive Psychology, 9*, 333-349.

Houle, C. O. (1980). *Continuing learning in the professions*. San Francisco: Jossey-Bass.

Hunter, D. R. (1995). Aeronautical decision making: Historical results and a new paradigm. In R. Fuller, N. Johnston & N. McDonald (Eds.), *Human factors in Aviation Operations* (pp. 11-16). Aldershot: Avebury Aviation.

ICAO (1991). Training of operational personnel in Human Factors. *Human Factors Digest, 3*, Circular 227-AN/136.

Jensen, R. S. (1988). Creating a "1000 hr" pilot in 300 hours through judgment training. *Proceedings of the Workshop on Aviation Psychology*. Newcastle, Australia: University of Newcastle.

Jensen, R. S. (1995). *Pilot judgment and crew resource management*. Aldershot: Avebury Aviation.

Jensen, R. S., & Benel, R. A. (1977). *Judgment evaluation and instruction in civil pilot training*. University of Illinois, Aviation Research Laboratory.

Jensen, R. S., Chubb, G. P., Adrion-Kochan, J., Kirkbride, L. A., & Fisher, J. (1995). Aeronautical decision making in general aviation: New intervention strategies. In R. Fuller, N. Johnston & N. McDonald (Eds.), *Human factors in aviation operations* (pp. 5-10). Aldershot: Avebury Aviation.

Johnston, N. (1993). Integrating human factors training into ab initio airline pilot curricula. *ICAO Journal, 48*(7), 14-17.

Kidd, J. R. (1973). *How adults learn*. New York: Association Press.

Knowles, M. (1980). *The modern practice of adult education: From pedagogy to andragogy*. Chicago: Association Press/Follett Publishing Co.

Kolb, D. (1984). *Experiential learning: Experience in the source of learning and development*. New Jersey: Prentice Hall.

Lynch, M. (1988). *Time for decision*. Sydney: Gower Northgate Training.

Montgomery, J. R. (1992). *The development, application, and implications of a strategy for reflective learning from experience*. Ann Arbor, MI: University Microfilms International.

Moore, P. (1991, August). *Approaches to learning in aviation: Does the difficulty of the topic make a difference?* Paper presented at the Human Factors in South Pacific Aviation Symposium, New Zealand Psychology Society's Annual Conference, Palmerston North, New Zealand (pp 1-14).

Moore, P., & Telfer, R. A. (1993). A comparative analysis of airline pilots' approaches to learning. *Journal of Aviation/Aerospace Education and Research, 4*(3), 17-23.

National Transportation Safety Board. (1990). *Aircraft Accident Report - United Airlines Flight 232, McDonnell Douglas DC-10-10, Sioux Gateway Airport, Sioux City, Iowa, July 19, 1989*. Springfield, VA: National Technical Information Service.

O'Lone, G., Ott, J., Scott, W., Dornheim, M., Henderson, B., Hughes, D., Kernstock, N., Philips, E., Brown, D., Proctor, P., Kandebo, S., Fotos, C., McKenna, J., & Mecham, M. (1989). Ageing aircraft issue presents major challenge to industry. *Aviation Week and Space Technology*, July 24 1989.

Petrin, D., & Young, J. (1991). Cockpit resource management: As vital in a trainer as in a jetliner. *FAA Aviation News*.

Pinet, J., & Enders, J. H. (1994, December). Human factors in aviation: A consolidated approach. *Flight Safety Digest* (pp. 7-12).

Ross, M. J. (1992). *The integrated commercial flying school: An evaluation of commercial flight training and flight simulator effectiveness*. Unpublished doctoral dissertation, University of Newcastle, Newcastle, Australia.

Sams, T. (1989, April). *Developing cockpit resource management training curricula for ab intio airline pilot training*. Paper presented at the 5th International Symposium on Aviation Psychology, Columbus, Ohio, (pp. 1-7).

Schmidt, H. G. (1983). Problem-Based Learning: rationale and description. *Medical Education, 17*, 11-16.

Schön, D. A. (1987). *Educating the reflective practitioner: Toward a new design for teaching and learning in the professions*. San Francisco: Jossey-Bass Publishers.

Schwartz, P. (1991) Persevering with problem-based learning. In D. J. Boud & G. Feletti (Eds.), *op.cit.* (pp. 65-71).

Smith, G. M. (1993, April). Self-analysis of LOFT as a strategy for learning CRM in undergraduate flight training. In R. S. Jensen & D. Neumeister (Eds.), *Proceedings of the 7th International Symposium on Aviation Psychology* (pp. 533-537). Columbus, OH: The Ohio State University, The Department of Aviation.

Smith, G. M. (1995). Active-learning strategies in undergraduate CRM flight training. In N. Johnston, R. Fuller, & N. McDonald (Eds.), *Aviation psychology: Training and selection* (pp. 17-22). Aldershot: Avebury Aviation.

Sparks-Langer, G. M., Simmons, J. M., Pasch, M., Colton, A., & Starko, A. (1990). Reflective pedagogical thinking: How can we promote it and measure it? *Journal of Teacher Education, 41*(4), 23-32.

Stockhausen, L. & Creedy, D. (1994). Journal writing: Untapped potential for reflection and consolidation. In S. E. Chen, R. Cowdroy, A. Kingsland & M. Ostwald (Eds.), *Reflections on Problem Based Learning* (pp. 73-86). Sydney: Australian Problem Based Learning Network.

Telfer, R. A., & Ashman, A. (1986). *Pilot judgement training - An Australian validation study*. The University of Newcastle, Newcastle, Australia.

Templer, P. R. (1981, Autumn). Effective training in industry. *Royal Air Force Education Bulletin, 19*, 47-54.

Todd, S. 1991, Preparing tertiary teacher for problem-based learning. In D. J. Boud & G. Feletti (Eds.), *The challenge of problem-based learning* (pp. 130-136). Kogan Page Limited, London.

Transport Canada. (1996a). *Human factors for aviation: Advanced Handbook*. Ottawa, ON: Transport Canada.

Transport Canada. (1996b). *Human factors for aviation: Basic Handbook.* Ottawa, ON: Transport Canada.

Transport Canada. (1996c). *Human factors for aviation: Instructor's guide.* Ottawa, ON: Transport Canada.

Trollip, S. R. (1995). Issues in teaching human factors efficiently. In N. Johnston, R. Fuller, & N. McDonald (Eds.), *Aviation psychology: Training and selection* (pp. 252-257). Aldershot: Avebury Aviation.

Tulving, E. & Thomson, D. M. (1973) Encoding specificity and retrieval processes in episodic memory. *Psychological Review, 80,* 352-373.

Wiggins, M. W. (1996). A computer-based approach to human factors education. In B. J. Hayward & A. R. Lowe (Eds.), *Applied aviation psychology: Achievement, change and challenge* (pp. 201-208). Aldershot: Ashgate.

Wiggins, M., & O'Hare, D. (1993). A skill-based approach to training aeronautical decision-making. In R. A. Telfer (Ed.), *Aviation instruction and training* (pp. 430-475). Aldershot: Ashgate.

Wiggins, M., O'Hare, D., Jensen, R., & Guilkey, J. (1997). The design, development and evaluation of a computer-based pilot judgement tutoring system for weather-related decisions. In R. S. Jensen & L. Rakovan (Eds.), *Proceedings of the 9th International Symposium on Aviation Psychology* (pp. 761-765). Columbus, OH: The Ohio State University, Department of Aviation.

Wiggins, M., Taylor, S., & Nendick, M. (1996). A computer-based application for the development of pre-flight decision-making skills amongst ab-initio pilots. *Proceedings of the 3rd ICAO Global Flight Safety and Human Factors Symposium (CIRC 226-AN/158)* (pp. 229-235). Montreal: International Civil Aviation Organisation.

Willems, J. (1981) Problem-based (group) teaching: a cognitive science approach to using available knowledge. *Instructional Science, 10,* 5-21.

Zeichner, K. M., & Liston, D. P. (1987, February). Teaching student teachers to reflect. *Harvard Educational Review, 57*(1), 23-48.

Zonnefeld, M. (1993, September). Cockpit resource management: Instructors and students should work as a team. *Flight Training* (p. 51).

6 Flying Light Aircraft: The Aircraft Control Problem and Psychomotor Skill Development

Don Harris

Overview

It is often said that flying an aeroplane is as easy as riding a bicycle. After even just a brief consideration of the psychomotor skills involved in cycling it would be more accurate to suggest that flying an aircraft is as difficult as riding a bicycle. At the end of most undergraduate courses in human skilled performance it should be occurring to the more talented undergraduate psychology majors that simultaneously walking and talking is quite a major feat of skill development and motor co-ordination. This thought will never have troubled their less talented classmates as it is unlikely that they will have mastered either of these skills in isolation, never mind in combination. The fact that pilots can successfully develop the skills required to conduct a light aircraft around the skies does not mean that the task of flying is a great deal simpler than psychologists make out. What it does suggest, though, is that the human psychomotor system is a great deal more complex and adaptable than one could ever imagine on first acquaintance.

This chapter is divided into two major sections. The first section deals with the control problem itself. This includes a brief examination of the dynamics of light aircraft flight (from an engineering perspective) followed by some consideration of the human control problem imposed on the pilot by the aircraft's dynamic behaviour. The second section is concerned with the problem of psychomotor skill acquisition. This section is itself divided into two major sub-sections, the first dealing with the nature of human psychomotor skills and the second addressing the manner in which they may be trained effectively.

Whenever addressing training in an aviation context, it is unavoidable that the use of flight simulation should be addressed. In this chapter, the consideration of flight simulation is limited to very low cost simulation, as this is the only form that is economically viable in the training of the general aviation pilot. Finally, the chapter concludes with a brief discussion of the operational implications of the topics discussed and some suggestions concerning topics that should be placed on the agenda for future research. However, let us start at the root of the problem with a consideration of the basic control problems encountered when flying a relatively technologically-simple, light aircraft.

The Control Problem

Aircraft dynamics are complex, and as a result, present complex control problems for the pilot. Somewhat perversely, it is the simplicity of light aircraft flight control systems that cause the complex nature of the control problem faced by the pilot. For once, the technologically sophisticated fly-by-wire (FBW) systems found in modern civil transport and military aircraft actually simplify the control task faced by the pilot by reducing the complexity of the control problem. As a result, we are faced with the problem of the least experienced pilots (often *ab initio* trainees) having to come to terms with a relatively complex control problem, when the most experienced pilots have a simple control problem. This would at first seem contrary to all the basic premises of human centred design, however as we shall see, this may not be such a bad thing.

Unless the pilot is performing aerobatics, the flight control problem is conceptually quite a simple one. The pilot is faced with a two dimensional compensatory tracking task (although as we will see later, this is not necessarily the way in which the flight task is performed; whenever a human being is involved in a system there is often a distinction between what the task *requires* of the operator and how the human being *responds* to that requirement). The pilot must maintain the desired vertical flight profile while maintaining the desired track over the face of the Earth. S/he will also need to pay attention to airspeed, which can be regarded as another aspect of the horizontal tracking task, (to simplify the problem for this discussion, the consideration of time will be omitted). The flight task is a compensatory tracking task (as opposed to a pursuit tracking task) as the control objectives of the pilot are simply to minimise errors. For example, in the simplest case the pilot may want to minimise the error between his/her present altitude and the desired altitude at that present moment in time. In a more complex case, the aim may be to attain and maintain a desired rate of climb (or descent). In this latter case, this requires the integration of airspeed and aircraft pitch attitude to attain the desired vertical speed. The flight task can be made even more complex if it is required to fly a prescribed vertical profile at a given vertical speed (e.g., when performing a standard instrument departure). This requires the integration of airspeed and aircraft pitch attitude at the appropriate point in three-dimensional space.

Several things should become apparent from the above. Firstly, what an error is depends upon the parameters within which you specify the task. Performance (and hence training objectives and training methods) can only be specified with respect to the parameters of the task. Secondly, although the aim of the flying task as described here is to minimise error, it is not always obvious to the pilot what the error is or how large it is. Unlike driving, where the error can easily be observed by the driver (e.g., the car is in either the centre of the lane or it isn't), flight path error cannot generally be seen by the pilot when flying under VFR (visual flight rules), although it can be determined from the instruments during instrument flight. Finally, what may not be apparent from the above description of the flight path task, is the fact that although the ultimate aim of the pilot is to minimise error in their desired flight path, the one parameter that the pilot has no *direct* control over is actually the flight path of their aircraft. Some readers may be astounded by this last observation. Nevertheless, with a brief consideration of aircraft control (both from an engineering and a pilot's perspective), it will be seen that this is true. Pilots control the flight path of their aircraft via a series of surrogates.

The Dynamics of Light Aircraft with Conventional Flight Control Systems

Every pilot commences their flying career in a light (general aviation) aircraft with 'conventional' dynamics. A conventional aircraft is termed an 'angle of attack' demand system in the long term and a 'pitch rate' demand system in the shorter term. When pilots move their stick (or column) in the fore/aft plane, (and hence are controlling the deflection of the elevator), they are commanding a new angle of attack, but they will initially see a change in pitch rate, which results in an increase or decrease in pitch attitude, and subsequently causes the aircraft to climb or descend. The initial pitch rate usually takes a second or two to build up to its steady state value. This steady state, pitch rate value will remain constant as long as the aircraft maintains airspeed (however this will not happen unless there is a concomitant increase in throttle opening and there is enough power available). A constant pitch rate response (i.e., constant stick position), combined with a constant airspeed will result in the aircraft performing a circular 'loop-the-loop' manoeuvre. If airspeed is allowed to decay, as can happen in most low-powered, light aircraft, the stick must be brought further back to maintain pitch rate.

The angle of attack can be thought of as the amount of lift that a wing can produce, independent of airspeed. Aircraft pitch attitude is the summation of flight path angle and angle of attack. Within the angle of attack limits for the wing, and assuming a constant airspeed, lift will increase as angle of attack increases. When the pilot makes an input to their aircraft's elevator, s/he is essentially controlling the amount of lift that the wing is producing, *not* flight path angle. However, it is this later parameter that is the ultimate objective of the control exercise. This is discussed further in the following section. The complexity of the pilot's control problem further begins to become apparent when it is considered that the required angle of attack to maintain straight and level flight will depend upon many parameters including airspeed; aircraft mass; altitude; power setting; and aircraft configuration (i.e., flap setting). Many of these parameters have a non-linear effect on the flight path control problem and also have complex interactions with other parameters.

If the pilot maintains a pitch input on the stick without changing the throttle position, the long term dynamics of the aircraft take over (i.e., the angle of attack response). If the pilot holds a constant pitching stick position, the aircraft will eventually adopt a new, constant angle of attack. During this long term response, the initial short term, pitch rate response will cease and the long-term response characteristics will take over. This will be observed as an oscillation in pitch attitude, airspeed and flight path, with each oscillation taking in the region of 30 to 60 seconds. This is called the phugoid and is characterised by a constant angle of attack.

The short term response (pitch rate) can be considered as the manoeuvring response. However, the longer term, angle of attack response, is also required so that the pilot can predictably stop the manoeuvre. Nevertheless, irrespective of the manner in which the aircraft's response is described in engineering terms, it is the pitch rate, short-term response which is most obvious to the pilot. Using pitch rate to control vertical flight path is a typical example of a hierarchical control system, (see the following section).

An idealised representation of a conventional aircraft's response to a step pitch input to the elevator can be seen in Figure 6.1. Two factors should be noted that are of importance to the discussion in the sections that follow. Firstly, the initial response to the elevator input is non-linear. Secondly, flight path lags behind the pitch attitude of the aircraft. Both of these factors can cause control difficulties for the human operator.

Certain other aspects of aircraft dynamics are not optimal from a human control viewpoint. It will not have escaped the notice of pilots that control response changes as a function of airspeed. Pitch rate response is much crisper at higher airspeeds (i.e., it is another non-linear control parameter) due to the time taken to reach the steady state angle of attack decreasing as the airspeed increases. Additionally, at low airspeeds, angle of attack must be increased simply to maintain lift. This causes pitch attitude increasingly to diverge from flight path as airspeed decreases. It will be seen in the following section when the serial pilot model is described, this has important implications for aircraft control.

Aircraft primary axis flight controls are also cross-coupled. A change in one parameter will effect other parameters. The most obvious example of this is the cross-coupling between speed and pitch. As speed increases, the aircraft will climb as lift over the wings increases, unless the trimmed angle of attack is reduced. Similarly, if the aircraft is pitched up without an increase in thrust, airspeed will decay. In terms of lateral control, rolling the aircraft will also result in a concomitant downward pitching moment as some lift is lost across the relatively slower-moving wing on the inside of the turn. Yaw inputs also exhibit cross coupling with both the roll and pitching axes of motion. These are the most obvious examples of control cross-coupling that pilots will be familiar with. However, there are more extreme examples of cross-coupling that every pilot encounters without even thinking about them. These will be described over.

Little of the above description of aircraft control dynamics will be unfamiliar to pilots. The essential message, though, is that as a result of these factors, the control strategies that must be employed by pilots are far more complex than those utilised by drivers, for example. The nature of the control problem also goes beyond many of the classical, laboratory-based studies of human skilled performance.

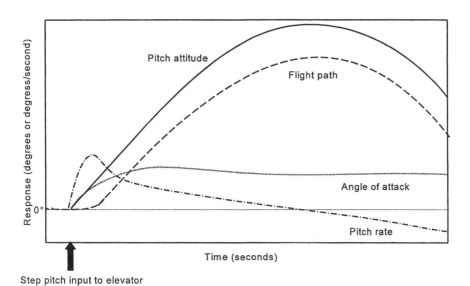

Figure 6.1   **Idealised aircraft response (angle of attack command system) to a step input to the elevator**
Source: Adapted from Field (1995)

Human Control of Dynamic Systems

As stated previously, the control of an aircraft is best described as a hierarchical control problem. For the sake of simplicity and ease of description, most of what follows will concentrate on control in the vertical plane. Nevertheless, most of the problems and concepts described apply equally in the horizontal plane.

The basic nature of a hierarchical control problem is that the parameter which needs to be controlled, in the case of an aircraft its flight path, can only be controlled indirectly via other lower order parameters (e.g., pitch rate). The hierarchical nature of altitude control is illustrated in Figure 6.2. Hierarchical control systems are also likely to be highly cross-coupled. Several examples of cross-coupling between the major controls found in an aircraft have already been described. However, an aircraft's controls also exhibit even tighter cross-coupling than that previously described. In a conventional-technology light aircraft it is impossible to change the pitch attitude of an aircraft without affecting either its vertical position or airspeed. The next generation thrust-vectoring combat aircraft with FBW flight control systems have de-coupled these control effects. The aircraft's pitch angle can be altered without it climbing or descending. However, it is unlikely that we will see thrust vectoring in general aviation aircraft in the foreseeable future, so these issues need not worry us here.

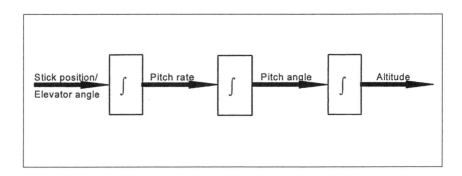

Figure 6.2   **Hierarchical nature of altitude control**

The most common model of pilot control is the series pilot model. It has already been noted that the control of an aircraft can be characterised as a compensatory tracking task, where the objective is to minimise any errors from the desired three-dimensional flight path. It has also been illustrated that the one thing that the pilot cannot directly control is flight path. This is done through a series of surrogates (pitch rate in the short term and angle of attack in the longer term). Finally, there is also one further problem. The pilot cannot actually observe either flight path or angle of attack directly. S/he can only see the aircraft's pitch attitude. In the series pilot model, the flight control problem is decomposed into a short term and a long term problem. The short term problem is one of the control of pitch attitude. Pitch attitude is used here as the control parameter as the pilot cannot actually observe angle of attack, which is actually the parameter that has the effect on flight path and is the parameter that the pilot has direct control over. As a 'rule of thumb', you cannot control a parameter that you cannot observe. The longer term control problem is one of flight path or altitude control.

The short term (pitch attitude) control problem can be considered to be 'nested' within the longer term (flight path angle or ultimately altitude) control problem. This is illustrated in Figure 6.3. What the pilot is doing is attempting to 'close' the inner control loop (i.e., the attitude control problem) in an attempt to control the 'real' problem, that of altitude/flight path angle control. In the horizontal plane, the pilot is dealing with a similar control problem, only in this case aircraft roll attitude is being used as a first approximation (inner loop parameter) in an attempt to control the aircraft's heading, and subsequently its track across the face of the Earth. As in the vertical control problem, the hapless pilot again has no direct control over the aircraft's heading. S/he only has control over roll attitude. Note again the cross-coupling of controls: roll attitude cannot the changed without effecting a change in heading and *vice versa*. Pilots of larger aircraft actually have direct control over the flight path problem via their autopilot or flight management systems. They command an altitude (or flight level) and track, and the aircraft selects the correct pitch attitude and heading to attain the desired flight path parameters. The pilot flying manually approaches the problem in reverse.

Some would argue that the pilot can actually observe flight path with reference to the cockpit instruments, in this case the altimeter, vertical speed indicator and horizontal situation indicator (HSI). Note that if the flight task is characterised as one of minimising flight path error, the attitude indicator actually tells pilots little about the direction and/or magnitude of their error. However, while these instruments will aid in the task of controlling flight path, any pilot who has undertaken any instrument flying will know that this is not easy. Irrespective of the format of the information provided, perhaps the characteristic of these instruments that makes them most difficult to use are the lags inherent in them. Lags in any aspect of the pilot-aircraft interface make control difficult. The manual control of any system (in this case the psychomotor control of an aircraft) cannot be separated from issues in the display of information. Display design is outside the bounds of this discussion, however, in a round about way it does neatly lead us into a description of the basic properties of control systems that can make them easy (or difficult) for the human operator to use. However, prior to engaging in this discussion, it is worth briefly describing the measurement of error in a dynamic, tracking task.

Three major types of error may be identified. The first, and simplest measure is the constant error, which is the signed, mean difference of the flight path obtained from the flight path desired. For example, over a ten nautical mile straight and level sector, taking deviations from desired altitude at half-mile intervals, you may observe that the pilot was on average 50 feet below the desired track. The absolute error is a variation on the constant error, however, in this case the sign of the deviation is removed before calculating the mean deviation. The second major error type is the variable error, which is the standard deviation of the constant error. This indicates how 'smoothly' a pilot is flying.

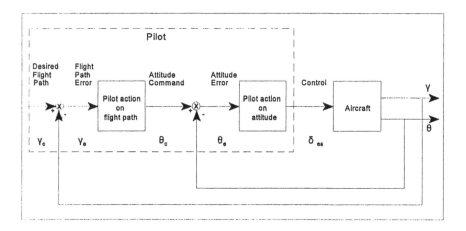

Figure 6.3   **The series pilot model illustrating a response in pitch. γ is the flight path angle; θ represents pitch attitude; γ_c is the commanded flight path angle; γ_e is the error between commanded and actual flight path; θ_c is the commanded pitch attitude; θ_e is pitch attitude error; δ_{es} represents a pitching input to the stick made by the pilot**
Source: Adapted from Field (1995)

One of the most common measurements of performance is the third main type of error measure, namely the root mean square error (RMSE). This is the square root of the mean of the squared error scores. This is perhaps the most popular measure of tracking performance. However, as Hubbard (1987) notes, it has several problems associated with it. In Figure 6.4(a), the points on the semicircle all have the same RMSE value, but actually represent totally different performances. On the y axis is a plot of variable error; on the x axis is a plot of constant error. Consider these as deviations from desired altitude. A pilot on the right hand side of the figure (pilot 1) is flying smoothly (low variable error), but is consistently high; one on the left (pilot 2) is again flying smoothly but is consistently low. A pilot in the centre of the diagram (pilot 3), on average, is flying at the desired height, but is frequently above and below the assigned altitude and is making gross control corrections to minimise the error. The first two pilots described are minimising velocity error at the expense of positional error as shown in Figure 6.4(b). The third pilot is minimising positional error at the expense of velocity error as shown in Figure 6.4(c).

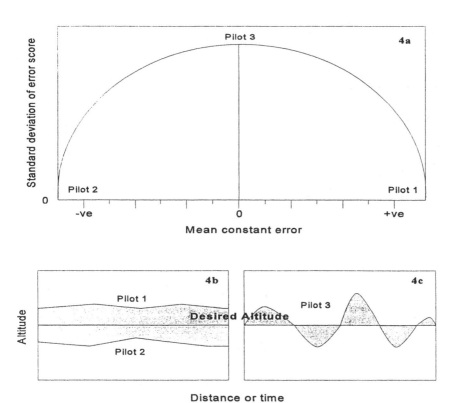

Figure 6.4   The relationship between root mean square error (RMSE), constant error and variable error (standard deviation)
Source: (a) is adapted from Hubbard (1987). (b) and (c) are adapted from Wickens (1992)

Returning to the characteristics of the types of control systems used for closed-loop, tracking-type tasks, Wickens (1992) identifies three aspects of system dynamics; gain, time delays and control order, that affect operator performance, either directly or indirectly, via a fourth factor, stability.

Control gain is also referred to as sensitivity or control: response ratio. In a high gain control, a small input by the operator will result in a large system output. The relationship between operator performance and gain is best described as an inverted 'U'. Performance suffers if gain is either too high or too low. This observation can be explained quite simply with reference to a general model of human control strategy. Any control movement consists of two distinct phases: a gross movement phase and one (or more) iterations where the operator subsequently 'adjusts' their control inputs until exactly the desired output is obtained (e.g., Woodworth, 1899). A high gain control minimises the amount of time required to make the initial gross movement, but makes the subsequent adjustment far more difficult and results in poorer performance, expressed in either speed or accuracy of response. A lower gain control decreases the adjustment time and improves accuracy, but increases the time required to complete the initial gross movement phase. The gain of any control system is a compromise between these two conflicting requirements, (see Figure 6.5).

Most general aviation (GA) and commercial aircraft have quite low control gains. Military aircraft have much higher control gains. This reflects the different levels of agility of the aircraft, which is a product of their different uses. Military aircraft are designed to undertake extreme combat manoeuvres, which require rapid roll and pitch rates, thus emphasis in the control system is placed upon optimising the initial phase of the control task. This can make the approach and landing phase difficult, especially in the later phases where precise control is required, which is difficult with a high gain control system. However, in several modern FBW combat aircraft the control gains are reduced when the undercarriage is lowered. GA and commercial aircraft have much lower control gains as they have no requirement to undertake rapid, gross manoeuvres. Furthermore, a very high gain control in an aircraft that cannot actually generate very high pitch and roll rates can make it seem (to the pilot) even more 'sluggish' in its responses.

Lower gain systems seem to be beneficial to performance in initial flight training (Young, 1994). In this study, low gain control systems produced better performance than high gain systems by minimising variable error. No differences were observed in constant error. High gain systems can also err towards instability, especially when the system also exhibits large control lags. As a result of control power being a product of the mass of air per unit time flowing over the aircraft's control surfaces, larger deflections are required for the same degree of aircraft response at lower speeds than at higher speeds (i.e., the control gain does not remain constant: it is a product of airspeed, where a higher airspeed has the effect of increasing gain). As aerodynamic pressure takes longer to build up at lower speeds, greater control lag and lower gain is evident to the pilot as speed decreases.

Some authors (e.g., Sanders & McCormick, 1987) have suggested that joystick-type controls exhibit almost no travel time (as noted in Figure 6.5, this line is flat and almost horizontal for all control gains), as a large deflection can be made almost as swiftly as a small deflection, thus the gain of these controls should be optimised for their adjustment phase. Yoke-type controls do not exhibit this 'flat line' response in travel time. Young (1994) observed that *ab initio* pilots performed better on a simulated approach and landing task when using a yoke control (especially if it had low gain) than when using a stick of the same gain. Again, this performance benefit was as a result of inducing lower variable error. To re-iterate, gain is the ratio between control system input (e.g., to roll an aircraft, the stick deflection in degrees or the stick force required), to system output (roll rate in degrees per second). Gain alone, cannot account for the better performance observed in *ab initio* pilots using a yoke. Therefore, the speed at which control inputs can be made must be a product of another aspect of the pilot/flight control system.

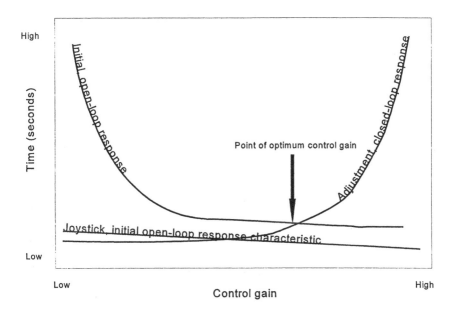

Figure 6.5   **Idealised method for the optimisation of a generic, manual control's gain for initial, open-loop control movements and subsequent, closed-loop adjustments**
          Source: Adapted from Sanders and McCormack (1987)

Gross control inputs cannot be made as quickly using a yoke as when using a stick, even if the gains of the two systems are the same. It can be suggested that the control design, to a certain extent, dictates the pilot's bandwidth. Bandwidth refers to the number of corrective control actions that can be made per second. Bandwidth is usually expressed as a frequency. In a predictable task, the upper limit to human bandwidth is in the region of 2-3 Hz (Pew, 1974). However, in an unpredictable task this upper bandwidth limit falls to 0.5-1Hz (Eklind & Sprague, 1961). The upper limit to a human operator's bandwidth is dictated by the speed of the human information processing system, but this can be modified (downwards) by the design of the control interface. This is essentially what was happening in the study reported by Young (1994). Participants using a yoke as the primary flight control were flying more smoothly as a result of the design of the control inceptor discouraging 'over-controlling', thereby promoting smoother flying.

Time delays take two forms: Control lags and display lags. (For any control engineers reading this chapter, it should be noted that in this case the terms 'lag' and 'delay' are being used synonymously.) For most private pilots it is the former that imposes the most frequent problem. However, referring back to a previous section, the latter also causes pilots problems when flying on instruments. As it is, the human operator cannot respond to flight path errors immediately. There will always be a lag between an error in the aircraft's flight path being detected and responded to. As will be seen in the following sub-section, this lag will be determined (to some degree) by the human response to the control order of the system. If there is also a lag in the system *per se,* performance will deteriorate even further. If the aircraft exhibits oscillatory behaviour (in the long term, phugoid response) it is possible for the pilot to make control responses 180 degrees out of phase at certain pitch oscillation frequencies in an attempt to control them. This is a direct result of lags in the control system and the human operator's response time to counteract them. When this is the case, instead of nullifying the uncommanded oscillatory pitching, the pilot's actions will actually reinforce the oscillations (pilot induced oscillations, or PIO). The pilot-vehicle system is more prone to instability when long lags are associated with a high gain control. This is partially obviated in the landing phase by a reduction in gain, which is a product of the lower airspeed at landing.

Human operators are particularly poor at anticipating what the future state of a system will be when control inputs and system outputs are separated appreciably in time. The greater the separation of input and response, and/or the greater the inertia in the system (as in sailing a super-tanker or, to a lesser extent, flying a Boeing 747), the more difficult the control problem becomes. With experience, the pilots develop a mental model of the aircraft's dynamics, which means that they can anticipate the likely size of effect of a control input without waiting for feedback about the subsequent size of the error. Their control inputs become open-loop. Pilots often refer to being 'ahead' of the aircraft on a good day, and 'behind' an aircraft on a bad day. This is discussed in greater detail in the following section on psychomotor skill development. The anticipatory problem faced by the human operator of a system is further compounded by its control order. The higher the control order, the greater the demands that are placed upon the operator's mental model of the system in an attempt to predict the likely outcome of a control input.

Control order refers to the manner in which the system output changes in response to a control input. In a 'zero-order' control system, control position directly changes the position of the controlled element. A reasonable analogy for this is the operator of a spot-light tracking an actor on a stage. Zero-order control systems are also known as 'position' controls. In a 'first-order' control, a control input alters the rate of change of the controlled element, and hence also its position. A classic example of this type of control is the accelerator in a car. Finally, a 'second-order' control changes the position of the controlled element through altering its rate of acceleration. The stick in an aircraft has elements of the lower two control orders. The one that the pilot will be most aware of is the first-order, short-period response where an input to the stick alters the rate of roll or pitch of the aircraft. In the longer term, the stick has zero-order control properties (it will be recalled that the response of the aircraft in the longer term is an angle of attack command system), where the stick position corresponds to angle of attack. However, it is unlikely that a pilot will choose to hold these stick forces for any length of time, so this positional response is usually 'trimmed out', so s/he is unlikely to be aware of this property. Zero- and first-order systems are less prone to becoming unstable than second-order systems. This is partly because human beings exhibit shorter initial response times to both zero- and first-order systems (in the order of 150-300 ms) than to second-order systems (400-500 ms) as well as their being easier to control (McRuer & Jex, 1967).

Optimisation of aircraft control systems There is little that can be done about control order or cross-coupling in a basic training aircraft using conventional, mechanical flight control systems. Control lag can be reduced by giving the aircraft larger, more powerful, control surfaces. This will also have the effect of increasing the rate at which the aircraft can perform rolling and pitching manoeuvres. These improvements in the system response would initially seem to be beneficial for improving pilot performance. This may not be the case though. The gain in the system also needs to be considered. If the control gain is too high, the aircraft may appear 'nervous' and unstable to trainee pilots and will increase the speed at which they can get into 'unusual' attitudes. Decreasing control lags by increasing manoeuvring power may need to be offset by decreasing control gain. Decreasing the control gain will also have the benefit of encouraging smoother flying, but not to the detriment of maintaining the desired heading or vertical flight path. Smoother flying can also be encouraged by effectively decreasing the bandwidth of the trainee pilot by utilising a yoke rather than a joystick type of primary axis flight control. It is perhaps no accident that high performance aircraft (e.g., military and aerobatic aircraft) have evolved in the way they have, using a joystick as the primary axes control inceptor, whereas lower performance aircraft more commonly use a yoke-type of control. Control system parameters should be *optimised* for the aircraft-type and the task at hand, not *maximised*.

Psychomotor Skill Development

Learning how to control the flight path of an aircraft using the stick, rudder and throttle is an example of psychomotor skill development It has been suggested on several occasions that this task can be characterised as a compensatory tracking task. A second characteristic of the task is that it is a continuous, closed-loop task. In a closed-loop task, human controllers are continually aiming to minimise the error component by attending to feedback about their performance and acting accordingly. This can be contrasted with open-loop tasks, in which the operator is not, and cannot, be dependent upon continuous feedback for performance. A typical example of this is striking a golf ball. No matter how hard a golfer tries to influence the flight of the ball after it has been struck, they will usually find that their efforts are to no avail. Furthermore, a golfer selects a certain type of swing to play a shot. This may be a slight variation on their normal swing, but the essential characteristic is that it is pre-selected prior to executing the shot. Once the swing is commenced it becomes 'fully automatic' and the motor programme employed is executed without reference to feedback.

Various stages have been proposed for the acquisition of a skill. Fitts and Posner (1967) suggested that trainees progressed through three stages when acquiring a skill: the cognitive, associative and autonomous phases. In the cognitive phase trainees are required to consciously attend to the cues which guide them concerning when the task should be performed, the individual components of the task and their inter-relationships, and the feedback about their performance. These cues may be from components of the task or task environment, or they may be in the form of guidance from their instructor or training system. In the associative phase, the trainee's actions begin to be guided directly by environmental cues but still some cognitive attention is required. In the final phase, the autonomous phase, task performance has become automatic and no longer requires conscious attention.

Anderson (1982) suggested that there were two stages to skill development, declarative and procedural, which roughly correspond to Fitts and Posner's cognitive and autonomous phases. Similarly, Rasmussen (1986) proposed that skill development progressed through three phases: knowledge-based, rule-based and skill-based. These also roughly correspond to Fitts and Posner's three stages. In all cases, the development of skilled performance is characterised by its becoming less available to consciousness and occupying less cognitive capacity. Motor skills also become less dependent upon external sources of feedback. Jaeger, Agarwal and Gottlieb (1980) proposed a three-stage developmental model of human skilled performance of tracking skills. They suggested that the student initially learns the directional relationships, followed by the timings and finally the spatial relationships. Wightman and Lintern (1985) suggested that in the development of three-dimensional tracking skills, such as those that a pilot is required to develop, a fourth and final factor, that of control coordination, was the final aspect to develop.

Poulton (1957) characterised the environments in which skilled performance took place as being either open or closed. Taking the latter first, a closed environment is a predictable environment. Human output (control) requirements are always identical for a given input. Aviation, however, takes place in an open environment, where a great deal of flexibility in skilled performance is required on the part of the pilot. The control requirements placed upon a pilot on a turbulent day to maintain altitude and track are quite different to those placed upon him/her on a calm day. This distinction becomes more pertinent when the differences between motor programs and motor schema are considered.

It may be beginning to seem that all aspects of the psychomotor performance can be broken down into two categories: open loop versus closed loop skills; open and closed environments; conscious versus automatic task performance, etc. An old colleague once observed that there were two types of people; those who broke everything down into two categories and those who did not! It is, in fact, a gross over-simplification to suggest that all of the above aspects of human skilled performance fall into convenient dichotomies; they do not. These descriptors are best thought of as representing ends of continua.

Up to this point it has been convenient, especially for the discussion of the aircraft dynamics, to regard flying as a closed-loop, compensatory tracking task. While this is what the task requires, and is also the manner in which an autoflight system tackles the problem (i.e., minimising the flight path error until it tends towards zero), it is not actually the way in which a skilled pilot performs the task. McRuer and Jex (1968) in their 'successive organisation of perception' model of tracking behaviour, proposed that skilled operators could potentially operate in three modes, the highest of which could only be achieved with extensive practise. They suggested that novices operated at the lowest level, which was a pure compensatory control strategy. In this mode, the operator simply minimises any error. However, this results in relatively poor control performance, as the response times of the human operator, control lags and display lags will all play a part. With increasing practise it becomes possible to operate in a pursuit mode, responding directly to the control input, thereby improving performance by avoiding many of the lags in the system. At the highest level, the operator is responding in a pre-cognitive mode. This is an entirely open-loop behaviour that operates independently of any feedback.

A highly skilled pilot can exhibit all of these modes of tracking behaviour, and will need to as flying, in the terminology of Poulton (1957), is very much an open environment. Flight path also always shows an appreciable lag in time behind the aircraft's angle of attack (see Figure 6.1), and hence the pilot's control inputs. Operating in a pre-cognitive mode allows the pilot to minimise or avoid many of the lags inherent in an aircraft's flight control system. However, the pilot's reactions to the disturbing effects of turbulence must be compensatory, because the effects of turbulence on flight path cannot be predicted. The experienced pilot compensating for a side-wind on final approach will exhibit pursuit tracking behaviour. Although the accuracy of the rudder application will become apparent as the pilot continues the approach, initially the pilot will put in an amount of rudder which 'feels about right'. Rotating the aircraft to the appropriate pitch attitude on take-off or in the flare, may be considered as an examples of the pre-cognitive mode of control behaviour.

As Wickens (1992) has noted, many real-world human control activities cannot be described simply as either open-loop or closed-loop activities. Piloting an aircraft falls into this category. When performing an instrument landing system (ILS) approach, in the initial stages of training the pilot will be operating in a compensatory fashion, attempting to minimise the glide slope and localiser error with reference to the ILS display. With experience, upon observing a flight path error, the pilot will put in an input of 'about the right size' (an initial open-loop control input) which will bring the aircraft (approximately) back on the desired flight path. This initial (large) input will then be modified with additional smaller inputs until the error is zeroed. Similarly, when the pilot wishes to execute a left hand turn, s/he makes a gross, open-loop left deflection of the stick to initiate the aircraft rolling in the desired direction at about the right rate of roll. When the rate of roll initiated has resulted in approximately the desired bank angle, the stick is centred. This is then followed by a series of closed-loop adjustments to obtain and maintain exactly the required bank angle. A similar process will be undertaken to roll out of the turn onto the desired heading (the pilot's surrogate control parameter for track).

This large movement followed by smaller movements is characteristic of many human control actions. As long ago as 1899, Woodworth noted that the initial large movement in an aiming task was 'ballistic' and not reliant upon feedback (what we now refer to as open-loop), followed by smaller control movements which relied on feedback for accuracy. As skill levels increase, the initial movement becomes larger and more accurate and the smaller control movements decrease both in size and number. Furthermore, the time between the phases of movement also decreases. Because the initial movement is open-loop, it is not dependent upon extrinsic feedback. Feedback is required, however, if the operator is to develop this initial 'ballistic' open-loop skill component.

Furthermore, momentarily referring back to the first section of this chapter, when these two movement phases are considered, it can be seen why it is important that control gains are optimised. The novice pilot flies in a compensatory tracking manner, placing emphasis less on the ballistic (open-loop) control movement phase and more on the adjustment (closed-loop), feedback-dependent phase. The novice pilot also does not undertake gross flight manoeuvres, concentrating on the basics. As a result, the novice pilot needs a lower gain control optimised for the adjustment phase. The military fast-jet pilot will, however, have amassed many more flight hours and therefore will employ a more open-loop (or 'precognitive' in the terminology of McRuer and Jex) control strategy, partly as a result of experience and partly as a result of the necessities imposed by the air combat environment. As a result, s/he will require a control optimised for the initial, ballistic control movement phase, (i.e., a higher gain control).

One of the basic tenets of this chapter has been that most pilots' control inputs have two components to them: an initial, open-loop response, followed by a closed-loop response. Until the mid-1970s, the generally-held theory was that open-loop motor skills were held as some sort of 'template' of related muscular movements, known as a motor programme, defined in space, sequence and timing (Adams, 1971). Unlike dynamic closed-loop tasks, these motor programs are not dependent upon visual feedback for their execution. They are, however, dependent upon proprioceptive (intrinsic) feedback for their performance.

Adams (1971) suggested that a motor programme had two distinct memory components associated with it: the memory trace and the perceptual trace. The former component selects and initiates the appropriate motor programme at an appropriate point in time from the person's 'library' of motor skills. The latter memory component evaluates the outcome. As the sequence of movements 'stored' in a motor programme is executed, kinaesthetic and/or proprioceptive feedback from the muscles and joints is compared against the programme's 'template', to guide performance. Visual feedback may also be used, but this is not totally necessary. Typical examples of motor programs include hitting a golf ball, tying a shoe lace and articulating words. The decision to initiate a motor programme is conscious. The actual execution of the programme is automatic and not open to conscious inspection. Adams identified two stages in the development of a motor programme. In the initial verbal-motor stage, trainees require extrinsic verbal or visual feedback on their performance. After a while, as the perceptual trace develops, trainees begin to be able to evaluate their own performance. Eventually, they can do this simply from the 'feel' of their actions (intrinsic feedback).

Wickens (1992) identified four attributes of a motor programme: they are produced as a result of high levels of practice; they demand few attentional resources for their execution; only a single response is required to select the programme; and finally, the result of the programme will be extremely consistent. It is on the last attribute, though, that motor programme theories fail at a general level. Psychological text books are festooned with 'hoary old chestnuts' of examples to illustrate certain points. The main problem with motor programme theory is best illustrated with reference to one of these time-honoured greats. Consider signing your signature, a typical open-loop motor programme. Your signature will appear 'the same' whether you are signing the strip on the back of your credit card or writing it in metre high letters on a blackboard. However, both processes use totally different muscle groups, so therefore both processes cannot belong to the same motor programme.

Schmidt (1975) in his motor schema theory suggested that what remained invariant was not the *process* by which a response was produced, but the sequence of, the relative proportions of, and the timings of execution pertaining to the *product*. Consider performing a gentle and a medium rate left hand turn: stick over, followed by a little rudder to keep the turn balanced and a little back pressure on the stick to keep the nose from dropping. The relative timing of the movements required and the relative control deflections are the same in both cases. The two turns differ only in the size of the control inputs required. Motor schemas can be considered as a more generalised form of a motor programme.

Motor schemas have two sub-components within them, a recall memory component and a recognition component (Stelmach, 1982). The recall component specifies the parameters for the initial, open-loop ballistic control movement. The recognition component acts as a template against which (intrinsic) feedback can be compared to check on performance. After the initial ballistic control movement has been completed, the recognition schema may also play a part in mediating the effects of extrinsic feedback.

Proctor and Dutta (1995) outlined several components of information that were required to form a motor schema. These included: the movement parameters (relative size, duration, and sequence) in the schema; the knowledge of the appropriateness of the outcome of these actions; and the environmental conditions prior to the execution of the movement. With reference to these information types, it can begin to be understood how the motor schemas associated with flight skills develop.

Movement parameters How do pilots develop the complex patterns of control movements required to control an aircraft? The answer would seem to be that the most effective way of demonstrating the inter-related, cross-coupled movements required is by using the 'traditional' technique of 'follow-through', where trainees rest their hands and feet on their controls, holding them loosely, while the instructor demonstrates the required manipulations. But why should this be effective? As mentioned two paragraphs previously, the proprioceptive and kinaesthetic systems play an important part in the execution of open-loop skills. However, there is also experimental evidence that suggests the development of closed-loop skills can be enhanced by physically guiding the hands of participants learning a new tracking skill (e.g., Williams & Rodney, 1978; Holding & Macrae, 1966). When learning to fly an aircraft, the instructor is basically trying to transmit complex information to the student pilot concerning the direction, frequency, amplitude and harmonisation of the inputs to the various controls. Spoken language cannot easily convey the complexity and precision of the control movements required, if indeed these highly skilled behaviours are capable of being verbalised, so the use of the kinaesthetic and proprioceptive systems would seem desirable.

In a study conducted by Rees and Harris (1995), 20 *ab initio* student pilots flew a series of simulated approaches using either linked or unlinked primary flight controls. In the linked condition, where control deflections on one side of the 'cockpit' were mirrored in the deflections of the other pilot's controls, the trainees were taught by the instructor using verbal guidance and 'follow-through'. The training regime was exactly the same for the trainees in the unlinked condition, however they only received verbal instructions since the instructor's control deflections now did not cause the trainee pilot's controls to move. The results indicated that those pilots who performed the task in the unlinked configuration showed higher variable errors for track keeping and also around the optimum descent profile, (i.e., they flew less smoothly) than those in the linked condition. The participants in the unlinked controls group also showed little evidence of any improvement in performance over the ten trials, unlike the participants in the linked group, who exhibited a decrease in variable error as trials progressed. The physical linkage of the student's and instructor's primary flight controls would seem to be an important channel of communication that enhances skill development in trainee pilots.

Not only is the kinaesthetic channel an important channel for communication, it is also the *correct* channel to use when trying to communicate subtleties in control strategy required to fly a light aircraft. Several cognitive psychologists have suggested that there are at least two separate identifiable components to human working memory, (e.g., Baddeley, 1990; Wickens, 1984). One component deals with verbal/linguistic information and the other is concerned with spatial information. Annett (1979) argued that verbal and motor memory systems *per se* were quite distinct but there was an action-language bridge that linked them. Wickens, Sandry, and Vidulich (1983) demonstrated that there was an optimal association between information code and the component of working memory utilised (spatial or verbal). Verbal information should address the verbal component of working memory, and spatial information should address the spatial component.

With reference to Multiple Resource Theory (Wickens, 1984), it can be demonstrated that instruction using linked and unlinked primary flight controls utilises different information processing routes and resources. In both cases, the response required of the student pilot is a spatial response (i.e., guiding the aircraft in three-dimensional space using control displacements). Using linked controls, the information is transmitted from the instructor using the kinaesthetic modality, whereas in the unlinked control condition the information is communicated via the auditory and/or visual modality. In the unlinked case, the instructor's directions are only transmitted verbally. This somewhat imprecise verbal information (e.g., 'left a bit'; 'back a little more on the stick') then requires translating into a spatial code, before the required manual response can be elicited. This is clearly not an optimal arrangement in human information processing terms.

Using the framework proposed by Annett (1979), the action-language bridge needs to be crossed, which appears to be a somewhat inefficient mechanism for learning. The transmission of information from instructor to trainee is better arranged in the linked-controls condition, subsequently resulting in superior performance.

Knowledge of the appropriateness of the outcome Unless information concerning the appropriateness of a control input is given to student pilots, they will never know the difference between a correct and an incorrect control strategy. This relates to the second type of information required to form a motor schema, knowledge of the appropriateness of the outcome. It has been known for nearly a century that knowledge of results (KR) is an important component in learning and training. However, using 'follow through' as an instructional technique actually provides knowledge of required performance (if the trainees are following through on their instructor's control inputs) or knowledge of their own performance (if the instructors are monitoring them via this medium).

Knowledge of performance (KP) is not just concerned with providing information about the outcome of a series of control inputs, as is KR (e.g., putting the aircraft down on the centreline of the runway). KP provides knowledge about the manner in which the final outcome was achieved: was the process correct or not? This feedback can come in two forms: knowledge of the forces required during a task (kinetic feedback) or knowledge of the spatial and temporal aspects of the task (kinematic feedback). Proctor and Dutta (1995) argued that KP was superior to KR for training the multiple degrees-of-freedom tracking tasks found in real life, such as flying an aircraft.

In 'traditional' flight training though, where trainees receive directions from their instructors in the air, KR and/or KP is sometimes difficult to provide, either in a usable or a durable format. The only sources of information about whether a task has been completed correctly or not are the flying instructor's 'on the spot' comments (which cannot be referred back to later), or his/her observations in a de-briefing session after the flight (and are thus temporally removed from the flight task). The enhanced feedback in the form of KR that flight simulators can provide (e.g., replay facilities or plots of flight path error) can help out in these respects. Simulators can also provide augmented, real-time feedback. However, simulation as an instructional tool is rarely utilised in the course of Private Pilot's Licence (PPL) training. Perhaps this should not be so.

Before continuing with this discussion, as a brief aside, even the most cursory overview of the literature will reveal that the majority of flight skills research (including some of the author's own work) has concentrated on the approach and landing phase. This is not too surprising as it is the phase of flight that places greatest demands on the pilot in terms of both the amount and the precision of the control requirements. It is, however, a little odd, in that it is the phase of flight that *does not* employ a 'typical' control strategy. When flying an approach in a light aircraft, the pilot is flying close to the 'back side' of the drag curve. Therefore, airspeed is controlled by fore/aft deflections of the stick and rate of descent (vertical flight path) is controlled on the throttle. This is not the typical control strategy used by pilots 95 per cent of the time. Nevertheless, the psychological motor schema development mechanisms hold good irrespective of the control strategy employed. Hopefully, by the end of the next paragraph it will begin to become apparent why this apparent brief digression from the main theme was required.

One of the basic problems faced by the *ab initio* pilot flying an approach is to establish what the correct approach path is. The pilot must know what the correct flight path is before he or she can apply the correct control inputs to obtain the desired result. They must know how big their flight path error is (if indeed there is one) before they can perform a control action of the required type and magnitude. As pitch attitude, angle of attack and flight path diverge at low airspeeds this is not an inconsiderable problem. It will be recalled that the ultimate objective of the pilot is to control flight path. However, this is done through the surrogate (inner-loop) parameter of pitch attitude.

Lintern, Roscoe, Koonce, and Segal (1989) suggested one approach to solving this instructional problem would be to train *ab initio* flight students in a flight simulator with augmented feedback. In this case, the augmented feedback was either in the form of a 'tunnel in the sky', which marked the correct three-dimensional flight path, or as a velocity vector showing their predicted position. The students who were trained with augmented feedback showed a positive transfer of training to actual flight performance, requiring fewer landings prior to going solo than the control group.

Augmented feedback, though, should be used with some caution. Not all studies which used augmented feedback to train psychomotor skills have been successful. In an attempt to train air-to-air gunnery skills, Goldstein and Rittenhouse (1954) used augmented feedback on a gunnery simulator. In this case, the experimental group received a tone when they were on-target. The control group received no such augmented feedback. In the final assessment of performance for both groups where no feedback was present, the experimental group, which had received the tone when they were on target, performed considerably worse. It was concluded that the tone was being used as a task cue by the experimental group, rather than as a source of information concerning their performance.

Knowledge of environmental conditions required prior to the execution of a control movement Proctor and Dutta's last requirement for learning a motor task is a knowledge of the conditions prior to the execution of a control movement. Put another way, how does the pilot know when s/he should make a control input and which one should it be? Dennis and Harris (1998) demonstrated that the development of flying skills could be enhanced without using a set of representative flight controls. In this study, two groups of ab initio student pilots were given training on a flight simulation package running on a desktop computer prior to performing some basic flight manoeuvres in the air. One group interacted with the computer using a representative set of flight controls. The other group used only the computer's cursor and function keys.

It was impossible to distinguish between the groups in terms of their in-flight performance (track, altitude and airspeed deviations in straight and level flight; altitude and heading deviations when maintaining and exiting turns). However both groups exhibited superior performance compared to a control group who had no computer-based training of any type. Interacting with the flight simulation package using only the computer's keyboard cannot in any way mimic the physical control actions required to fly a light aircraft, nevertheless, it can help the trainee decide when a control movement should be initiated and terminated.

For example, when rolling out of a turn, the *ab initio* pilot needs to know when (in degrees) to start putting in a control deflection opposite to the one that initiated the turn in order to roll out onto the desired heading. *When* to initiate the control input (the environmental conditions) can simply be trained by pressing a cursor key. The size and coordination of the inputs cannot be trained using just the computer's keyboard as the primary flight controls. Training on the PC-based flight simulator provides trainees with a moderately realistic preview of the pace of the task; it affords them the ability to begin to develop the cognitive template of what the required performance on the task looks like, thereby aiding in the development of their motor schema and also familiarises them with some of the control interactions, for example between roll and pitch. In the words of the Cranfield College of Aeronautics' Chief Test Pilot, '... it helps to get their [the trainee pilot's] clock up to speed'.

What was also interesting in this study was that although both groups trained on the computer performed equally well and out-performed the control group, when the workload data were analysed, it was observed that the group trained using the representative set of flight controls was under considerably lower workload than the other two groups. The *ab initio* pilots trained on the PC using the computer's keyboard could perform up to the standard of their peers who used a representative set of flight controls, but only at a cost of incurring higher levels of workload.

While equipping the computer with a set of controls did not help to train psychomotor skills *per se,* as their movement (kinematic) characteristics and feel (kinetic) characteristics did not represent those found in the aircraft, it did aid in developing three aspects of the student pilot's motor schema: the movement parameters, knowledge of the appropriateness of the outcome; and knowledge of environmental conditions required prior to the execution of a control movement. Controlling the 'aircraft' using the computer's keyboard addressed only one (knowledge of environmental conditions) or possibly two (knowledge of the appropriateness of the outcome) of the parameters required to develop a motor schema, hence the higher workload experienced in flight by the group trained this way.

Training Schedules

Learning to fly an aircraft is an excellent example of part-task training. In part-task training, the student learns individual aspects of the task before compiling them into the whole task. Wightman and Lintern (1985) identified three types of part-task training: segmentation, fractionation and simplification. Initial flight training, in general, utilises a great deal of the first two techniques. The training of psychomotor skills for flight has typically only used the first and last of these part-task techniques, with varying degrees of success.

Segmentation involves decomposing a task into discrete, serial components in a task (for example take-off, climbing and straight and level flight). Each of these components is then trained in isolation (not necessarily in the order in which that occur in the actual task) before being assembled into the whole task. Consider the first few lessons in most PPL syllabi. The first lesson covers straight and level flight. The second lesson usually combines teaching climbs and descents (e.g., the student pilot takes over shortly after take-off), punctuated by periods of straight and level flight. By the third lesson, the student may be conducting the take-off, before continuing to practice climbs and descents, and straight and level flight, while also being taught new manoeuvres. This can be considered as an example of backwards chaining (see Figure 6.6). Like most examples, this one begins to break down if examined too closely, however, it does serve to illustrate the principle.

As can be seen in the other examples of segmentation illustrated in Figure 6.6, not only is it important that the basic flight skills are taught first before the more advanced manoeuvres, the sequence of events when flying also dictates a backward chaining approach; you cannot just keep taking-off on your first lesson!

Fractionation Not only can the flight task be segmented serially, it can also be fractionated. If serial segmentation is conceptualised as breaking down the flight task into its serial components, (i.e., along the *y* axis), fractionation can be thought of as decomposing the flight task into its parallel component tasks (distributed up and down the *x* axis). These individual parallel components in the task can then be trained separately. The various methods of fractionation are illustrated in Figure 6.7. The three classic parallel tasks that any pilot must perform are 'aviate', 'navigate' and 'communicate'. Most trainee pilots have enough problems just completing the 'aviate' component of the task for the first few lessons. As their first solo flight approaches, the 'communicate' aspect of the flight task is introduced in parallel. Finally, student pilots will be expected to perform the 'navigate' component as well. This is a classic example of repetitive part training, as is illustrated in Figure 6.7.

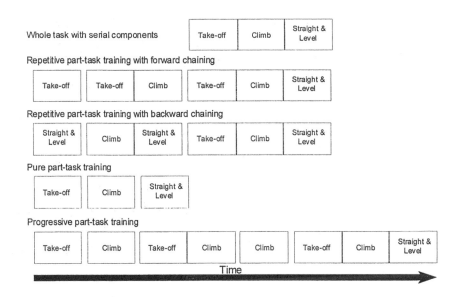

Figure 6.6   Segmented training schedule for initial stages of flight training
Source: Adapted from Proctor and Dutta (1995)

Simplification is the last of the part-task training techniques described by Wightman and Lintern (1985). This is difficult to describe diagrammatically, but can be explained quite simply with reference to a simple example. When learning to ride a two-wheeled bicycle, most parents simplify the task for their children by fitting stabilisers. Perhaps fortunately, this part-task training approach is not easily implemented in initial flight training. As we shall see in a little while, evidence would suggest that simplification is not desirable when attempting to acquire the motor skills associated with flying.

Part-Task Training for Psychomotor Skills in Aviation

Wightman and Lintern (1985) conducted a meta-analysis of the results of studies to assess the effectiveness of the various types of part-task training for the acquisition of the motor skills involved in flying. Many of the studies reviewed used a 'transfer of training' paradigm. This type of experiment involves training one group in the conventional manner (e.g., in the aircraft) and training a second, experimental group, in a different way (e.g., in a simulator or using a part-task training approach), before transferring to the conventional training programme. The usual measure taken is the time to reach a criterion level of performance on the whole task. If the experimental group takes less time to reach the criterion than the control group, there is said to be a positive 'transfer of training'. This does not necessarily mean that the training is more effective, as if they save three hours during the conventional training programme time but spend six hours in the simulator during the pre-training period, then these students have actually spent three hours longer on the training course. If, however, the cost of the pre-training is less than the cost of the training time saved on the conventional programme, it still may be worthwhile in terms of cost savings, despite taking longer. This is typically found in flight training when simulators are used. The cost of operating a flight simulator is considerably less than that of operating an aircraft, so the extra overall time in the flight training programme may be justified on this basis.

Segmentation can only reasonably be used as a part-task training technique when undertaken on a flight simulator. Most flight-based studies utilising segmentation have concentrated on the approach and landing phase, typically with the greatest degree of repetition being placed upon the very final stages of approach and the landing flare. All the studies reviewed by Wightman and Lintern (1985) showed positive transfer of training when this technique was employed. Segmentation through part-task training in a simulator allows repeated practice of particularly difficult sections of the flying task without interference from intervening aspects of the whole task that would be unavoidable when training in the aircraft itself. For example landing practise can be undertake repeatedly without having to take-off again and fly another circuit.

Implementation of fractionation in a motor control task such as controlling an aircraft is somewhat difficult to undertake in anything other than a simulator. Fractionation in flight training, as previously discussed, tends to be implemented on the age old basis of 'aviate, navigate, communicate'. Fractionation within each of these components is rare. Wightman and Lintern (1985) describe the work of Briggs and Waters (1958) and Adams and Hufford (1961) who attempted to train students in an approach and landing task using this part-task approach. In the former study, trainees were initially trained by teaching and practising the pitching aspect and the rolling aspect of the task separately. It was concluded that there was some positive transfer of training to the whole task, but it was not high. The amount of training transfer decreased as the degree of interaction of components in the sub-tasks increased.

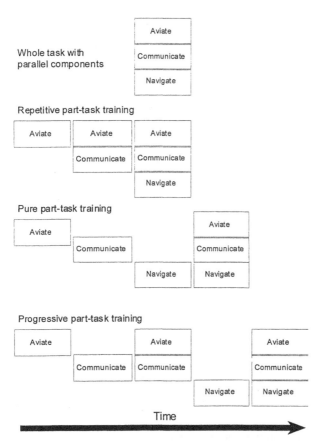

Figure 6.7 **Fractionated training schedule for initial stages of flight training**
Source: Adapted from Proctor and Dutta (1995)

The Adams and Hufford (1961) study used a different approach to fractionation. In this case it was attempted to pre-train the perceptual element of the task by having trainees view a series of pre-programmed exemplar landings. It was observed that this group showed no significant difference in performance to a control group who did not undergo the perceptual pre-training. In short, fractionation would appear to be neither a viable nor desirable way to train motor control flight skills.

Simplification would not appear to be a viable technique for the training of the motor skills required to fly a light aircraft, although it can be argued that commencing training in a light aircraft before transferring to a more complex aircraft is an excellent example of simplification that works. However, in the review by Wightman and Lintern (1985), it was concluded that there was little empirical evidence that simplification of the flight control problem provided any benefits compared to full-task training. However, there also appears to be evidence that simplification may actually be detrimental to the acquisition of flight skills. In a later paper by Lintern, Roscoe and Sivier (1989) student pilots were trained in a flight simulator on an approach and landing task in which either they trained on the full task, or the control problem was simplified by reducing the control order on the primary axis flight controls to approximate that of zero-order (positional) controls. It was observed that pilots trained in the latter condition performed considerably worse during transfer trials than did pilots trained in the simulations which mimicked the normal flight dynamics of a light aircraft.

In the trials conducted by Young (1994), volunteers conducted ten trials to approach and landing. In this study the flight task was simplified in a different way. During the first five trials, the trainees did not control the throttle of the aircraft (hence did not control the rate of descent). This was done for them by the instructor. From the sixth trial onwards, they were also required to manipulate the throttle. Irrespective of the control condition, performance by the end of the tenth trial (as measured by deviations from localiser and glide slope) had never recovered to the level observed at the end of the fifth trial.

Perhaps these observations are not too surprising. It has been known for many years that when a trainee is attempting to master a complex, closely co-ordinated task with many control interactions (such as flying an aircraft), it is better to train the task as a whole rather than individually in its component parts, (e.g., McGuigan & MacCaslin, 1955; Briggs & Waters, 1958; Naylor & Briggs, 1963). Miller, Pribram, and Gallanter (1960) argued that while each individual muscular movement in a larger motor unit could be acquired in isolation, the main difficulty was in combining these individual tactical details into the larger super-set of co-ordinated movements.

Operational Implications and Recommendations

In an odd sort of way, there are almost no operational implications or recommendations for the training of the psychomotor skills required to fly an aircraft. In other situations, Ergonomists have often found that when attempting to improve many traditional hand tools, they cannot. As the tools are used over a protracted period of time, improvements are made by the users to compensate for their inadequacies. Eventually, it becomes impossible to make any further improvements to the design. This is the ultimate form of user-centred design. The same has happened over 80 years of training pilots to fly. The basics have not changed: fly the aircraft; know where you are; communicate with the other users of the sky. When all else fails, the first priority is to fly the aircraft. This latter aspect of the flying task is what psychomotor skill development is about. The techniques used to teach and fly an aircraft have developed to such a point through the pragmatic evolution of instructional practises by instructors, that they probably cannot be improved. Perhaps the most important issue is to ensure that the content of the PPL syllabi across the world are not eroded in order to produce cost and/or time savings. Psychomotor skills take time to develop, and the more complex the skill, the longer it takes to acquire to a point where it becomes automatic.

There is some potential to improve the quality of PPL training through the use of low cost, personal computer (PC) based flight simulation (see Figure 6.8). It has been suggested that the acquisition of flight skills can be improved even when non-representative controls are used on the computer. The use of PCs for flight training has the advantage of being able to place instruction on the ground and increase the amount of air-time available for practice, rather than having to teach and practice in the aircraft, which is both wasteful of time and is also not an optimal learning environment. PCs offer a cost-effective means of providing flight simulation facilities to flying clubs and smaller flight training schools who cannot afford a larger flight training device. Not only will these devices aid in the procurement of control skills, they can also be used in the development of the more cognitive aspects of the flight task, for example navigation and instrument flying.

The US Federal Aviation Administration is now showing increasing interest in the capabilities of these packages (referred to as PCATDs—PC-based aviation training devices) with a view to qualifying them as approved training devices for selected procedures under part 141 of the Federal Aviation Regulations (McLaughlin, 1994; Williams & Blanchard, 1995; Williams, 1996). In general, irrespective of what new devices are being developed to train ab initio pilots, the emphasis should be on improving the quality of instruction, not on reducing the time of instruction. Time in the air at the controls of an aircraft should not be reduced under any circumstances. There is potential to add to it, however, at the controls of a simulator. The only way to develop proficiency in a psychomotor skill is through practice. The more practice, the better. Most readers will agree that it doesn't require a psychologist to come to this conclusion!

Research Agenda

Low fidelity simulators can be highly effective in the initial stages of flight training. Without a doubt, the usage of PCs to train pilots at all levels will increase. The only questions that remain are which tasks are suitable for training on a PC (assuming the current levels of control and display fidelity) and which are not. This research is well underway and as discussed in the previous section, regulations are being incorporated to include PC-based flight simulators into certain parts of the training curriculum.

While there is little immediate research required on the training technology aspects of psychomotor skill acquisition, there are some serious questions to be asked about potential control configurations in future flight training aircraft. It cannot be too long before a manufacturer decides to produce a flight training aircraft with FBW flight controls. This will probably not be an aircraft for initial flight training, as FBW will be prohibitively expensive for aircraft in this category. However, it is not outwith the bounds of probability that the next generation of lead-in trainers for military fast jets (e.g., aircraft in the same class as the Shorts Tucano and the Pilatus PC9) will employ this technology. There will not necessarily be a requirement for the aircraft's primary axis flight controls to be linked within the cockpit (see the control configurations described by Rees & Harris, 1995). While students in these aircraft will not be *ab initio* trainees, they will be learning advanced air combat manoeuvres techniques, which are often taught using 'follow through'. The effect of the deletion of the linkage between the instructor's and the student's flight controls would have to be carefully evaluated to ensure that there were no negative implications for flight training before it could be sanctioned.

Figure 6.8   A typical low-cost PC-based flight simulation

Moreover, with an advanced FBW trainer, there would be no reason to implement the same flight control laws as those found in a conventional technology flight training aircraft. Most FBW fast jets employ a pure pitch rate command system, rather than an angle of attack/pitch rate system, as found in conventional light aircraft. Employing a pure pitch rate command system would seem desirable in a lead-in trainer to a fast jet. However, it does require a slightly different control strategy on the part of the pilot. This would seem to be another excellent reason to maintain a cross-cockpit linkage of the primary flight controls if this transition between these types of aircraft control system is to be completed successfully.

Furthermore, the whole of the psychomotor skills learning process should be seen as a system in which the instructor plays a vital role in providing feedback to students about the appropriateness of their control inputs. The linkage of the flight controls not only provides the student pilot with a learning channel from the instructor, it also provides the instructors with a feedback/monitoring channel from their students about the appropriateness of their control inputs. Without this feedback of control inputs to the instructor, training effectiveness could be adversely effected. So far, the deletion of the cross-cockpit linkage of the controls has only been evaluated from the point of view of the student pilot. If a system's perspective is taken, it also needs to be evaluated from the point of view of the instructor.

In a similar vein, the role of 'feel' through the flight controls needs to be assessed. Every student is taught to recognise the onset of the stall from the buffeting encountered and the 'sloppy' feel of the stick and poor response of the aircraft to control inputs. With the event of FBW flight training aircraft, these sources of feedback could be removed, possibly to the detriment of training.

There is an old adage: 'if it ain't broken, don't fix it'. In many respects, the whole system of initial flight training (in terms of motor skill development) has evolved to point at which it can be said that 'it ain't broken'. There is a risk though, that in an attempt to improve aircraft, many desirable characteristics of the cockpit will be deleted to the detriment of training. It is vital that this does not happen.

Acknowledgements

I would like to thank Jim Gautrey for his help and advice in preparing this chapter, especially the section on the flight dynamics of light aircraft. I would also like to thank Fiona Smith for commenting on, and proof reading the drafts of this chapter. I am most grateful to both of them.

References

Adams, J. A. (1971). A closed-loop theory of motor learning. *Journal of Motor Behavior, 3,* 111-150.

Adams, J. A., & Hufford, L. E. (1961). *Effects of programmed perceptual training on the learning of contact landing skills* (NAVTRADEVCEN 247-3). Port Washington, NY: US Naval Training Center.

Anderson, J. R. (1982). Acquisition of cognitive skill. *Psychological Review, 89,* 369-406.

Annett, J. (1979). Memory for skill. In L. Wankel & R. B. Wilberg (Eds.), *Applied problems in memory* (pp. 215-247). London: Academic Press.

Baddeley, A. (1990). *Human memory: theory and practice*. Hillsdale, NJ: Lawrence Erlbaum Associates.

Briggs, G. E., & Waters, L. K. (1958). Training transfer as a function of component interaction. *Journal of Experimental Psychology, 56*, 492-500.

Dennis, K. A., & Harris, D. (1998). Computer-based simulation as an adjunct to *ab initio* flight training. *The International Journal of Aviation Psychology, 8*, 261-276.

Eklind, J. I., & Sprague, L. T. (1961). Transmission of information in simple manual control systems. *IEEE Transactions on Human Factors in Electronics,* HFE-2 (pp. 58-60).

Field, E. (1995). *Flying qualities of large aircraft: Precognitive or compensatory?* Unpublished doctoral dissertation, College of Aeronautics, Cranfield University.

Fitts, P. M., & Posner, M. I. (1967). *Human performance*. Belmont, CA: Brooks Cole.

Goldstein, M., & Rittenhouse, C. H. (1954). Knowledge of results in the acquisition and transfer of a gunnery skill. *Journal of Experimental Psychology, 48*, 187-196.

Holding, D. H., & Macrae, A. W. (1966). Rate and force of guidance in perceptual-motor tasks with reversed or random spatial correspondence. *Ergonomics, 9*, 289-296.

Hubbard, D. C. (1987). Inadequacy of root mean square error as a performance measure. In R. S. Jensen (Ed.), *Proceedings of the 4th International Symposium on Aviation Psychology* (pp. 698-704). Columbus: Ohio State University.

Jaeger, R. J., Agarwal, G. C., & Gottlieb, G. L. (1980). Predictor operator in pursuit and compensatory tracking. *Human Factors, 22,* 497-506.

Lintern, G., Roscoe, S. N., Koonce, J. M., & Segal, L. D. (1989). Transfer of landing skills in beginning flight training. In R. S. Jensen (Ed.) *Proceedings of the 5th International Symposium on Aviation Psychology* (pp. 128-133). Columbus, OH: Ohio State University.

Lintern, G., Roscoe, S. N., & Sivier, J. E. (1989). Display principles, control dynamics, and environmental factors in pilot performance and transfer of training. In R. S. Jensen (Ed.), *Proceedings of the 5th International Symposium on Aviation Psychology* (pp. 134-148). Columbus, OH: Ohio State University.

McLaughlin, K. (1994). Need for recognition of PCATDs. In K. W. Williams (Ed.), *Summary proceedings of the joint industry-FAA conference on the development and use of PC-based aviation training devices* (p. 3). Washington, DC: Department of Transportation, Federal Aviation Administration Report DOT/FAA/AM-94/25.

McGuigan, F. J., & MacCaslin, E. F. (1955). Whole and part methods in learning a perceptual motor skill. *American Journal of Psychology, 68*, 658-661.

McRuer, D. T., & Jex, H. R. (1968). A review of quasi-linear pilot models. *IEEE Transactions on Human Factors in Electronics, 8*, 240.

Miller, G. A., Pribram, K. H., & Gallanter, E. (1960). *Plans and the structure of behaviour*. London: Holt.

Naylor, J. C., & Briggs, J. E. (1963). Effects of task complexity and task organisation on the relative efficiency of part and whole training methods. *Journal of Experimental Psychology, 65,* 217-224.

Pew, R. W. (1974). Human perceptual-motor performance. In B. Kantowitz (Ed.), *Human information processing: Tutorials in performance and cognition* (pp. 178-218). Hillsdale, NJ: Erlbaum Associates.

Poulton, E. C.(1957). On prediction in skilled movements. *Psychological Bulletin, 54,* 467-478.

Proctor, R. W., & Dutta, A. (1995). *Skill acquisition and human performance.* Thousand Oaks, CA: Sage.

Rasmussen, J. (1986). *Information processing and human-machine interaction: an approach to cognitive engineering.* Amsterdam: North-Holland.

Sanders, M. S., & McCormick E. J. (1987). *Human factors in engineering and design* (6th ed.). New York, NY: McGraw-Hill.

Schmidt, R. A. (1975). A schema theory of discrete motor skill learning. *Psychological Review, 82,* 225-260.

Stelmach, G. E. (1982). Information processing framework for understanding human motor behaviour. In J. A. Kelso (Ed.), *Human motor behaviour—an introduction* (pp. 12-32). London: Lawrence Erlbaum Associates.

Wickens, C. D. (1984). Processing resources in attention. In R. Parasuraman and R. Davies (Eds.), *Varieties of attention* (pp. 63-101). New York, NY: Academic Press.

Wickens, C. D. (1992). *Engineering psychology and human performance* (2nd ed.). New York, NY: Harper Collins.

Wickens, C. D., Sandry, D., & Vidulich, M. (1983). Compatibility and resource competition between modalities of input, output, and central processing. *Human Factors, 25,* 533-543.

Williams, I. D., & Rodney, M. (1978). Intrinsic feedback, interpolation and the closed-loop theory. *Journal of Motor Behavior, 3,* 205-212.

Williams, K. W. (1996). *Qualification guidelines for personal computer-based aviation training devices: instrument training.* Washington, DC: Department of Transportation, Federal Aviation Administration Report DOT/FAA/AM-96/8.

Williams, K. W., & Blanchard, R. E. (1995). *Development of qualification guidelines for personal computer-based aviation training devices.* Washington, DC: Department of Transportation, Federal Aviation Administration Report DOT/FAA/AM-95/6.

Woodworth, R. S. (1899). The accuracy of voluntary movement. *Psychological Review, 3 (Monograph Supplement),* 1-119.

Young, A. (1993). *An Investigation into the Effect of Control Design and Sensitivity on the Performance of a Simulated Approach and Landing Task.* Unpublished master's thesis. College of Aeronautics, Cranfield University.

Part 4
Computer-Based Training in General Aviation

7 The Development of Computer-Assisted Learning (CAL) Systems for General Aviation

Mark Wiggins

Introduction

Errors involving failures in decision-making and information processing contribute significantly to aircraft accident and incident rates involving general aviation aircraft (O'Hare, Wiggins, Batt, & Morrison, 1994). This has prompted the development of a series of training initiatives which have been designed to focus upon a variety of cognitive skills including decision-making (Jensen, 1992; Lester, Diehl, & Buch, 1985), risk management (Brecke, 1982), and fatigue and stress management (Adams & McConkey, 1990). These initiatives have formed the basis of training programs which have been implemented to the extent that a significant proportion of general aviation pilots are now familiar with concepts such as "Hazardous Thoughts" and "IMSAFE" (Buch, 1984; Harris, 1994; Holt, Boehm-Davis, Fitzgerald, Matyuf, Baughman, & Littman, 1991; Lester et al., 1985). Moreover, subjective responses from pilots have suggested that there is a considerable level of face validity involved in the application of these concepts within general aviation (Buch, 1984; Lester et al., 1985; Telfer, 1988).

However, despite the apparent success of these training initiatives, empirical analyses have indicated that it is particularly difficult for learners to transfer skills such as decision-making from a training scenario to the operational environment (Holt et al., 1991; Wiggins & O'Hare, 1993). This suggests that while pilots may understand the broad significance of concepts such as decision-making, they may lack the strategies necessary to develop and implement these types of skills as part of their operational performance.

Theoretical support for a distinction between "understanding" and "implementation" can be derived from Jensen's (1992) notion that the application of skilled performance within the aviation domain, involves both a motivational component and a skills-based component. In the majority of cases, however, cognitive training strategies within general aviation have focused upon the motivational component of cognitive skills, to the exclusion of the skills-based component. Consequently, while pilots may develop an understanding of the inappropriate motivational factors which are likely to impact upon their performance, they may lack the skills necessary to acquire, integrate and respond appropriately to information which may impact upon the conduct of the flight.

According to Casner (1994) and Lintern (1995), the development of skilled cognitive performance requires a training system which situates the particular skill within the environment in which it is expected to be applied. From a practical perspective, however, it must be acknowledged that the learning environment within aviation is limited to some extent, by training costs, safety concerns, and environmental factors such as noise and heat. An alternative approach to cognitive skills development may, therefore, involve the development of computer-based systems which are capable of incorporating both the cognitive basis for skill acquisition, and the relevant cues and demands which exist within the operational environment. Combined with strategies which focus upon the motivational component of cognitive skills, computer-based training systems may provide a mechanism to reduce the proportion of general aviation accidents which result from cognitive failures.

Computer-Assisted Learning

Computer-assisted learning (CAL) is one of a number of Computer-Based Training (CBT) systems, and one in which the aim is to facilitate, rather than direct the process of learning. There are a number of advantages associated with the application of CAL within the aviation environment including cost effectiveness, flexibility, the opportunity for part-task training, student-centred learning and the potential for the development of a detailed mental model of the operational environment (Wiggins, 1996). As an educational strategy, it is intended to augment existing training systems in order to improve training efficiency, and/or to facilitate the acquisition of additional skills. CAL is not therefore, intended as a mechanism to usurp either the primary role of the flight instructor, or the role of the training organisation in flight instruction.

Despite the potential of computers as a training aid, it must be acknowledged that the advent of such systems was expected to yield far greater pedagogical returns than has currently been the case (See O'Hare, Chapter 7, this volume). This has been due to a combination of factors including the development of systems which failed to meet the particular needs of end-users, and/or the application of systems within the environment which did not facilitate the process of learning.

The significance of the end-user during systems development is illustrated through the experience of *SimuFlite*, a corporate pilot training company located in the United States. Established in 1981, computer-based training systems were developed as a core component of *SimuFlite's* ground school training strategy. While the package was a relative success initially, users quickly lost interested in the system, and *SimuFlite* was faced with a 50-60% client retention rate (Bovier, 1993). Following a detailed investigation by the company, it was revealed that although the system was somewhat flexible in terms of its training capability, it failed to engage the user following repeated interactions. In addition, pilots were critical of the lack of human interaction within the training programme, and the lack of depth within the system. According to Bovier (1993), these difficulties could be traced to the failure to consult the needs of the end-users during the initial stage of systems development.

Consistent with the growing appreciation of the needs of the end-user, there has been a concerted shift in recent years, from computer-based systems which simply provide declarative knowledge, such as that derived from textbooks, to systems which are designed to facilitate the development of more applied, procedural knowledge (Mattoon, 1994). This is consistent with the notion that human performance within a complex, uncertain environment is dependent upon both the accumulation of declarative information associated with a domain, and the knowledge and the skills necessary to apply this information within the environment. For example, the declarative knowledge that an aircraft has a capacity for eight quarts of oil does little to assist pilots, unless it is associated with the procedural knowledge necessary to respond appropriately in the case of a reduction in oil pressure. Wiggins and O'Hare (1993) assert that the development of this combination of knowledge is an essential requirement for the acquisition of skilled performance within the aviation domain.

While the focus on skill-based development is the ideal, it should be noted that the training system itself must be allied with the appropriate pedagogical technique for each particular environment. For example, one of the main criticisms associated with *SimuFlite's* approach to computer-based training was the lack of human interaction involved in the programme (Bovier, 1993). These pilots were familiar with a pedagogical strategy in which they had access to a facilitator who would provide feedback and relate the information to the operational environment. This highlights the significance of the application of an appropriate pedagogical strategy in facilitating the acceptance of CAL amongst users.

CAL and General Aviation

In most cases, the nature of general aviation operations is such that pilots are operating in a geographical area with which they may be unfamiliar; they often lack the type-specific training evident in airline operations; they typically do not have the benefits associated with multi-crew operations or advanced technology; they are more likely to confront convective meteorological activity en-route; and they generally lack the breadth of task-oriented experience afforded by airline operations. Consequently, general aviation pilots are far more vulnerable than airline pilots, to the precursors of human factors errors and the consequences of those errors.

In addition, one of the major difficulties associated with general aviation operations lies in the potential for skill degradation due to either inactivity, or changes in the type of aircraft or the environment within which the pilot operates. Moreover, this degradation in skilled performance appears to be more significant and more rapid for cognitive skills than it is for perceptual-motor skills (Childs, Spears, & Prophet, 1983). The result is that while general aviation pilots may retain the skills necessary to physically control the aircraft, they may lack those skills such as decision-making, risk assessment, information acquisition and problem-solving, which are essential components associated with the safe operation of an aircraft.

On the basis of these arguments, general aviation is a domain in which CAL has the potential to effect a significant, qualitative change in terms of the both skill acquisition, and skill retention during periods of inactivity. Moreover, it has the potential for application either as an initial training platform, or for the purposes of re-current training at relatively minimal expense.

Systems Design and Development

The development of CAL begins with the identification of a target group for whom a particular programme is designed. One of the mistakes associated with many approaches to CAL is to define this target group too widely (Bovier, 1993), and therefore risk the development of a package which is both too large to meet the specific needs of users, and too cumbersome to be used on a regular basis. Moreover, the nature of expertise is such that it is relatively specific to a particular type of environment (Cohen, 1993; Logan, 1988). For example, an expert trans-oceanic airline pilot would not necessary be considered an expert when flying a light aircraft in mountainous terrain, and in deteriorating weather conditions.

On the basis of this argument, defining the target audience for a piece of training software can be difficult, although software developers generally find it easier to narrow, rather than broaden the context for a particular piece of software. In terms of weather-related decision-making, for example, it might be argued that the skills involved in one type of meteorological environment may be quite different to the skills required in another. This may be due to differences in the characteristics of meteorological phenomena, differences in topography, and or differences in aircraft capabilities. Consequently, a software programme might focus upon problem-solving during particular classes or types of meteorological phenomena, rather than attempt to facilitate the acquisition of weather-related decision-making skills in general. This perspective is consistent with most conceptualisations of the nature of skill acquisition (Anderson, 1993; Leake, 1996) in which expertise (or the capability to generalise) is assumed to derive from repeated attempts to master performance within a particular situation, and the subsequent generalisation of these skills to other tasks. For example, a pilot may learn to fly using a particular type of aircraft, and having reached a high level of performance, may be able to apply these same skills when operating quite different aircraft.

Having defined the target group, it behoves the system designer to undertake a detailed analysis of potential users, both in terms of their skill-based requirements, and in terms of their expectations concerning the nature and the format of the instructional system. This process can be accomplished using one of a number of knowledge acquisition techniques referred to collectively as Cognitive Task Analysis (CTA). The primary object of these knowledge acquisition strategies is to capture in as much detail as possible, the cognitive, perceptual and behavioural features involved in the performance of a task. This provides the basis for an understanding of the domain, and therefore facilitates the development of an effective training environment. A range of strategies are available including critical incident techniques (Flanagan, 1954), concept mapping (Kaempf, Wolf, Thordsen, & Klein, 1992), cognitive interviews (Cooke, 1994), process tracing (Wiggins & O'Hare, 1995), protocol analysis (Carroll & Johnson, 1990), and cognitive graph analysis (Cooke, 1994)

A significant part of the process of systems design involves a consideration of the type of reasoning the user is expected to engage in during the course of skill acquisition. In general, perspectives regarding human reasoning can be divided into two distinct camps: production-based reasoning and case-based reasoning. The former is based upon the assumption that reasoning involves the development of an association between declarative and procedural knowledge associated with a domain. Whereas declarative knowledge involves factual information (e.g., knowledge of the optimal manoeuvring speed of an aircraft), procedural knowledge involves the implementation of declarative knowledge to influence some change in the environment or in understanding (e.g., the capability to manipulate the aircraft to maintain the optimal manoeuvring speed) (Anderson, 1993). The notion can be expressed as an IF-THEN rule/production, where the initial component represents a condition (declaration) under which a subsequent outcome or strategy (procedure) is applied.

According to this model, the link between declarative and procedural knowledge occurs when both the conditions and outcomes occur simultaneously within working memory. This association is generalised subsequently, such that inferences can be applied to relatively novel situations. From a training perspective, the main implications of this approach, include the requirement to ensure that the learner is practically engaged in the performance of the task; the provision of rapid, task-oriented feedback; and the development of a training scenario which embodies the conditions and actions evident within the operational environment.

Unlike production-based reasoning, case-based reasoning (CBR) is based upon the assumption that the repeated interaction with the operational environment results in the acquisition and retention of an extensive series of cases or examples. This repertoire of cases is recalled subsequently as the basis for mapping and solving similar problems both rapidly and accurately. According to Leake (1996), case-based reasoning involves either the classification of a particular situation according to previously acquired examples, or the adaptation and implementation of a problem-solving strategy which has been applied during similar problems in the past. In each case, the reasoning process involves situation assessment, case retrieval and an assessment of the similarity of the present case to previous examples.

From a training perspective, the emphasis involves building upon the existing knowledge of the learner. The aim is to develop sufficient cases, such that the learner can compare and contrast situations and perceive slight differences which might exist between examples. This process, however, is dependent upon an efficient and effective knowledge classification system which enables the process of generalisation between otherwise independent cases (Schank, 1996). Therefore, the goal of the instructional designer within a CBR framework, is to facilitate the acquisition of skills necessary to classify problems according to their similarities in terms of goals, environmental characteristics, and/or the type of planning involved.

Production-based and case-based models of reasoning differ in many respects, and these differences will reflect upon the nature of the instructional systems under development. A summary of the implications of these models for instructional systems design is provided in Table 7.1. In terms of the efficacy of one type of reasoning over another, there is empirical support for each of the approaches, and it is difficult to advocate one type of reasoning as generally more effective than the other. There is, however, some evidence to suggest that reasoning may differ, depending upon the cognitive demands associated with a task.

Table 7.1   A summary of implications of production-based reasoning and case-based reasoning for the design of instructional systems

Production-Based Reasoning	Case-Based Reasoning
Task Involvement	Task Involvement
Cue Identification	Problem Classification
Outcome/ Action-Oriented	Goal
Immediate Feedback	Environment
	Plan
	Case Retrieval
	Immediate Feedback
	Similarity Assessment/ Evaluation

The successful transfer of skilled performance from the training environment to the operational environment can be problematic. In the case of skills such as decision-making, the problem may lie in the failure to consider the demands imposed upon these strategies in the operational environment. In the case of decision-making, for example, general aviation pilots are often required to formulate and evaluate options based upon complex and relatively uncertain information. Moreover, this process is often time-critical, while the consequences of a poor decision can be significant. Failing to take account of these cognitive demands during training, may diminish the extent to which the skills developed during this phase may be transferred to the operational environment (Woods, 1988).

The cognitive demands associated with a particular task can be derived from the cognitive analysis of end-users and experts within the domain. Within the aviation environment, the cognitive demands differ considerably, depending upon the type of operation. Hence, there is often a need for task-specific training systems. The flexibility associated with CAL, however, is such that it has the facility to incorporate either a case-based or a production-based training system within an environment which is consistent with the demands associated with the operational performance. It thereby provides an opportunity for the transfer of skills necessary to cope effectively with the variety of demands imposed by the operational environment.

Human Factors Principles

Computer-based systems are essentially a communication platform between the software designer and the user. This communication is effected through the software interface: the point within the system at which the psychological intentions of the user are translated into physical instructions (Redmond-Pyle & Moore, 1995). Bridging the gap between the user and the designer are what Norman (1990) suggests metaphorically, are the "Gulfs of execution and evaluation". These concepts refer to the need to consider the cognitive processes (psychological intentions) through which a user relates to a machine ('gulf of execution'), and through which a user receives feedback from the machine concerning the original intentions ('gulf of evaluation').

A part of the process of bridging the gulfs of execution and evaluation involves a detailed consideration of the user's mental model of the system as it applies within the operational environment. According to Redmond-Pyle and Moore (1995), the user's mental model of the training system is dependent upon a number of features including the interface design, the type of information included within the 'help' function, and the training which the user might receive. Norman (1990), however, notes that the most effective system design is one which incorporates a transparent or visible model of the system. Therefore, it should conform to a "common sense" approach which might be expected by the user within a particular type of environment. This results in relatively rapid learning, and enhanced performance, since the user is not spending valuable time learning to the use the system through a complex set of instructions.

According to Norman (1990), adherence to the user's existing mental model is one of seven principles of system design which are purported to improve user performance. The remaining six principles include:

- Simplifying the structure of tasks;
- Ensuring that functions are visible to the user at all times;
- Ensuring that user's perceive the relationship between their intentions and system responses;
- The application of perceptual constraints within the system;
- Designing for potential errors; and
- The standardisation of the system, both internally and externally.

Norman's (1990) principles were developed primarily as a result of empirical observation and anecdotal evidence, although they have now become somewhat of a "standard" to which software interface designs must adhere (Redmond-Pyle & Moore, 1995).

Interface Design and Evaluation

The relative efficacy of a piece of software is fundamentally dependent upon the development of an appropriate interface between the software and the user (see Figure 7.1). According to Bradford (1987), the types of information which might be presented, and the process through which this information is presented, requires the interaction between potential users and the software designer to ensure the development of an optimal product. The successful outcome of this interaction can be defined according to four dimensions including effectiveness, learning potential, flexibility, and attitude (Shackel, 1990).

The effectiveness of a system interface refers to the efficiency and the accuracy with which the user acquires and interprets task-related information (Shackel, 1990). The user must, therefore, possess both an accurate mental model of the system environment and an understanding of the strategies through which to acquire this information. This includes a basic understanding of any navigational aids which may be included as part of the interface. Generally, this dimension is evaluated using a performance-based criteria such as accuracy and/or response latency.

Allied to this dimension is the learning potential, or the rate at which the user acquires an understanding of the system sufficient to operate the programme effectively and efficiently (Shackel, 1990). This dimension examines both the level of internal and external consistency to ensure that the user can generalise from one segment to another, and one system to another. The greater the consistency, the greater the rate at which system-related skills are acquired (Andre & Wickens, 1992; Fisk & Gallini, 1989).

The flexibility of a system interface involves the extent to which it remains effective and efficient despite the different demands made by users. Although the system designer has a particular understanding of the process through which the system will be used, users may elect to use the system for a number of tasks other than those for which it was directly designed. The system interface should, therefore, be sufficiently flexible to meet the distinct and diverse demands which may be made by users without losing its overall capability. This dimension is typically evaluated subjectively, with potential users indicating how they would use the system.

Fundamentally, the use of a system is dependent upon the attitude of the user. Where a system interface is perceived as difficult and/or frustrating, the system is less likely to be used in the future (Redmond-Pyle & Moore, 1995). Similarly, if the information embodied within the system is considered of little value to the user, then there is less likelihood that the system will be re-accessed. Consequently, the attitude of the user must involve a consideration of both the information within a system, and the interface through which this information is acquired.

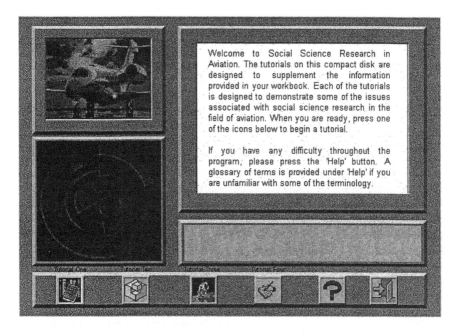

Figure 7.1   An example of a computer interface

System Validity

According to Wiggins (1996), one of the most significant factors associated with the introduction of CAL involves the development of suitable evaluative techniques. Most researchers and practitioners would argue that there is little point in developing knowledge if that knowledge is unable to be applied within the operational environment. Consequently, there is a need to establish the validity associated with training initiatives, and this involves a concentration upon both declarative and procedural knowledge, including the development of skills such as diagnosis, information acquisition, information integration, decision-making, and stress management.

The effective assessment of the acquisition of such skills involves the development of a criterion, against which the performance of users can be examined. Wiggins (1997) suggests that one mechanism through which to set a criterion is to compare the performance of experts and novices within a particular domain. In general, novices perform at a lower level, irrespective of the relative familiarity of a particular task. As task-related experience is acquired, performance approaches that of experts during familiar tasks, but reverts to novice-level performance during relatively unfamiliar tasks. Since the essence of expertise is the capability to generalise across similar situations, a performance criterion should be set to examine the extent to which this generalisation of performance occurs.

Having established a criterion, however, there are a number of strategies available through which the performance of users might be evaluated following training. The first stage of this process involves the identification of the potential users of the system, and a descriptive analysis of the population characteristics. Aspects such as the level of education, experience with computers, age, and occupation are all likely to influence the learning outcomes.

Subjective and/or objective evaluations of performance yield different types of information and, therefore, are more or less effective, depending upon the goals of the analysis. For most analyses, however, a combination of strategies is likely to be beneficial, since one set of data can be used to provide support or an explanation for the data arising from the other. A similar argument can be applied for the consideration of qualitative and/or quantitative forms of data. A mixed-method approach comprising a combination of qualitative and quantitative data is likely to yield the most comprehensive results. Moreover, one form of data often can be used to validate the results arising from the other. Finally, the nature of interface design is such that it is often possible for users to suggest alternative strategies, rather than simply highlight that a problem exists.

A Practical Example

The following is an example of the process through which these principles were applied during the development of a computer-assisted learning tool which was designed to facilitate the development of skilled weather-related decision-making.

Introduction

In-flight weather-related decision-making is a task that is well suited to CAL programs. The cognitive demands associated with in-flight weather-related decision-making are such that the situation is highly dynamic, complex, and may be time-constrained in many cases. Moreover, there are often serious, short-term consequences associated with an incorrect decision. CAL, however, has the facility to provide a safe and efficient learning environment, within which pilots can be taught to recognise and respond appropriately to a deterioration in the weather conditions during flight.

The most serious error associated with in-flight weather-related decision-making arises when a pilot, operating under Visual Flight Rules (VFR), violates the requirements for Visual Meteorological Conditions (VMC), and enters instrument conditions. This restricts both horizontal and vertical visibility, to the extent that the pilot must rely upon cockpit instruments for operational stability and control. Flying an aircraft using flight instruments is a highly skilled and highly demanding task which requires sustained practice for successful performance. VFR pilots are often ill-equipped for this task and, consequently, there is a strong potential for a loss of operational control of the aircraft. In the United States, for example, unintentional VFR flight into IMC accounted for 580 aircraft crashes between 1982 and 1993 (Aircraft Owners and Pilots Association, 1996). Moreover,

82% of these crashes resulted in fatalities, and it is, therefore, regarded as the deadliest of all weather-related aircraft accidents in terms of injuries.

Within the operational environment, effective in-flight weather-related decision-making requires: the synthesis of weather-related information both from outside the cockpit and the meteorological forecast obtained prior to flight; the development of a strategy necessary to avoid in-flight encounters with cloud; and the willingness to implement such a strategy. In addition, the pilot is expected to maintain appropriate control and management of the aircraft. This is a complex information processing task, made all the more difficult by the inexperience of many pilots.

Cognitive Task Analysis

The computer-based tutoring system in the present example was based upon the fundamental assumption that the acquisition of skills develops through an association between cues arising from the operational environment, and the various consequences associated with those cues. Therefore, it was designed to provide an environment which focused upon the perceptual, cognitive and affective demands which would normally occur during this type of decision-making process.

To determine the relative significance of the perceptual, cognitive and behavioural demands during in-flight weather-related decision-making, a cognitive interview protocol was designed, the information from which would provide the basis for a cognitive task analysis. The protocol was developed in-house, and was loosely based around research completed by Kaempf, Klein, and Thordsen (1991). The data arising from the cognitive interviews were transcribed and combined to form the basis of conceptual graphs for each of a number of meteorological characteristics which might be encountered by a pilot during flight. It was assumed that the identification of the various cognitive and procedural elements based upon interviews with experts, would facilitate the construction of an effective training environment for inexperienced decision-makers.

The major difficulty associated with this type of knowledge acquisition is the amount of data that is collected and the process through which it is managed. Conceptual Graph Analysis (CGA) is one mechanism through which the data arising from cognitive task analytic techniques can be modelled and integrated for subsequent application in the training domain. Originally developed within the computer science domain, CGA has been adapted to facilitate the representation of cognitive structures amongst subject-matter experts (Gordon, Schmierer, & Gill, 1993).

The CGA for in-flight weather-related decision-making was developed through a series of cognitive interviews with experts who were asked to recount a situation in which they had made what they regarded as a "poor" in-flight weather-related decision. During the course of the interview, experts were probed using a semi-structured protocol which was designed to ensure the identification of task-related cues and associated consequences.

Conceptual Graphs

The process of developing conceptual graphs is subject to a great deal of interpretation on the part of the analyst. Consequently, the conceptual graphs developed in the present study were constructed on the basis of keywords mentioned during interviews, and conformed to the principle of parsimony. The aim was to ensure that minimal interpretation was required on the part of the analyst, and therefore, the graphs would be capable of replication. These graphs were subsequently presented to the experts for comment and manipulation if necessary.

Ultimately, conceptual graphs must be sufficiently meaningful to provide the basis for the development of instructional systems. The application of an appropriate theoretical perspective is one mechanism to ensure the accurate and meaningful interpretation of data arising from cognitive interviews. The current study involved an assumption that expert decision-making in critical environments is primarily recognition-driven (Klein, 1989), and that this capability is developed through the repeated association between task-oriented cues and consequences (Wiggins & O'Hare, 1993). This production system model was an important theoretical basis upon which the interpretation and subsequent application of the conceptual graphs was based.

An example of a conceptual graph for stable, deteriorating conditions is provided in Figure 7.2. Note on the left of the figure, the concepts which are required in order to recognise and respond appropriately to stable meteorological conditions with a deteriorating cloud-base. This might be equated to declarative knowledge while the goal/action statements to the right of the diagram might equate to procedural knowledge. The relationships between cues and consequences have, therefore, devolved to elicit the precise cues necessary to recognise and respond to this particular type of meteorological phenomena.

The four cues identified in Figure 7.2 were used as the basis for the development of the Recognition stage of the In-Flight Weather-Related Tutoring System. This is an essential phase of the training process, since one of the key elements associated with inadequate in-flight weather-related decision-making lies in the failure to recognise deteriorations in weather conditions at the earliest possible stage of a flight.

At the initial stage of the programme, users are asked to identify a series of cues associated with deteriorations in the particular weather conditions under analysis (See Figure 7.3). In the case of stable, deteriorating conditions, for example, clearance from terrain, visibility, and clearance from the cloudbase are important determinants for the recognition of hazy conditions. The user has a number of sources of information available to facilitate the identification of these cues including two video clips, and an interview with an expert. Each video clip contains an in-flight encounter with the meteorological conditions under analysis. In addition, there is an opportunity for users to acquire additional information by passing the cursor over strategic parts of the image. This process accesses relevant information and assists the user to both identify the relevant cues, and develop an

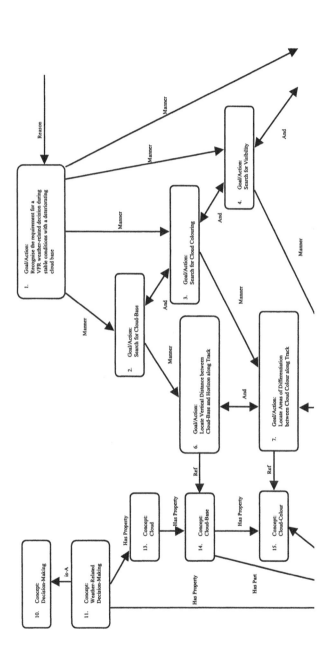

Figure 7.2 An example of a section of a conceptual graph for in-flight weather-related decision making illustrating the conceptual/declarative elements and their relationship to the goal/action elements associated with procedural knowledge.

understanding of the process through which the cues can be recognised and applied during flight.

Once the user has identified a cue, the keyword is typed into the textbox provided and a corresponding cue card is revealed. Additional information is provided to the user and each card can be re-examined by "clicking" the mouse on the appropriate tab. An expert interview is also provided to reinforce the consequences associated with the failure to recognise the deterioration in the dimensions of these cues.

Programme Development

The cognitive process involved in the identification of cues is, fundamentally, a problem-based exercise which structures the process which would otherwise occur within the operational environment. While the information provided is declarative in nature, the attentional and problem-solving resources which are brought to bear during this exercise are likely to result in the retention of this information within long-term memory.

The development of procedural knowledge is accomplished through a two stage comparative process, recognising that the process is dependent upon the capacity to identify a deterioration in cues, rather than any absolute value. The first of these stages allows users to access each of three sets of video clips pertaining to two stages of flight. The user is asked to compare each of three sets of video clips and determine the set during which the most significant deterioration occurred for a given cue (see Figure 7.4). Each set is identified by numeric value and the corresponding value is typed into a text box at the base of the screen. This reveals a cue card with feedback concerning the performance of the user.

The process of comparison enables users to practice the identification of the deterioration of a particular cue during a simulated flight. In many cases, the deterioration is sufficiently marked to warrant a reliance upon flight instruments had the situation been encountered within the operational environment. Consequently, the user is gaining experience relating the cue to an associated consequence: that is, procedural knowledge.

The second stage of the comparative process involves users' observation of each of the cues simultaneously, during a series of six video clips. These are designed to simulate the changing meteorological conditions which might occur during the course of a flight. The users' task is to examine the meteorological forecast and determine the point during the flight at which the conditions deteriorate below the conditions forecast. This is designed to act as a trigger for the initiation of a subsequent problem-solving process. The user receives feedback concerning the nature of their decision and has the opportunity to re-examine the cues and their associated strategies.

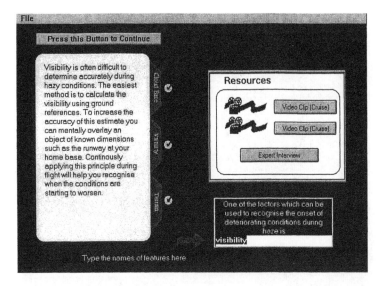

Figure 7.3   **An example of the computer screen during the identification phase in which users identify the task-oriented cues associated with the recognition of a particular weather phenomena during flight**

Operational Implications and Recommendations

From a conceptual perspective, CAL programs such as the one described, have the potential to significantly improve and maintain the cognitive performance of pilots operating within the general aviation environment. In particular, CAL has the capacity to provide a semi-structured learning system which embodies the cognitive demands and the cues which exist within the operational environment. This is designed to facilitate the acquisition of skilled performance at a rate which supersedes alternative training systems, and within a safe and responsive environment.

However, while there appears to be significant potential for CAL within general aviation, the implementation of these systems is limited by both a lack of resources for systems development, and a scepticism within the general aviation industry concerning the efficacy of computer systems to effect any meaningful change in operational performance. The latter is a reasonable concern, since very few training systems within the aviation industry have ever been subject to empirical validation within the operational environment.

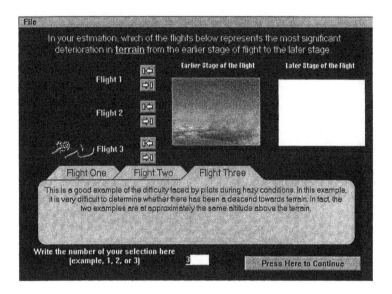

Figure 7.4   **An example of the computer screen during the comparative phase in which users compare two video clips for each of three flights based on a particular task-oriented cue (in this case, clearance from terrain)**

Although there is a responsibility on the part of the designer to ensure the efficacy of CAL prior to implementation, there is also a responsibility on the part of the educator to ensure that training systems are integrated into learning schedules at an appropriate level for the student. Integration also includes the provision for facilitation as a mechanism to provide both motivation for the student, and to link the training materials to the operational environment.

Like any other training material, computer-based systems are subject to the perceptions, personal goals, and personal preferences of system designers. Consequently, there is a strong potential for the development of a plethora of computer-based aids, few of which conform to the human factors principles of consistency and usability. Previous evidence suggests that this is likely to result in an unwillingness to use a system (Bovair, 1993), an increase in training time (Norman, 1990), and/or a potential reduction in the level of transfer to the operational environment. As a result, there may be a requirement in the near future to develop "best practice" standards for the design and development of computer-based training systems which are used within the aviation domain. This would not only ensure a level of consistency between designers, but it might also provide guidelines and objectives for novice systems designers which might not otherwise be available.

Research Agenda

Clearly, one the major limitations associated with computer-assisted learning involves the lack of precise principles and guidelines which might underpin the development of such systems. While there exists a number of mechanisms to evaluate and validate such systems, there are no specified criteria for the purposes of evaluation within the aviation environment. The development of such criteria will require consensus amongst educators, and the implementation of a research programme which examines the impact of various systems on human performance. within the operational domain.

Allied to the development of evaluative criteria is the notion of validity, or the extent to which a training system develops the knowledge and/or the skills intended. Moreover, evaluative techniques must progress beyond a simplistic analysis of perceptual-motor skills or declarative knowledge, and include a detailed consideration of the type and level of the cognitive skills acquired. This requires a research programme to develop standardised measurement protocols which can be applied efficiently and cost-effectively across a variety of domains.

Conclusion

While it might be tempting to perceive computer-based approaches as a type of panacea to improve a broad range of pilot standards and capabilities, it should be noted that unless qualitative improvements are made in terms of the process of education and training, they are unlikely to afford any advantage over alternative strategies. Therefore, computer-based training initiatives must be conceptualised as more than simply a mechanism through which to transfer. "paper and pencil" strategies to a computer-based format.

The major advantage associated with CAL is its capability to incorporate task appropriate cues within a safe and cost-effective learning environment. The application of appropriate human factors principles, coupled with a detailed understanding of the cognitive demands associated with a domain, must serve as a necessary pre-requisite for the development of these systems within the general aviation industry. In addition, there is a requirement for the development of standardised principles and competency-based objectives, against which to assess the efficacy of particular initiatives.

Combined, these strategies have the potential to facilitate the acquisition of cognitive skills amongst the general aviation pilot community. This is likely to correspond to an eventual reduction in the proportion of general aviation aircraft accidents that involve cognitive failures.

References

Adams, R. J., & McConkey, E. D. (1990). *Aeronautical decision-making for air ambulance administrators.* Washington, DC: Federal Aviation Administration (NTIS DOT/FAA/DS-88/8).

Aircraft Owners and Pilots Association. (1996*). Safety review: General aviation weather accidents.* Frederick, MD: Author.

Anderson, J. R. (1993). *Rules of the mind.* Hillsdale, NJ: Lawrence Erlbaum.

André, A. D., & Wickens, C. D. (1992). Compatibility and consistency in display-control systems: Implications for aircraft decision aid design. *Human Factors, 34,* 639-653.

Bovier, C. (1993, August). How a high-tech training system crashed and burned. *Training,* 26-29.

Bradford, A. N. (1987). Presenting information on a computer screen: A decision process for the information planner. *Human Factors Society Bulletin, 30,* 1-3.

Brecke, F. H. (1982). Instructional design for aircrew judgement. *Aviation, Space, and Environmental Medicine, 53,* 951-957.

Buch, G. (1984). An investigation of the effectiveness of pilot judgement training. *Human Factors, 26,* 557-564.

Carroll, J. S., & Johnson, E. J. (1990). *Decision research: A field guide.* Newbury Park: Sage Publications.

Casner, S. (1994). Understanding the determinants of problem-solving behaviour in a complex environment. *Human Factors, 36,* 580-596.

Childs, J. M., Spears, W. D., & Prophet, W. W. (1983). *Private pilot flight skill retention 8, 16, and 24 months following certification* (NTIS DOT/FAA/CT-83/34). Washington, DC: Federal Aviation Administration.

Cohen, M. S. (1993). The naturalistic basis of decision biases. In G. A. Klein, J. Orasanu, R. Calderwood & C. E. Zsambok (Eds.), *Decision making in action: Models and methods* (pp. 51-99). Norwood, NJ: Ablex.

Cooke, N. (1994). Varieties of knowledge elicitation techniques. *The International Journal of Human-Computer Studies, 41,* 801-849.

Fisk, A. D., & Gallini, J. K. (1989). Training consistent components of tasks: Developing an instructional system based on automatic/controlled processing principles. *Human Factors, 31,* 453-463.

Flanagan, J. C. (1954). The critical incident technique. *Psychological Bulletin, 51,* 327-358.

Gordon, S. E., Schmierer, K. A., & Gill, R. T. (1993). Conceptual graph analysis: Knowledge acquisition for instructional system design. *Human Factors, 35,* 459-481.

Graesser, A. C., & Clark, L. C. (1985*). Structures and procedures of implicit knowledge.* Norwood, NJ: Ablex.

Graesser, A. C., & Franklin, S. P. (1990). Quest: A cognitive model of question answering. *Discourse Processes, 13,* 279-304.

Hammond, K. R., Hamm, R. M., Grassia, J., & Pearson, T. (1987). Direct comparison of the efficacy of intuitive and analytical cognition in expert judgement. *IEEE Transactions on Systems, Man, and Cybernetics, 17*, 753-770.

Harris, J. S. (1994). Improved aeronautical decision-making can reduce accidents. Flight Safety Foundation: *Helicopter Safety, 20*, 1-6.

Holt, R. W., Boehm-Davis, D. A., Fitzgerald, K. A., Matyuf, M. M., Baughman, W. A., & Littman, D. C. (1991). Behavioural validation of a hazardous thought pattern instrument. *Proceedings of the Human Factors Society 35th Annual Meeting* (pp. 77-81) San Diego, CA: Human Factors Society.

Jensen, R. S. (1992). Do pilots need human factors training. *CSERIAC Gateway, 3*, 1-4.

Kaempf, G. L., Klein, G. A., & Thordsen, M. L. (1991). *Applying recognition-primed decision to man-machine interface design* (NASA Technical Report NAS2-13359). Yellow Springs, OH: Klein Associates.

Kaempf, G. L., Wolf, S. P., Thordsen, M. L., & Klein, G. (1992). *Decision-making in the Aegis combat information centre* (Naval Command, Control and Ocean Surveillance Centre Report N66001-90-C-6023). San Diego, CA: Klein Associates.

Klein, G. A. (1989). Recognition-primed decisions (RPD). *Advances in Man-Machine Systems, 5*, 47-92.

Leake, D. B. (1996). CBR in context: the present and the future. In D. B. Leake (Ed.), *Case-based reasoning: Experiences, lessons, and future directions* (pp. 3-30). Menlo Park, CA: AAAI Press.

Lester, L. F., Diehl, A., & Buch, G. (1985). Private pilot judgement training in flight school settings: A demonstration project. In R. S. Jensen (Ed.), *Proceedings of the 3rd Symposium on Aviation Psychology* (pp. 353-365). Columbus, OH: Ohio State University.

Lintern, G. (1995). Flight instruction: The challenge from situated cognition. *The International Journal of Aviation Psychology, 5*, 327-350.

Logan, G. D. (1988). Toward an instance theory of automatisation. *Psychological Review, 95*, 492-527.

Mattoon, J. S. (1994). Designing instructional simulations: Effects of instructional control and type of training task on developing display-interpretation skills. *The International Journal of Aviation Psychology, 4*, 189-210.

Means, B. (1993). Cognitive task analysis as a basis for instructional design. In M. Rabinowitz (Ed.), *Cognitive science foundations of instruction* (pp. 97-118). Hillsdale, NJ: Lawrence Erlbaum.

National Transportation Safety Board. (1996). *Annual review of aircraft accident data, U.S. General Aviation, calender year 1994* (NTIS NTSB/ARG-96/01). Washington, DC: Author.

Norman, D. A. (1990). *The design of everyday things.* New York, NY: Doubleday.

O'Hare, D., Wiggins, M., Batt, R., & Morrison, D. (1994). Cognitive failure analysis for aircraft accident investigation. *Ergonomics, 37*, 1855-1869.

Orasanu, J. (1994). Temporal factors in aviation decision-making. *Proceedings of the Human Factors and Ergonomics Society 38th Annual Meeting* (pp. 935-939). San Diego, CA: Human Factors and Ergonomics Society.

Redding, R. E., & Seamster, T. L. (1994). Cognitive task analysis in air traffic controller and aviation crew training. In N. Johnston, N. McDonald & R. Fuller (Eds.), *Aviation psychology in practice* (pp. 190-222). Aldershot, UK: Avebury.

Redmond-Pyle, D., & Moore, A. (1995). *Graphical user interface design and evaluation.* London, UK: Prentice Hall.

Schank, R. C. (1996). Goal-based scenarios: Case-based reasoning meets learning by doing. In D. B. Leake (Ed.), *Case-based reasoning: Experiences, lessons, and future directions* (pp. 295-348). Menlo Park, CA: AAAI Press.

Shackel, B. (1990). Human factors and useability. In J. Preece & L. Keller (Eds.), *Human-computer interaction* (pp. 27-41). New York, NY: Prentice Hall.

Shanteau, J. (1988). Psychological characteristics and strategies of expert decision-makers. *Acta Psychologia, 68,* 203-215.

Telfer, R. (1988). Pilot decision-making and judgement. In R. S. Jensen (Ed.), *Aviation psychology* (pp. 154-175). Aldershot, UK: Gower.

Wiggins, M. (1996). A computer-based approach to human factors education. In B. J. Hayward & A. R. Lowe (Eds.), *Applied aviation psychology: Achievement, change and challenge* (pp. 201-208). Aldershot, UK: Ashgate.

Wiggins, M., & O'Hare, D. (1993). A skill-based approach to aeronautical decision-making. In R. Telfer (Ed.), *Aviation instruction and training* (pp. 430-475). Aldershot: Ashgate.

Wiggins, M. W. & O'Hare, D. (1995). Expertise in aeronautical weather-related decision-making: A cross-sectional analysis of general aviation pilots. *Journal of Experimental Psychology: Applied, 1,* 305-320.

Wiggins, M. W. (1997). Expertise and cognitive skills development for ab-initio pilots. In R. A. Telfer & P. J. Moore (Eds.), *Aviation training: Learners, instruction and organisation* (pp. 54-66). Aldershot, UK: Avebury Aviation.

Woods, D. D. (1988). Coping with complexity: The psychology of human behaviour in complex systems. In L. P. Goodstein, H. B. Andersen & S. E. Olsen (Eds.), *Tasks, errors and mental models* (pp. 128-148). London: Taylor & Francis Ltd.

8   A Pilot for all Seasons: Beyond Simulation

David O'Hare and Richard Batt

All there is to thinking ... is seeing something noticeable which makes you see something you weren't noticing which makes you see something that isn't even visible (Norman Maclean—A River Runs Through It)

Introduction

In recent years extensive investments in information technology have taken place in organisations involved in education and training. For example, more than a quarter of U.S. college courses now involve use of email or internet resources, and more than 10% use some form of multimedia resources. The size of this investment is estimated at over US$70 billion in the past decade and a half (Geoghegan, 1994). Detractors of this headlong rush down the superhighways of information technology have criticised the tendency to use technology as a cheap substitute for human instruction without first determining that the use of IT will increase the quality of the training provided.

It has become increasingly clear that the introduction of IT can result in highly unsatisfactory outcomes. The most spectacular and well-publicised example was the debacle involving the London Ambulance Service (Beynon-Davies, 1995). The computer-aided despatch system, introduced to automate much of the ambulance despatchers task, failed on 26 October 1992. Claims were made that the system had despatched 20-30 Londoners to a premature death as ambulances arrived up to three hours late. This is not an isolated example: a recent survey of 45 leading IT experts in the U.K. found that "on average interviewees estimated that around 80-90% of investments in new technology fail to meet all their objectives" (Clegg, Axtell, Damodaran, Farby, Hull, Lloyd-Jones, Nicholls, Sell, & Tomlinson, 1997, p. 855).

The aviation industry naturally tends to be at the forefront of innovation and technological development and has been a leader in using simulation and computer-based training (CBT) for many years. Currently there is much interest in the potential of Virtual Reality (VR) training. Despite much hype and enthusiasm for VR as a training technology, there is a severe lack of published data to demonstrate the value of VR in acquiring or maintaining skills. Most studies have reported either subjective ratings of training effectiveness (e.g., Loftin, Savely, Benedetti, Culbert, Pusch, Jones, Lucas, Muratore, Menninger, Engleberg, Kenney, Nguyen, Saito, & Voss, 1997) or incidence of sickness and nausea (e.g., Regan, 1997) rather than performance measures.

One of the challenges as we head into the next millennium is to ensure that the best possible training is available to the maximum number of people throughout the aviation industry. This entails extending the availability of synthetic experience (e.g., simulation and CBT) beyond those groups who have traditionally been exposed to it (i.e., military and air transport pilots). As noted in the Introduction to this volume, general aviation pilots, who collectively fly more hours, and experience much higher accident rates than commercial air transport pilots, stand to gain the most from the widespread application of more effective training technologies.

Early approaches to CBT followed the programmed instruction model of learning developed by behavioural psychologists in the 1950s (Brock, 1997). The principal advantages were that the learner could dictate the pace of instruction and immediate feedback could be provided. Current CBT is virtually synonymous with 'multimedia system' in which a CD-ROM equipped computer delivers instructional material involving some combination of text, images, animation, sound, and full-motion video. Increasingly important are multimedia presentations delivered via the World Wide Web (WWW). There are certainly some impressive examples of multimedia CBT on the market. At least they are impressive in terms of their ability to attract attention and engage the learner's interest for a period of time. Delivering high quality multimedia CBT in aviation would, at first sight, seem to be mainly a matter of finding a good multimedia designer to package up the content material.

Hammers are good for knocking in nails. Software designers are good for designing multimedia software. Neither of these are necessarily good for producing effective CBT (although hammers can come in handy for delivering 'percussive maintenance' on recalcitrant machinery!). It is essential for CBT development to be driven by a specification of (i) what is to be learnt and (ii) what conditions are required to enable the learner to develop the necessary skills. In contrast, much training development has been driven by the desire to make use of emerging technologies.

It is readily apparent that much multimedia design is driven by this desire to make use of 'cutting edge' technologies. It is common to find Web pages and CBT systems dripping with fancy graphics and excessive visual clutter. What is pertinent ('Figure') can easily become Obscured or overpowered by the visually distracting 'Ground'. Lost in a 'FOG' in otherwords! Such multimedia CBT may well be less effective than conventional training, and the time and money spent on development might have been better invested elsewhere.

It is all too easy to find examples which support the claim that "All the computer does is provide an efficient means for bad instruction to be distributed" (Brock, 1997, p. 579). If we are to avoid falling into the trap of simply making bad instruction more widely available, a number of fundamental issues in the design of CBT must be addressed.

What is to be Learnt?

We all know exactly what pilots do and what makes a good pilot don't we? Surprisingly, the past fifty years of aviation psychology has produced very little in the way of detailed job analyses of being a pilot (Damos, 1996). Our failure to determine exactly what constitutes a 'good' pilot is reflected in the inability of current selection tests to predict the distant criteria of operational performance (Roscoe & North, 1980). We can certainly provide reasonable predictions of success at the early stages of a pilot's career, but not her or his longer term prospects. Improvements in this field are being made with the help of more cognitively complex measures of 'higher-order' cognitive abilities such as planning and prioritising, rather than 'lower-level' abilities such as reaction-time and short-term memory. The greater success of these measures in predicting advanced performance (O'Hare, 1997) indicates that the key elements of what all pilots do, and good pilots do best, lies in the area of higher-order cognitive skills. Like a fine timepiece, the simplicity of the observable movement of the hands belies the complexity of the hidden mechanism. Since we cannot dis-assemble pilots like watches, we have to study closely from the outside, drawing carefully reasoned inferences about the underlying processes.

The studies that Mark Wiggins and I have conducted with expert pilots (e.g., Wiggins & O'Hare, 1995) have been remarkably consistent with the results from studies of other experts reported in the cognitive science literature. Experts have a greater depth of knowledge in their field, they navigate through that knowledge in different ways from novices, they continually monitor and update their 'situational awareness', they are more aware of their limitations, and most importantly they notice cues and make distinctions that novices fail to see or recognise.

For example, the instrument scanning patterns of expert pilots are different from those of novice pilots (Bellenkes, Wickens, & Kramer, 1997). A dozen flight instructors with an average of 80 hours instrument time were compared with a dozen student pilots (average 1 hr instrument time) on a range of instrument manoeuvres in a simulated task. A display of simulated instruments was presented on a 19-in computer monitor. Participants made control inputs through a sidearm joystick. The scanning patterns of participants was measured with a head-mounted eye tracker. The experts were able to extract information from flight displays more rapidly and were better attuned to the inter-relationships between parameters (e.g., altitude deviations with heading changes). The authors go on to suggest that these findings can be used as the basis of "more efficient, theory-based, pilot training" (Bellenkes et al., 1997, p. 578).

It turns out that, as in the above study, the basis of many impressive judgmental and decision making skills lies in the increasingly well-attuned perceptions of the expert to subtle environmental features. The skilled fly-fisher 'reads' the water and knows where the trout will be (Maclean, 1976); chicken-sexers can sex 1000 chicks per hour at over 98% accuracy (Biederman & Shiffrar, 1987); chess-masters can instantly recognise the pattern of pieces in a game and generate the appropriate move even when playing under 'blitz' conditions with only 6 seconds per move (Chase & Simon, 1973; Calderwood, Klein, & Crandall, 1988); expert firefighters rapidly size up a situation from the immediately available perceptual information (Klein, 1989); experienced radiologists are able to make complex diagnostic decisions from single fixations of chest x-rays viewed for only one-fifth of a second (Kundel & Nodine, 1975). The bottom line is this: the superior ability of the expert to generate appropriate and effective judgements and decisions does not arise from better memory skills or more powerful decision making strategies, but rather from enhanced perceptual-recognitional skills.

In contrast to this story of astonishing expert performance, research on considered, analytical expert judgement shows quite a different picture. Many studies have shown that experts' decisions about psychiatric outcomes, criminal behaviours, financial problems etc. can be modelled by very simple linear combinations of a few central variables (Dawes, 1988). In one particularly stunning demonstration, the weight of psychiatric patients' files predicted four important criteria (work, health, family, and rehospitalisation) much better than the expert judgements of any of the staff psychiatrists, social workers, nurses or occupational therapists (Lasky, Hover, Smith, Bostian, Duffendack, & Nord, 1959).

The accuracy of expert judgements in predicting outcomes in most areas has generally been found to be quite poor. For example, the basic unstructured employment interview results in decisions which are based on a small number of variables such as the likeability of the interviewee and their perceived similarity to the interviewer. Unsurprisingly, these decisions bear virtually no relationship to the subsequent job performance of the candidates (Arvey & Campion, 1982).

In aviation, it is common to divide piloting skills into two groups: perceptual-motor skills involved in tasks such as landing, and 'headwork' skills involving judgement, decision-making, 'airmanship'. The perceptual-motor skills are trained through repeated exposure to subtle variations in environmental conditions—one practices landings over and over, with and without flaps, with and without cross-wind, etc. Feedback is immediate and relatively unambiguous. The result is the (eventual) development of well-calibrated perceptual skills. With increasing experience and currency, errors due to mishandling the aircraft become fewer and fewer (O'Hare, Wiggins, Batt, & Morrison, 1994).

The 'headwork' or decision and judgement skills have received quite a different treatment. In the past they have not been formally trained at all. Novices were supposed to acquire sound judgement merely through observation and experience. The prevalence of decision making errors in fatal accidents (Jensen & Benel, 1977) led to a realisation that these skills are too important to be left to chance and resulted in a change of attitude towards formal training in 'headwork' skills. Now, decision making and judgement skills are commonly subject to formal training.

However, the approach has been to treat decision making as a conscious and highly analytical process in which one must be aware of biases (e.g., so-called 'hazardous attitudes') and limitations (e.g., failure to consider sufficient options). Training in decision making has thus sought to overcome these problems by increasing awareness of decision making biases and limitations. Pilots are taught to adopt a careful analytical decision making approach, perhaps aided by the use of some acronym, such as 'DECIDE' (Clarke, 1986).

Unfortunately, since as we have shown, the essence of expert skilled performance is in making critical perceptual distinctions which the novice simply cannot make, the approach of trying to improve pilot decision making by training the analytical process itself is doomed to failure. Since the expert pilot "recognises how to react in a situation" without engaging in deep, analytical activity (Klein, 1991), then it follows that the first goal of training should be to enable the novice to make the appropriate discriminations (Charness, 1989). As shown below, computer-based training (CBT) can be used in innovative and effective ways to support the development of the required skills.

Facilitating the Development of Skilled Decision Making

Table 8.1 outlines the four essential stages in developing a training system for 'headwork' or cognitive skills. Some of the issues are discussed in more detail by Mark Wiggins in Chapter Seven of the present volume, which outlines the development of a prototype CBT programme for weather-related decision making.

Table 8.1   Four steps in developing skilled performance

1. Identify critical constraints on decisions
2. Develop simple problem environments
3. Develop challenging problem environments
4. Evaluate/Test

Step 1: Task Analysis.

The first step is to identify the critical cues and differentiations which support decision making in the task under consideration. "The student must learn what to look for, how to identify important cues, how to make critical discriminations" (Fitts, 1965). This is by no means easy. Even in apparently straightforward situations like the approach to landing, there may be disagreement as to what constitutes the critical cues (Lintern & Liu, 1991). One approach to this issue is to use some form of cognitive task analysis (CTA) to probe expert knowledge. This is the approach which has been most widely used in the analysis of aviation expertise (e.g., Seamster, Redding, Cannon, Ryder, & Purcell, 1993). There are a number of specific methods used in CTA, both verbal (e.g., the Critical Decision Method of Klein, Calderwood, & MacGregor 1989) and non-verbal (e.g., the analysis of similarity judgements of key concepts in the task domain). A common feature of many CTAs is the comparison between expert and less-experienced practitioners.

The alternative approach involves a detailed analysis of the task environment itself to uncover the critical constraints which determine possible action. This approach is referred to as cognitive work analysis or CWA (Vicente, 1995), or ecological task analysis or ETA (Kirlik, Walker, Fisk, & Nagel, 1996). Instead of focusing on what people do, the CWA/ETA approach focuses on the environment (broadly construed to include physical, task, structural and organisational aspects) and the limitations and constraints which are imposed on people's behaviour.

Further discussions of these methodologies are beyond the scope of the present chapter. Suffice it to say that this initial step constitutes the critical part of the whole project since it lays the foundation for the subsequent training programme. Unfortunately there are very few people in the academic or training communities who are familiar with these techniques of cognitive engineering. To its credit, the Australian Research Council (1996) has recognised the urgent national need for more trained scientists in this area; called for the establishment of a national centre for human factors research; and cited plans by the Defence Science and Technology Organisation (DSTO) to double its number of human factors scientists by 2000. Other countries with aspirations involving high participation rates in technologically advanced areas might be wise to follow the Australian example.

Step 2: Seeing the Invisible.

Once they have been identified, it is necessary to provide an environment within which trainees can learn the critical perceptual differentiations and practice the required actions to each cue. This environment should be dynamic and interactive. It has often been pointed out that the cockpit of a light aircraft provides a poor learning environment—noisy, cramped, and expensive to run. In addition to these practical drawbacks, the actual environment of flight can present such a plethora of stimulation that the learner may have considerable difficulty in determining the critical cues for any given flight manoeuvre.

Computer-based training can provide an improved delivery platform which overcomes most of these drawbacks, but certain key requirements must be met. These are best illustrated with respect to a prototype CBT programme for weather-related decision making developed by Mark Wiggins in collaboration with researchers at The Ohio State University and the U.S. Federal Aviation Administration (see Chapter 7).

Trainees must first be exposed to examples containing the critical perceptual cues. This is done by viewing and analysing a series of in-flight video clips and interviews with experts. There is also the opportunity to gain further information by 'clicking' on parts of the image. This provides the learner with essential 'declarative' or conceptual knowledge about the important cues. Because the learner is actively engaged with the material it is more likely to become part of the learner's long term memory than would be the case if the learner was simply given a list to learn containing the critical cues.

Active engagement is taken a step further in the next part of the CBT where the learner is required to make active judgements involving which of three flights shows the greatest deterioration in terms of a specified cue (see Figure 7.4) and also to determine at which point in a sequence of video clips from a single flight the conditions deteriorated from those forecast. The aim in this stage is for the learner to develop 'procedural' skill from the declarative knowledge obtained earlier. Procedural skill involves the rapid execution of actions associated with particular conditions. As Anderson has pointed out, "... in a system in which one can only learn skills by doing them, the importance of formal instruction diminishes and the importance of practice increases" (Anderson, 1987, p. 204).

The essential condition for the development of procedural skill is that there is an invariant or constant relationship between a condition and the consequent action. However, the trainee must eventually be able to deal with novel or unusual situations. In the past, it has been assumed that this capacity was developed through 'experience'. Ernest Gann (1961) provides an excellent description of this 'seasoning' process as he began his commercial flying career under the guidance of veteran captains. The essence of this apprenticeship was exposure to a wide range of challenging situations under the guidance of a knowledgeable expert. Assigned to fly the DC-2 route between Newark, Buffalo and Syracuse, Gann is pitied by his contemporaries "flying fast and high on longer routes between greater cities ... I have been sent into exile ... the planes are antiques ... the Captains notorious taskmasters, and the weather always distressing. They do concede the route offers superb training. So began my true apprenticeship" (Gann, 1961, pp. 61-62)

We can no longer rely on an apprenticeship model for aviation training. For one thing, with the upward flow of pilots from GA into the airlines, flight instructors in GA are likely to be relatively inexperienced themselves. To address this lack of real seasoning amongst GA flight instructors, we must learn to use computer technology more creatively to provide an 'artificial seasoning' process.

Step 3: Artificial Seasoning.

The ability to respond to varying circumstances is developed by providing the learner with increasingly challenging situations and examples. In the skill acquisition literature this is referred to as the tuning phase (Anderson, 1982). The learner becomes able to extract generalities from previous experiences and thus can apply these general principles to new situations. The learner can also become more discriminating, which involves narrowing the range of application of a principle. We need not be concerned with the details of these processes, and in any case this takes us into relatively uncharted areas of cognitive science. Many fundamental questions about learning from experience have yet to be answered. In the meantime, we have to apply what we do know to the development of training systems.

It has often been noted that one year of experience in a busy control tower/hospital/police station etc. is worth ten years experience in a quiet backwater (e.g., Klein & Hoffman, 1993). Thus 'time-on-the-job' is a poor measure of expertise. Twenty years of experience may be just that, or it may be one year of experience twenty times over. In situations where feedback is poor or delayed, then learning may be slow or even non-existent. This problem is likely to be particularly marked in complex and dynamically changing environments. Hence, experience, per se, does not equate to expertise; the important thing is what is learned as a result of that experience (Shanteau, 1988).

Unfortunately, in aviation 'time-on-the-job' is still used as a necessary (but not sufficient) criteria for issuing licences and qualifications. We wouldn't be happy with issuing a CPL to someone with 100 hours total time because they would be 'lacking in depth of experience'. We haven't yet been sufficiently innovative or thoughtful about using CBT to support the development of experience. We are still tied to a flight simulator model of CBT. As Schneider (1985) points out, "If training-programme developers blindly make computers perform the same type of simulation activities that were previously done with simulators, there is no reason to expect training efficiency to improve" (Schneider, 1985, p. 285).

Unfortunately, flight simulation has been side-tracked by controversy over the fidelity issue. The crucial question however, is not whether the simulation is 'real' enough, but whether it makes visible the essential and necessary constraints on performance. It has been shown that teaching novices to land an airplane can be done more efficiently with a simplified representation of the landing scene which makes it easier to perceive the critical constraints (Lintern & Liu, 1991). Simulations can also be more effective if they are not in real-time. 'Above Real-Time Training' or ARTT (Miller, Stanney, Guckenberger, & Guckenberger, 1997) has been used to train fast-jet test pilots and air traffic controllers. Such techniques could be extended to a wider range of tasks in aviation training.

For example, novice pilots tend to have a rather static understanding of meteorology. Time compression techniques might be used to enhance the understanding of synoptic charts and dynamic weather patterns. These are currently used on some television weather reports and the visual impact of seeing the movement of cloud patterns ahead of and behind frontal bands is certainly quite impressive. For training purposes however, it would be essential to ensure that the cues being made more visible are actually the ones crucial for effective task performance. This requires that Step 1—the analysis of expert performance or an investigation of inherent task structure—has been adequately performed. Having a software developer produce something that looks good is no guarantee that the critical task constraints have been identified!

Vicarious Experiences

Another approach to speeding up the acquisition of experience is the provision of what Klein and Hoffman (1993) refer to as 'vicarious experiences'. These include incident accounts and stories of relevant events. Klein and Hoffman refer to studies which show that carefully described incidents can be helpful in expanding the expertise of newly qualified staff. Schank (1996) has described in detail a framework for designing training based on the acquisition of new cases. According to Schank, "learning is the accumulation and indexing of cases and thinking is the finding and consideration of an old case to use for decision making about a new case" (Schank, 1996, p. 299). In the initial stages it may be sufficient for the learner to be acquiring and storing as many new experiences (cases) as possible. Later on, it is necessary for the learner to experience failures of expectation as previous cases turn out to be inadequate for dealing with current circumstances.

At this stage it is necessary for the learner to be encouraged to actively reflect on the reasons why the old cases need to be modified to account for the current experience. Henley, Anderson and Wiggins outlined the process of reflective learning in their chapter on flight instruction (Chapter 5) in the present volume. Schank outlines four principles which govern the creation of a case-based training system. The overarching principle of such a training system is that: "one can only learn from experience if the experience one is having is strongly related to an experience one has already had" (Schank, 1996, p. 343).

The first principle is that cases should be presented only when the learner has an 'expectation, question, or curiosity' about some issue. There must be some prior experience which arouses the need for the subsequent example, otherwise it will be quickly forgotten. The second principle is that there should be a progression of cases which create clear expectations about appropriate behaviours. In practice, it may be quite difficult to determine what this progression of cases should be. The third principle is that cases are best absorbed following some real-life experience. If the real-life experience is memorable, then the associated cases will be as well. Finally, cases must be followed by appropriate actions. Otherwise, learners may simply be learning to 'parrot right answers independent of real visceral decision making' (p. 347). Our emphasis on the pairing of conditions with actions (Step 2) is strongly consistent with this principle.

Aviation of course contains a vast repository of accident and incident reports as well as 'I-learned-about-flying-from that' stories. If experience involves encounters with a wide range of challenging events, then we could surely speed up the acquisition of experience by using CBT to involve the learner in new events. No one has yet fully exploited the principles described above to create a case-based CBT system for aviation instruction and training. A number of case-based decision aids have been developed for use in aviation. These include PROSPER, a system for choosing wing-section aerofoils (Whitaker, Stottler, Henke, & King, 1990); CASSIOPEE, a system designed to support troubleshooting of CFM 56-3 aircraft engines on B737 aircraft (Heider, 1996); CASELINE, for B747-400 fault diagnosis and rectification (Dattani, Magaldi, & Bramer, 1996); and BASIS, the British Airways Safety Information System which aggregates information from operational safety reports, maintenance quality reports, human factors information, and flight data recorder data from all flight operations (McGregor & Hopfl, 1993).

We have just begun to research the effectiveness of a simple case-based CBT system on performance in a laboratory-based pilot decision making task. The experiment was designed to determine if exposure to case-based information alone would positively influence decisions made in a simulated flight. Following the theory presented by Schank (1996) it might be expected that there would be no positive effect of simply presenting cases to learners who have no prior expectations regarding suitable actions in a weather-related decision making task. A second group of participants were therefore given a prior grounding in the basic rules of weather-related decision making before being exposed to the case-based training. The design of the study is shown in Figure 8.2 The results of our initial investigation are outlined in the following section.

Group 1	No Training	Simulated Flight Task
Group 2	Cases Only Training	Simulated Flight Task
Group 3	Rules Training + Cases Training	Simulated Flight Task

Figure 8.2   Experimental design of the preliminary case-based training study

The ICARUS Experiment

Using a specially developed simulation called ICARUS (see Figure 8.3), Richard Batt and I have recently found that trainees exposed to both a set of rules and a range of case examples performed better on a weather-related decision making task than participants who received no training or who were exposed to the examples alone. Participants in this study were flight naive undergraduates who were given a simulated 'search-and-rescue' scenario to follow. ICARUS was designed to represent a 'real' GA panel but to require no previous flight skills to control the aircraft. All commands are issued by clicking on the appropriate arrow key on the 'flight director'. By breaking the direct link between control input and outcome, the simulation was far more stable than commercially available PC flight simulators, and participants were able to control the aircraft without any prior practice. The ICARUS simulation was developed to measure actual behaviour in a simulated setting rather than factual knowledge or stated intentions.

Fifty-seven university undergraduates took part in the study. Participants were divided into three groups. The control group received no preliminary training. The third group was given a brief CBT course (Rules Training) consisting of a series of screens containing both text and illustrations describing the dangers of flying into IMC, including advice about being trapped by a lowering cloudbase and rising terrain. The material was drawn from a range of pilot training manuals and reference books including Buck (1988), Collins (1978), Job (1994), and Wagtendonk (1994). The material stressed the need for good aeronautical decision making and the timely recognition of deteriorating conditions. Strategies such as making a 180 degree turn were explained.

Both training groups received a case-based CBT training module containing four illustrated examples (see Figure 8.4) of New Zealand weather-related decision making accidents. The case histories on which the material was based were drawn from a range of flight safety publications from different countries, including 'Flight Safety' (New Zealand CAA); 'Asia-Pacific Air Safety' (Bureau of Air Safety Investigation, Australia); 'Aviation Safety Reflexions' (Transport Safety Board, Canada); and GASIL (UK CAA).

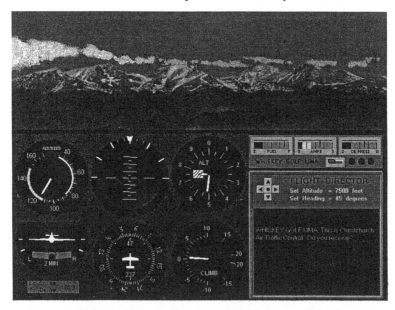

Figure 8.3 **The ICARUS simulation developed by Richard Batt at The University of Otago to investigate the development of experience through exposure to examples**

The aircraft was carrying out a private VFR flight from Wellington to Christchurch. Weather conditions were poor with cloud cover across most of Nelson-Marlborough and Canterbury. A squall-line passing across the coast near Kaikoura resulted in areas of driving rain, cloud in excess of 5/8's cover down to 500 feet, with lower patches, and visibility down to 500 metres.

Screen "EXBW3A"   Setup   Exit

Figure 8.4 **A screen from the case-based CBT module**

Following this brief period of training, participants were shown how to operate the ICARUS simulation and sent on a cross-country flight during which they came across a crashed aircraft which was unable to communicate with ATC directly but could relay calls through the participant's aircraft. This provided the motivation to continue the flight for as long as possible. Participants were required to maintain assigned altitudes and headings, which were changed from time to time by air traffic control. If they allowed these values to fall outside acceptable limits then both visual and auditory warnings were presented. As the simulation progressed, the visual scene would gradually fill with cloud. At the start of the flight there was only 6% of the scene occluded by cloud. We were interested in seeing how far in to deteriorating weather participants would progress before turning back. This was measured by the percentage of cloud cover at the time the decision was made to discontinue the flight.

The results are shown in Figure 8.3. We found that participants whose training included both a rule-based framework as well as specific case-based examples of weather-related crashes turned back very much earlier than the other participants. This is in accordance with Schanks's (1996) theory described earlier. According to this perspective, presenting cases 'before their time' i.e., without a prior background of expectations, will not be effective. These data show that participants were only influenced in their decision making when they received the case-based information after receiving a basic grounding in the rules of weather-related decision making. Unfortunately, due to practical constraints, it was not possible to include a 'rules only' training condition to control for any effects which learning the rules alone might have. Nevertheless, the fact that the findings are clearly in accord with theoretical predictions is highly encouraging.

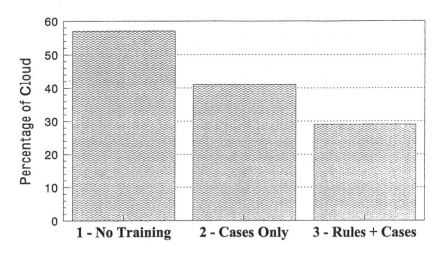

Figure 8.3   Results of the ICARUS experiment

These preliminary results are broadly consistent with findings emerging from similar lines of research elsewhere. Kirlik, Walker, Fisk, and Nagel (1997) trained participants to make judgements in a simulation of a game of American football. Participants who had to rely solely on abstract knowledge of the rules were much slower than participants whose training was enhanced by being given concrete perceptual examples. The value of experience then, both in 'real-life' and in these laboratory studies may be to provide concrete, visible examples of abstract rules and propositions. As well as making the rules more memorable, examples are fundamental in 'tuning' our experience by presenting the learner with a myriad of variations and subtleties. It is essential that this tuning stage in the development of expertise be preceded by the development of the necessary skills in cue perception and differentiation described in Step 2 above.

Step 4: Evaluation and Testing.

If we are to improve on 'time-on-the-job' measures then it will be necessary to develop new means of measuring the development of expertise. These will also be necessary for evaluating the success of CBT or other methods designed to facilitate the acquisition of expertise. Klein and Hoffman (1993) summarise possible approaches in three categories: performance measures, knowledge differences, and developmental milestones. Further discussion of these issues in an aviation context is provided by Wiggins (1996).

Conclusions

Training is an activity designed to prepare people for performance. Effective performance in many skills is based on the ability to recognise situations on the basis of subtle cues and distinctions. In aviation, as in other areas, there needs to be a greater emphasis on analysing expert performance and task structure to identify these critical constraints. The results can form the basis of CBT to enhance the ability of the novice to pay attention to the critical cues and constraints which govern action in a wide variety of circumstances. As well as sharpening up their situational awareness, the CBT should be designed so as to facilitate the development of procedural skill—learning by doing. Little, if any, CBT in aviation training has been designed in this way. The CBT developed by Mark Wiggins as part of the FAA's efforts to improve pilot decision making (see Chapter 7) is an indication of what can be done.

Someone who has learned to make crucial distinctions in their field has certainly begun to acquire the necessary basis of expertise. To become a fully-fledged expert traditionally requires many years of further experience in sharpening and tuning one's skills. We need to develop ways of speeding up and systematising this process. I have labelled this the development of 'artificial seasoning'. We have evidence that involvement in simulated experiences can increase the effectiveness of basic training, but we do not yet know enough about learning from examples to be able to prescribe exactly how this aspect of training should be developed. If we can side-step the question of simulator fidelity and concentrate our efforts on the creative exploitation of CBT to provide new ways of gaining experience, then we can take aviation education beyond 2000 with some confidence.

Future Directions

As Roscoe (1980) has clearly observed, the field of aviation psychology can logically be defined in terms of two complementary areas of activity. The first involves the design of systems—hardware, software, procedures and processes. The second involves the selection and training of individuals to operate those systems: "... the former serves to reduce the need for the latter; the latter completes the job left undone by the former" (Roscoe, 1980, p. 3). The existence of extensive training requirements for relatively simple tasks is a sure sign of deficient design which fails to match human requirements. Even in more complex systems, such as aviation, it is important to recognise the primary role of systems design in making tasks manageable and safe. Some of the technological advances which have simplified and improved the reliability of performance in airline cockpits are likely to be extended to general aviation in the next millennium (see Beringer, Chapter 10).

The potential influence of a case-based approach to training is matched by the potential of case-based designs for aiding pilot decision making. The speed and confidence of expert decision making is related to the ability of the expert to retrieve a suitable prior experience to assist with formulating a response to the present situation. Given a large enough stock of previous experiences, and a well-organised memory system which allows for the efficient retrieval of prior experiences, the expert decision maker does not have to invent new solutions to problems on the spot. The novice, however, lacking an appropriately indexed knowledge base must engage in effortful, conscious, deliberate problem solving under pressure.

If the novice had access to a knowledge base of the kind which underlies expert performance, then this might provide a valuable decision making aid. As Whitaker et al (1990) note: "... one of a CBR system's strengths lies in its ability to prompt the user with a better memory retrieval system. CBR systems retrieve information systematically and provide it to the user. This information can be targeted to match critical attributes of the current problem and hence provide information most useful to the operator's immediate needs" (Whitaker et al., 1990, p. 314). The most obvious use for such a system would be as an aid to pre-flight planning and decision making, although the possibility of in-flight decision aiding deserves to be explored. The concept of a cockpit decision support system is not new. For example, the Pilot's Associate (Banks & Lizza, 1991) was developed to aid military pilots' decision making.

Although this, and other chapters in the present volume, have been primarily concerned with issues involved in training, it is important to bear Roscoe's dictum in mind. That is to say, that training serves largely to complete the job left undone by design. For example, no amount of training in how to operate computers would have produced the same benefits as were achieved by the move from syntax driven systems to point-and-click graphical interfaces. If pilots are left to perform extremely challenging flight tasks in highly uncertain environments with limited information, then no amount of improved training can be expected to compensate for these intrinsic deficiencies. On the other hand, an integrated approach to both the design of the next generation of general aviation aircraft, and the training of the pilots who will operate in them, may yield substantial benefits in terms of enhanced safety and efficiency of operation.

References

Anderson, J. R. (1982). Acquisition of cognitive skill. *Psychological Review, 89*, 369-406.

Anderson, J. R. (1987). Skill acquisition: Compilation of weak-method problem solutions. *Psychological Review, 94*, 192-210.

Arvey, R. D., & Campion, J. E. (1982). The employment interview: A summary and review of recent research. *Personnel Psychology, 35*, 281-322.

Australian Research Council (1996). *Psychological science in Australia*. Canberra: Australian Government Publishing Service.

Banks, S. B., & Lizza, C. S. (1991, June). Pilot's Associate: A cooperative knowledge-based system application. *IEEE Expert,* 18-29.

Bellenkes, A. H., Wickens, C. D., & Kramer, A. F. (1997). Visual scanning and pilot expertise: The role of attentional flexibility and mental model development. *Aviation, Space, and Environmental Medicine, 68*, 569-579.

Biederman, I., & Shiffrar, M. M. (1987). Sexing day-old chicks: A case study and expert systems analysis of a difficult perceptual-learning task. *Journal of Experimental Psychology: Learning, Memory, and Cognition, 13*, 640-645.

Beynon-Davies, P. (1995). Information systems 'failure': The case of the London Ambulance Service's computer aided despatch Project. *European Journal of Information Systems, 4*, 171-184.

Brock, J. F. (1997). Computer-based instruction. In G. Salvendy (Ed.), *Handbook of human factors and ergonomics, second edition* (pp. 578-593). New York: Wiley.

Buck, R. N. (1988). *Weather flying*. New York: Macmillan.

Calderwood, R., Klein, G. A., & Crandall, B. W. (1988). Time pressure, skill, and move quality in chess. *American Journal of Psychology, 101*, 481-493.

Charness, N. (1989). Expertise in chess and bridge. In D. Klahr & K. Kotovsky (Eds.), *Complex information processing: The impact of Herbert A. Simon* (pp. 183-208). Hillsdale, NJ: Lawrence Erlbaum.

Chase, W. G., & Simon, H. A. (1973). Perception in chess. *Cognitive Psychology, 4*, 55-81.

Clarke, R. (1986). *A new approach to training pilots in aeronautical decision making*. Frederick, MD: AOPA Air Safety Foundation.

Clegg, C., Axtell, C., Damodaran, L., Farby, B., Hull, R., Lloyd-Jones, R., Nicholls, J., Sell, R., & Tomlinson, C. (1997). Information technology: A study of performance and the role of human and organizational factors. *Ergonomics, 40*, 851-871.

Collins, R. L. (1978). *Flying safely*. London: Black.

Dattani, I., Magaldi, R. V., & Bramer, M. A. (1996). A review and evaluation of the application of case-based reasoning (CBR) technology to aircraft maintenance. *Research and development in expert systems, XIII* (ES96) (pp. 189-203). Oxford, UK: SGES Publications.

Dawes, R. M. (1988). *Rational choice in an uncertain world*. Orlando, FL: Harcourt Brace Jovanovich.

Damos, D. L. (1996). Pilot selection batteries: Shortcomings and perspectives. *The International Journal of Aviation Psychology, 6*, 199-209.

Fitts, P. M. (1965). Factors in complex skill training. In R. Glaser (Ed.), *Training research and education*.

Gann, E. K. (1961). *Fate is the hunter*. London: Hodder & Stoughton.

Geoghegan, W. H. (1994). *Whatever happened to instructional technology?* Paper presented at the 22nd Annual Conference of the International Business Schools Computing Association.

Heider, R. (1996). Troubleshooting CFM 56-3 engines for the Boeing 737 using CBR and data-mining. In I. Smith & B. Faltings (Eds.), *Advances in case-based reasoning: Proceedings of the 3rd European workshop on CBR, EWCBR-96*. Berlin: Springer-Verlag.

Jensen, R. S., & Benel, R. A. (1977). *Judgment evaluation and instruction in civil pilot training* (Final Report FAA-RD-78-24). Springfield, VA: National Technical Information Service.

Job, M. (1994). *The old and the bold: Who learnt about flying that way*. Melbourne, Australia: Iona.

Kirlik, A., Walker, N., Fisk, A. D., & Nagel, K. (1996). Supporting perception in the service of dynamic decision making. *Human Factors, 38*, 288-299.

Klein, G. (1989). Recognition-primed decisions. In W. B. Rouse (Ed.), *Advances in Man-Machine Systems Research, 5* (pp. 47-92). Greenwich, CT: JAI Press.

Klein, G. A. (1991). Models of skilled decision making. *Proceedings of the Human Factors Society 35th Annual Meeting*. Santa Monica, CA: Human Factors Society.

Klein, G. A., Calderwoood, R., & MacGregor, D. (1989). Critical decision method for eliciting knowledge. *IEEE Transactions on Systems, Man, and Cybernetics, 19*, 462-472.

Klein, G., & Hoffman, R. R. (1993). Seeing the invisible: Perceptual-cognitive aspects of expertise. In M. Rabinowitz (Ed.), *Cognitive science foundations of instruction* (pp. 203-226). Hillsdale, NJ: Lawrence Erlbaum.

Kundel, H. L., & Nodine, C. F. (1975). Interpreting chest radiographs without visual search. *Radiology, 116*, 527-532.

Lasky, J. J., Hover, G. L., Smith, P. A., Bostian, D. W., Duffendack, S. C., & Nord, C. L. (1959). Post-hospital adjustment as predicted by psychiatric patients and by their staff. *Journal of Consulting Psychology, 23*, 213-218.

Lintern, G., & Liu, Y-T. (1991). Explicit and implicit horizons for simulated landing approaches. *Human Factors, 33*, 401-417.

Loftin, R. B., Savely, R. T., Benedetti, R., Culbert, C., Pusch, L., Jones, R., Lucas, P., Muratore, J., Menninger, M., Engleberg, M., Kenney, P., Nguyen, L., Saito, T., & Voss, M. (1997). Virtual environment technology in training: Results from the Hubble space telescope mission of 1993. In R. J. Seidel & P. R. Chatelier (Eds.), *Virtual reality, training's future?* (pp. 93-103). New York: Plenum.

MacGregor, C., & Hopfl, H. (1993). A commitment to change: Safety management in British Airways. *Disaster Prevention and Management, 2*, 6-13.

Mclean, N. (1976). *A river runs through it.* Chicago: University of Chicago Press.

Miller, L., Stanney, K., Guckenberger, D., & Guckenberger, E. (1997, December). Above real-time training. *Ergonomics in Design*, 21-24.

O'Hare, D. (1997). Cognitive ability determinants of elite pilot performance. *Human Factors, 39*, 540-552.

O'Hare, D., Wiggins, M., Batt, R., & Morrison, D. (1994). Cognitive failure analysis for aircraft accident investigation. *Ergonomics, 37*, 1855-1869.

Regan, C. (1997). Some effects of using virtual reality technology: Data and suggestions. In R. J. Seidel & P. R. Chatelier (Eds.), *Virtual reality, training's future?* (pp. 77-83). New York: Plenum.

Roscoe, S. N., & North, R. A. (1980). Prediction of pilot performance. In S. N. Roscoe (Ed.), *Aviation psychology* (pp. 127-133). Ames: The Iowa State University Press.

Schank, R. (1996). Goal-based scenarios: Case-based reasoning meets learning by doing. In D. B. Leake (Ed.), *Case-based reasoning: Experiences, lessons, and future directions* (pp. 295-347). Menlo Park, CA: AAAI Press.

Seamster, T. L., Redding, R. E., Cannon, J. R., Ryder, J. M., & Purcell, J. A. (1993). Cognitive task analysis of expertise in air traffic control. *The International Journal of Aviation Psychology, 3*, 257-283.

Shanteau, J. (1988). Psychological characteristics and strategies of expert decision makers. *Acta Psychologica, 68*, 203-215.

Vicente, K. J. (1995). Task analyis, cognitive task analysis, cognitive work analysis: What's the difference? *Proceedings of the Human Factors and Ergonomics Society 39th Annual Meeting* (pp. 534-537). Santa Monica, CA: Human Factors and Ergonomics Society.

Wagtendonk, W. (1994). *Weather to fly for recreational pilots.* Tauranga, NZ: Author.

Whitaker, L. A., Stottler, R. H., Henke, A., & King, J. A. (1990). Case-based reasoning: Taming the similarity heuristic. *Proceedings of the Human Factors Society 34th Annual Meeting* (pp. 312-315). Santa Monica, CA: Human Factors Society.

Wiggins, M. (1996). A computer-based approach to human factors education. In B. J. Hayward & A. R. Lowe (Eds.), *Applied aviation psychology: Achievement, change, and challenges* (pp. 201-208). Aldershot: Ashgate.

Wiggins, M., & O'Hare, D. (1995). Expertise in aeronautical weather-related decision making: A cross-sectional analysis of general aviation pilots. *Journal of Experimental Psychology: Applied, 1*, 305-320.

Part 5
New Technology and General Aviation

9 Lost in Space: Warning, Warning, Satellite Navigation

Ruth M. Heron and Michael D. Nendick

Introduction

Why, it may be asked, should a lost-in-space warning about satellite navigation be issued? Undeniably, the highly touted Global Navigation Satellite System (GNSS) represents a quantum leap towards improvement in navigation for both corporate and general aviation (GA) pilots. Moreover, the foundation of GNSS, the U.S. Global Positioning System (GPS) comprising 24 earth-orbiting satellites, will surely bring into the flight deck an advanced world of avionics promising unprecedented capability for providing pilots with precise navigation information in all phases of flight. Indeed, GPS is already improving GA en route flight under both visual flight rules (VFR) and instrument flight rules (IFR), as well as the navigation of IFR nonprecision approaches. Eventually, as Johns (1997) points out, through augmentation systems such as ground-based local and wide area differential GPS, satellite navigation is expected to provide for cost-effective flying of precision approaches.

Clearly, though, if the potential benefits of GPS are to be fully realised by GA pilots, the receivers they use in order to avail themselves of this technology must be ergonomically sound in all respects. However, so keen were rival manufacturers to get their products to market that, early in the history of these units, a multitude of models emerged with a rapidity that could not be matched by regulatory authorities attempting to establish standards for their design and operation. Moreover, no ergonomics expertise appears to have been involved in their development (nor in the development of Loran-C predecessors whose hardware and software were often the basis for GPS designs)—an omission the more lamentable in light of the much greater complexity of GPS avionics vis-à-vis that of the VOR and ADF systems they are intended to replace. Little wonder, then, that various ergonomics/human factors problems with these new receivers have become manifest, many of which will likely be solved only as the new systems evolve. For precisely this reason, it must be stressed that, in the meantime, over-confidence during operation of either the small handheld VFR units used for en route navigation, or the more complex panel-mounted units designed and certified for IFR en route and approach navigation, may present a serious risk to aviation safety.

It follows, then, that the ergonomics features associated with using GPS equipment, along with their implications for aviation safety, are extremely important concerns for GA pilots, instructors, and small-scale operators. The intention of the authors is to illuminate the pilot performance problems relevant to these concerns. Beginning with a brief overview of GNSS developments, the authors then set out an ergonomics model and perspective within which subsequent descriptions of GPS receiver design deficiencies and their effects on pilot performance may be understood. Next, the chapter draws attention to pilot attitudes and behaviours associated with use of GPS receivers. Subsequently it discusses remedial measures and concludes with a summary and important messages for designers of avionics systems and those who use them.

Overview of GNSS Developments and Status

In the 1980s, members of the Future Air Navigation System (FANS) committee, formed by the International Civil Aviation Organization (ICAO), met to discuss ways of eliminating problems related to air traffic congestion, the inadequacy of HF communications, the increasing amount of traffic in non-radar areas, and financial difficulties that loomed on the horizon for airlines and civil aviation authorities around the world (SatNav Program Office, 1998). GNSS is the navigation part of the more comprehensive Communications, Navigation, Surveillance and Air Traffic Management (CNS/ATM) concept adopted by FANS to solve these problems and, thereby, to lead aviation into the 21st century (Galotti, 1997).

Heijl (1997) reports that one of the conclusions of FANS was that full implementation of the CNS/ATM concept would provide a new freedom to airspace users, a system he refers to as having 'VFR flexibility with IFR safety', but one known in the aviation community simply as "free flight". This freedom, he explains, will depend heavily upon the automation within CNS/ATM, such as that presently offered by GPS. Clearly, major changes to air traffic control (ATC) and GA operations are foreshadowed in this system which, in providing such freedom, allocates to the pilot the responsibility of maintaining appropriate aircraft separation. An important relevant component of CNS/ATM is Automatic Dependent Surveillance (ADS), the data link system that will provide ATC with information about the aircraft's position conveyed directly from the vehicle's navigation equipment, as well as with route information stored in its computers (SatNav Program Office, 1998).

Through its Satellite Operational Implementation Team (SOIT), the U.S. Federal Aviation Administration (FAA) has been working co-operatively with Nav Canada to have a unified satellite navigation system fully implemented in North America early in the 21st century, this to be done in phases that are feasible, given their technical, operational, and certification requirements (GNSS Working Group, 1993, June 23). Since the launching of the first series of operational satellites in 1979, the U.S. GPS constellation of satellites became fully operational in July 1995 and, as of May 3, 1997, a total of 26 satellites had been brought into service (SatNav Program Office, 1997). Launching of a follow-on generation of 33 satellites is planned for 2012, these now being built with advanced features allowing for calculation of ionospheric errors in real time and, hence, greater accuracy (SatNav Program Office, 1997). The total number of U.S. satellites could be greatly enlarged through integration with the Russian Global Orbiting Navigation Satellite System (GLONASS) which, in fully operational state, will consist of 24 satellites (GNSS Working Group, 1993, June 23).

Apart from the preparation and launching of satellites, work relevant to the design of approaches is also well advanced. R. Bowie, SatNav Program Manager, NAV CANADA, reports that, to date, virtually all VOR and NDB approaches in the U.S. have been overlain by GPS configurations and, of 1,901 T-configured stand-alones designed, 1,357 have been published (R. Bowie, personal communication, August 15, 1998). By the year 2001, the U.S. plans to have architecture for local area augmentation systems (LAAS) available, and to have the implementation of wide area augmentation systems (WAAS) well in progress (SatNav Program Office, 1998).

Complementing U.S. developments, Canada now has about 150 overlay GPS nonprecision approaches in use, along with 42 stand-alone approaches; NAV CANADA expects to have a total of 100-125 stand-alones published by the end of 1998 (R. Bowie, personal communication, August 15, 1998). Capability for augmented GPS precision approaches meeting high-level requirements of accuracy, integrity, and service availability is expected between 2005 and 2010 (GNSS Working Group, 1995, September 15).

In the interest of achieving a "seamless" global guidance system, the U.S. promised ICAO in 1996 that their GPS system would be provided as a basis for worldwide transition to a global satnav system; hence, U.S. policy is one that welcomes use of its system by other countries without user fees (SatNav Program Office, 1997). Accordingly, Japan and Europe are now proceeding with their systems, the MT-Sat Based Satellite Augmentation System (MSAS) and the European Geostationary Navigation Overlay System (EGNOS), respectively (R. Bowie, personal communication, August 15, 1998). Lawson-Smith (1997) reports that the Airservices Australia Strategic Plan for the period 1997 to 2012 calls for (i) en route IFR navigation and gradual introduction of nonprecision approaches between 1997 and 2000, (ii) implementation of trial systems for precision approaches between 2000 and 2005, and (iii) sole reliance on GNSS for en route navigation and nonprecision approaches with precision approaches introduced for "compliant aircraft" between 2005-2012.

On balance, satellite navigation will undoubtedly become the main available option for civil aviation worldwide. However, while the ICAO plan is to gradually remove all traditional ground-based aids, safety may dictate that, in some cases, GPS should be backed up by navigation systems using separate technology for redundancy and cross-checking. Whether such systems will utilise the Inertial Reference System (IRS) expected to be affordable even to GA operators, on-board terrain mapping radar, a skeleton VOR/DME, ILS/MLS, or some other system, is an issue for further deliberation. Of course most of these alternative systems, especially those of older vintage, will eventually become economically nonviable and, thus, disappear over time. It scarcely need be said that replacement systems should be technically and ergonomically sound, with safety the keynote of their design.

From the foregoing it will be clear that implementation of an entirely new communication/navigation aviation system is well underway, and that its immense complexity will change the art and craft of GA flying in a way that raises many ergonomics/human factors issues that have implications for aviation safety. Not the least of these is the design of GPS receivers. The section that follows provides a framework for understanding the ergonomics of these units and their effects on pilot performance.

An Ergonomics Perspective

Ergonomics, a term the authors of this chapter regard as synonymous with human factors, is concerned with how humans interact with machines. Synthesising information from a number of sources (Evarts, Shinoda, & Wise, 1984, Chapter 5; Eysenck & Keane, 1993; Kantowitz & Sorkin, 1987; Sanders & McCormick, 1992; Wickens & Flach, 1988), this section begins by describing a working model for a pilot/aircraft system—one, that is to say, that will be sufficient for an understanding of some of the basic ergonomics of the GA pilot's task prior to use of GPS. A later part of the section considers changes to the model necessary to accommodate use of a GPS receiver.

In the simplest form of the ergonomics theoretical model, the pilot is seen as operating in a closed loop pilot/aircraft system, processing information obtained through the senses from displays in the flight deck (as well as from the environment) and, on the basis of outcomes of this mental activity, executing control responses that change the aircraft's behaviour. The cycle begins again when results of this behaviour are fed back into the system. Sub-processes involved in the information processing that occurs between input to the brain and output of the control responses include attention to and detection of stimulus objects, brief sensory storage, pattern discrimination and recognition, reference to working and long-term memory, evaluation and decision-making and, finally, organisation of motor components in preparation of external responses. These sub-processes do not proceed in a linear fashion; rather they are interactive, characterised as the overall process is by many feedback and feedforward loops.

Adding to the complexity of the brain's task is the pilot's frequent reception of simultaneous input from several sensory modalities (e.g., visual, auditory, tactile, and kinaesthetic), the processing load being particularly heavy during the approach phase of flight. Normally, the brain of the experienced pilot will assign routinised motor-control parts of the flying task to lower levels where they can be run off seemingly without conscious effort, thus freeing up higher levels to meet whatever cognitive demands may arise. However, during the approach phase of flight, the speed and multiplicity of inputs can be so great—especially if ATC calls for an approach different from that planned—that the brain's well known capacity for parallel cognitive processing (Everts, Shinoda, & Wise, 1984, Chapter 5; Stokes, Wickens, & Kite, 1990) must be maximised. What must be considered also is that, remarkable as the brain's ability to process information is, it does have its limits—these being related, on the one hand, to how much it can process in a given unit of time and, on the other, to how efficient it can be when input is incompatible with human sensory, perceptual, or cognitive capabilities. Apropos of the latter, an important fact is that the pilot's information-processing workload is increased if input is either too fast or too slow, or if it is inaccurate, ambiguous, or degraded in some way.

When interpreted in terms of information-processing limitations, the May 1995 mid-air collision between a Fairchild Metro 23 and a Piper Navajo PA-31 provides a compelling illustration of how violation of some of these limitations can spell aviation disaster. According to the accident report by the Transportation Safety Board of Canada (1995), the former aircraft was inbound to, the latter outbound from, an airport in Ontario; and, because of the high degree of lateral accuracy for the GPS system each pilot appeared to have been using for guidance, separation between the two tracks was extremely narrow. There can be only speculation as to how much information processing was accomplished by these pilots prior to the impact of their aircraft—probably detection of something, perhaps perception (the stage of processing in which meaning is attributed to an object), and possibly—at least in the case of the Piper Navajo's pilot—even decision-making regarding evasive action. There can be certainty only with respect to the fact that neither pilot was able to process the information fast enough to mobilise and execute a satisfactory collision-avoidance response.

Time, although an important factor in this accident, may not have been the only one contributing to it. The above mentioned accident report indicates that, although there were no obstructions to visibility, the sky was overcast and the colours of the aircraft were white and light beige respectively. Under such conditions, impaired perception may have played a role; more explicitly, each pilot may have found the other's aircraft indiscriminable from the environment. Regardless of which factors were involved, it is clear that the accident underscores in a most dramatic way that there are inviolable limits to human information processing.

With reference again to the pilot/aircraft model described above, it is important to realise that, as implied earlier, it is really only rudimentary, just a bare-bones reflection of the GA pilot's real-life system and its relevant information-processing demands - even though that real-life pilot was being guided by pre-GPS navaids such as VOR and ADF, and hence had only to select a frequency to arrive at directional information. Consider, then, how inadequate this model must be if it is to accommodate the fact that the GA pilot using a GPS receiver must be able to programme this unit in a way that relates raw position and integrity data to an intended flight path, and to do so with both speed and accuracy. In view of this fact, the model put forth by Heron, Krolak, and Coyle (1997a) appears to be more suitable. These authors postulate that, when a GPS receiver becomes part of the GA pilot's flight deck, *another closed loop system is added to that for the pilot/aircraft system,* as indicated in Figure 9.1, thus imposing an additional information-processing burden on the pilot.

Before the nature of this additional workload is explored, a word must be said about a significant feature of theoretical ergonomics, namely, that, as the models outlined above imply, ergonomics is based upon the systems approach. In a system, every single component is related to every other component, and this interdependency of parts militates against adopting simplistic notions of causality to explain incidents or accidents, attributing them to 'pilot error'. Instead, without at all dismissing such influences on pilot performance as alertness, competence, and responsibility, the human factors specialist sees the pilot as unable to perform optimally unless all other parts of the system are also performing optimally. Accordingly, given a healthy, competent, and responsible pilot, system efficiency and accident/incident prevention will then be maximised only if the machine parts of the system are designed in a way that avoids challenging or violating human limitations.

The ergonomics models and concepts briefly outlined above will be applied in the next section to the discussion of some of the troubling design features of today's GPS receivers and the behaviours they may elicit from the GA pilots who must operate them.

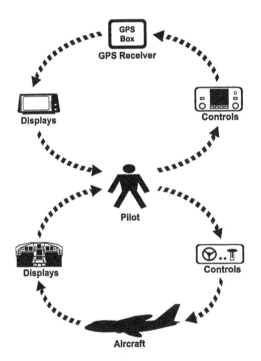

Figure 9.1   A GPS receiver in the flight deck adds another closed loop system to the pilot/aircraft system
Source: Heron et al. (1997a)

GPS Receiver Design

In treating the design deficiencies of GPS receivers, this section will ignore those relevant to systems interfaced with a flight management system (FMS); rather, it will confine itself to those prevalent with handheld models used for VFR flying and the panel-mounted models used for IFR flying. After all, when GPS was first added to the FMS, the situation for the pilot simply became one in which another sensor had been added to the computer. Most, if not all, of these pilots had already had extensive training in using the FMS and were practised in programming flight plans and approaches. Hence, they were used to being the human part of a double closed-loop system that would, if modelled, be the same as that in Figure 9.1, albeit with the GPS receiver box replaced by the FMS. However, prior to the appearance of GPS receivers, the GA pilot without an FMS had become practised in selecting frequencies, *not* in complicated computer flight programming.

Moreover, in the accounts below, less attention will be given to the "handhelds", as they are commonly called, than to the more sophisticated panel-mounted units. The rationale for this difference in emphasis is that only the latter can be certified under Technical Service Order (TSO) C129 for IFR flying (FAA, 1992) which, inasmuch as it imposes on the pilot a greater cognitive workload than does VFR—especially during approach and landing phases of flight—is associated with human factors problems that are greater in both variety and complexity.

Handheld GPS Receivers and VFR Navigation

General characteristics Although the databases of most of the better handheld receivers have incorporated all existing ground-based navaids, their lack of the waypoints for nonprecision approaches, along with other simplifications, render them uncertifiable under TSO C129. For this reason, these receivers should be used only for VFR en route navigation. Being small and light, these third-generation receivers can be comfortably held by the user in the manner shown in Figure 9.2. However these units have a number of other design features which, as explained in the paragraphs below, pose threats to aviation safety.

Cable connections Handhelds are designed not only for holding in the hand, but also for attachment to the control column by a yoke mount, or onto or on top of the instrument panel. Unfortunately, because of their portability and ready transferability from one aircraft to another, the attachments are likely to be impermanent, this increasing the possibility of insecure connections that leave power and antenna wires trailing hazardously around the cockpit. According to one National Transportation Safety Board (NTSB) (1997a) report, the pilot of a home-built Boccher P51 Mustang, which was destroyed when it struck trees on departure from an airport in the state of New York, told the FAA inspector that his handheld GPS receiver had fallen from the instrument panel during the take-off roll and jammed the flight controls. Although the inspector was unable to retrieve the receiver, the pilot's report gains plausibility in light of a similar accident (NTSB,

1998a) involving substantial damage to a Piper PA-18-160 which veered to the right during take-off from Talkeetna Airport in Alaska and landed in a ditch. In this case, the Piper's owner found the antenna cable of the GPS receiver lodged in the right parking brake; apparently it had not been connected and, hence, had slipped off the dash and jammed the brake.

Signal reception A major factor underlying poor GPS signal reception is poor antenna configuration. Signal interruption may be the result of an insecurely connected antenna cable, a condition highly relevant to handheld and portable receivers, because of their easily dislodged plug-in connections. However the location of the antenna is also an important factor. As UHF signals cannot penetrate the hull of the aircraft, reliable GPS reception cannot be expected unless the receiver is connected to an external antenna with an unimpeded view of the sky and sufficiently distant from other aerials to prevent signal interference. St. George and Nendick's (1997) observation of short-term loss of GPS navigation as a result of signal breakthrough, even from normal air traffic, points up the concern about this problem.

Figure 9.2   Typical third-generation VFR handheld GPS receivers with map displays

Loss of power Another common cause of GPS information failure is loss of power to the receiver. Handheld receivers are generally powered by an AA battery pack, with rechargeable NiCad cells and connections to the aircraft-powered cigarette lighter as optional accessories. If connections to the aircraft source are insecure, the receiver will revert to battery power. However, if the battery is low, those systems boasting a warning indicator may deliver the message "Warning, Battery Low" only shortly before the display goes blank, thus leaving the pilot without navigation guidance from the receiver. Just such a mischance may have befallen the pilot of a Bellanca 14-13, who in the fall of 1994 was using a GPS receiver to navigate while on a personal flight to Tunica, Mississippi. About 10 minutes before arrival, the receiver batteries failed. Becoming disoriented, the pilot then used up his fuel trying to locate the airport, eventually making a forced landing into a casino parking lot where he crashed his airplane into some trees (NTSB, 1995a).

Some GPS handhelds do not have a low-battery warning indicator, in which case it is incumbent on the pilot to conduct a regular check of the power source in use, an exercise sometimes entailing an unduly time-consuming search for the relevant page in the receiver information system.

Controls and displays Controls on handhelds generally have either an array of push buttons resembling a touch-tone telephone pad for entry of alphanumeric data, or a limited number of push buttons for choosing functions. In either case, the controls are small and have very little tactile or auditory feedback, so that key-punch errors are easily made, especially by pilots who have large fingers or who wear gloves, and especially in turbulent conditions. Moreover, most keys have multiple functions, a property prompting St. George and Nendick (1997) to stress that the pilot must maintain awareness of the mode in which the receiver is operating.

As the very small dimensions of the display screens prevent simultaneous presentation of large amounts of data, information for a given category or mode must be spread over several screens. For example, the navigation (Nav) mode typically includes such data as present position, groundspeed, bearing to a waypoint, desired track, cross-track error, and track-angle error. However, to be readable at all, only a few of these can appear on one screen, so that several screens are needed to exhaust the store of information for that mode. Given this constraint, the ergonomics ideal would then demand that the most useful information appear on the first screen, with additional information presented on subsequent screens in descending order of importance. Nendick and St. George (1995) found that pilots regarded groundspeed, present position, distance to go, cross-track error, CDI-bar, and moving map as the most helpful units of information: a useful next research step for these investigators would be to establish the best presentation order for groups of these items.

The diminutive size of the screen of a handheld, coupled with the large amount of information to be displayed on it, dictates also that letters, symbols, and icons be exceedingly small. Primary symbol heights are, in fact, in the order of 4-5 mm, those for secondary symbols only 2-3 mm. Accordingly, numbers can be easily misread, especially under certain conditions of ambient illumination, while warning symbols may be so inconspicuous as to be missed entirely.

In the light of the results of Williams' (1998) simulator study, it might be surmised that the poor display properties of handhelds could be offset to some extent by an electronic moving map (see Figure 9.2); his subjects made fewer errors and took less time to judge the relative direction to nearest airport with a map display than with a text display listing airports along with bearing and distance. Certainly it is to the pilot's advantage to have a display showing the aircraft's location with reference to airports, navigation aids, and controlled airspace boundaries, eliminating as it does the need to calculate the difference between the aircraft's heading and the bearing of an airport in order to arrive at relevant orientation data. Indeed, many handhelds now offer not only moving maps, some with an adjustable scale and ability to declutter the screen, but also other graphical displays such as virtual horizontal situation indicators (HSIs).

However, while graphical enhancements, if used properly, undoubtedly have potential for assisting the GA pilot, they cannot remove the potential for error if either the remaining text screens or the graphical screens themselves have the property of poor readability. If the pilot has difficulty in detecting or discriminating letters and/or symbols on any GPS receiver screen, the sensory and perceptual phases of information processing will be subject to error or at least slowed down, thus delaying and possibly impairing the cognitive processes of evaluation and decision-making and, thereby, incurring a loss of situation awareness.

Summary While treatment of the ergonomics of handheld GPS receivers has been brief, it will be complemented with additional detail in the subsection on panel-mounted receivers which, as may be surmised, share many of their design deficiencies with these smaller units. Meanwhile, the material presented above should make it obvious that there are two basic types of human factors problems relevant to the design of handheld receivers—namely, those that can be minimised through careful pre-flight preparation on the part of the pilot, and those that arise primarily as the result of design features that strain the pilot's information processing capacities. With respect to the first category, for example, the jamming of controls by cables can be avoided, and potential for poor signal reception or reversion to battery power reduced, if the pilot ensures during the walk-around that all relevant cable connections are secure. Additionally, signal reception can be enhanced if the pilot connects to a properly configured antenna system, while power failure due to low batteries can be eliminated altogether if the pilot checks their storage level prior to flight.

However receiver properties such as the size and shape of controls and the size and density of symbols, letters, and digits appearing on the screen—design properties such as these, if not sensitively addressed by experienced human factors specialists, can leave the pilot at the mercy of poor ergonomics. In the first case, he/she will be subject to input errors, and, in the second, to an undue burden of information processing that leads to head-down time, loss of situation awareness, and possible accident.

More will be said about the effects of poor GPS receiver design on the pilot's information processing load as the discussion turns now to the ergonomics of the panel-mounted models used for IFR flying. Issues relating to pilot responsibility will be picked up for further attention in a later section.

Panel-Mounted Receivers and IFR Navigation

IFR certifiability Panel-mounted GPS receivers can be used for IFR navigation of both en route and nonprecision approach phases of flight, provided they meet relevant requisites of TSO C129, which was developed by the FAA and is accepted in Canada, Australia, New Zealand, the U.K., Germany, France and other countries. To be certified under TSO C129 for IFR en route flight (and, therefore, in Class A2), a GPS receiver must be equipped with Receiver Autonomous Integrity Monitoring (RAIM) capability. By identifying satellites broadcasting erroneous data, thereby enabling pilots to ignore such signals, RAIM adds a significant measure of accuracy and, hence, safety to GPS navigation (GNSS Working Group, 1993, September 15).

The GNSS Working Group (1993, December 10) sets out additional TSO C129 requirements for certification of a receiver to Class A1 which permits navigation of IFR nonprecision approaches. Such a receiver must have, in addition to RAIM, a database that contains the locations of all waypoints required to define the approach, these to be stored in precisely the same order as those appearing on the approach plate for the NDB or VOR approaches they overlie. Given this condition, the sensitivity of the course deviation indicator (CDI) will automatically change as the aircraft changes from en route to final approach, so that tracking accuracy is ensured and collision with obstacles is thereby avoided. A further database requirement is that the waypoint data must be identified by the currently published name of the relevant approach.

From the foregoing it will be obvious that both the programming and usage of panel-mounted receivers are much more involved than they are in the case of handheld units; it follows, unsurprisingly, that the human factors problems associated with their use are much more complex. A report prepared by the Safety and Security Directorate (1996) of Transport Canada indicates that attempts are being made to address some of these problems at the second stage of certifiability required in Canada, namely, the installation stage, at which time consideration is given to the pilot interface and workload, integration with existing aircraft systems, and the flight manual. However this report straightforwardly acknowledges the shortcomings of this process, citing, for instance, the unstructured nature of assessment methods and the nonspecificity of requirements for the various items under test. Extensive coverage of most of the remaining design-related human factors problems can be found in Heron et al. (1997a, 1997b). The presentation below draws selectively from these, adducing new information where relevant.

System architecture Figure 9.3 displays some typical third-generation IFR approved GPS receivers. Because an understanding of the ergonomics of these models will benefit from an appreciation of the architecture of a typical panel-mounted receiver, a representation of one is displayed schematically in Figure 9.4. As this figure shows, the information base of the unit is organised around a number of modes—

e.g., Nav, Database, Flight Plan, etc.—and, within each mode, there is a page for each of several topics. Both Nav and Emergency Search modes, for example, have eight topics and therefore eight pages, Database mode has six, System mode has four, and so on. Some pages may have sub-pages. Depressing a button with the appropriate label provides access to a given mode, while turning one of the knobs does so with respect to a given page. Turning another knob accesses a given sub-page. Although not included in the diagram, three other labelled buttons are available in the depicted model, namely, Sel (Select), Ent (Enter), and -D-> (Direct To). Other models may include buttons for special functions such as NRST (Nearest).

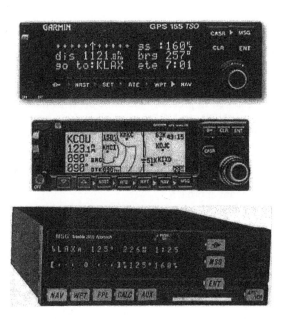

Figure 9.3   Typical third-generation TSO C129 IFR-certified panel-mounted receivers

Controls For the typical systems shown in Figure 9.3, controls consist of rectangular push buttons and round knobs with inner and outer components. Reporting on results of their study assessing menu formats and relevant operations of GPS receivers, Wreggit and Marsh II (1998) state that various physical properties of the push-buttons on the unit they used—e.g., diameter, displacement, and centre-to-centre spacing—were deemed to be acceptable.

However, while Wreggit and Marsh II's (1998) finding can likely be generalised to most of today's panel-mounted receivers, Heron et al. (1997a, 1997b) noted other features of keys that are far from ideal. For example, contours that are incompatible with finger contours are sometimes found, as well as the absence of any kind of feedback—tactile, auditory, or even visual. Poor contours invite keystrike errors; lack of feedback creates uncertainty about whether keystrikes—intended or otherwise—have indeed been made. Without an "undo" button to reverse recognised keystrike errors and to eliminate confusion arising from the uncertainty about whether keys have been struck or not, the eventually weary and frustrated pilot has no alternative but to go back to the beginning and start afresh.

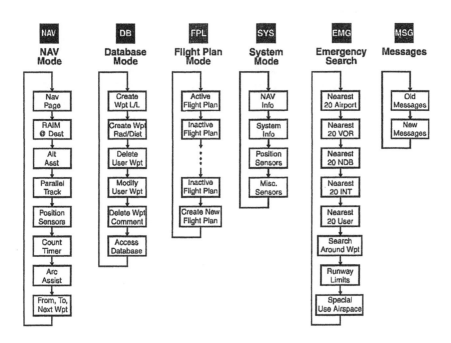

Figure 9.4   Architecture of a typical IFR GPS receiver system
Source: II Morrow, Salem, OR

With obvious sensitivity to the pilot's error/uncertainty predicament, Wreggit and Marsh II (1998) suggest including a button that takes the user back to a central location. However, although this notion has merit, it is important to realise that such a button would only *end* effects of existing error and uncertainty, not *remove* them. To be more explicit, such effects will have already translated into disruption of information processing in the pilot/receiver loop referred to earlier and, hence, into delay in the processing that must take place in the pilot/aircraft loop. Moreover, because this inefficient pilot/receiver behaviour (that of correcting errors and trying desperately to understand what input the receiver needs) is associated with excessive head-down time, loss of situation awareness, with all its potential for error, will likely already have been the unfortunate result. Accordingly, while the necessity of an undo button is unquestioned, and the utility of a 'take-back-to-centre' button unhesitatingly recognised, it must nevertheless be strongly emphasised that, if ergonomics efficiency is to be served, the primary requisite here is button and system design that is *preventive* in nature. In different words, what is needed first and foremost is design that will minimise to the greatest possible extent the probability of error and uncertainty and, thereby, the necessity of using either the undo or the take-back button.

Other problems with controls abound. Some buttons may have multiple functions as, for example, the Direct-To button on one model which, at one stage of its development, required one press for from-A-to-B navigation, and two to activate the flight plan. Similarly, knobs often have multiple functions. Both the left and right knobs on a commonly used receiver have different functions depending on whether the inner knob is in or out and whether the cursor is on or off. GPS receivers are, of course, extremely limited with respect to space and, for this reason, will call for design compromises. Nevertheless, it is a commonplace that multifunction controls have difficulty meeting the demands of high-level ergonomics efficiency, placing on the user as they do, the burden of committing to memory the specific details related to, and the responses required for, each function. The result is, again, a potential slowdown in information processing in both the pilot/receiver and pilot/aircraft systems, a possibility that could become a tense reality during the period leading to the high-workload approach and landing phases of flight. If the pilot is fatigued or under stress at this time, memory may fail altogether, in which case the implications for aviation safety would be serious indeed.

The labels on receiver buttons represent another control feature with implications for memory loading. Memory is always taxed when any aspect of a system is in disagreement with the "population stereotype", a term commonly used by ergonomists to mean roughly "what the majority of the population expects". Heron et al. (1997a) cite a number of receiver button labels that fall into this category. The Direct-To button has already been mentioned for its ergonomics sin of having two functions. It may be recalled that, while one of these was to control the en route flight path of an A-to-B destination and, therefore, is congruent with the meaning pilots would attach to the label, the other was to activate the flight plan. Inasmuch as the label makes not a jot of sense in respect of the latter function, it violates the population stereotype and, thereby, incurs immoderate load on memory. Similarly, use of the Select button to exit the flight plan—a requirement on one unit—is decidedly nonintuitive and, therefore, not easily remembered. Another example - one that is enormously frustrating to pilots—is the Hold button. When ATC calls for a hold at the Missed Approach Holding Waypoint (MAHWP), this button must be used to prevent the aircraft from autosequencing to the next approach waypoint.   However the meaning the pilot attaches to the word "hold" is just what ATC has called for, namely, "go into a holding pattern", and *not* the meaning the receiver system attaches to it, namely, "suspend the flight plan". This inconsistency with the population stereotype makes it difficult for the pilot to remember to take the receiver system out of its hold function when released by ATC from the hold pattern; but this he/she must do— and expediently—in order to proceed with the approach.

Operational difficulties Nendick and St. George (1996) report that their pilot subjects perceived workload to be significantly reduced with use of a GPS receiver, possibly because the requirements of reading maps and making mental navigation computations had been lessened. However these authors add that, while GPS did not appear to be encouraging more time scanning the instruments at the expense of looking outside, lookout had not increased as much as might have been expected, an outcome due possibly to additional time spent scanning the GPS unit.

Certainly, from an ergonomics perspective, there is much about the operation of current GPS receivers to keep the pilot's head down. Loading a nonprecision approach, for example, requires eight inputs on one model but, as Horne (1996) reports, may require as many as thirty on others. Without suitable prompts, and this may well be the case, the pilot must again rely on memory which, as mentioned earlier, is apt to fail during periods of stress or fatigue.

S. Coyle (personal communication, September 12, 1998) reports a number of human factors difficulties encountered while conducting a series of tests with a computer simulation of a particular model. He discovered, for example, that although the list of approach waypoints obtained prior to loading an approach is complete, the waypoints cannot be verified against the approach plate until the approach is loaded. If the verification process were to expose a discrepancy, the loaded approach would then, of course, need to be abandoned and re-loaded. Additionally, although in both the Approach List and the unit's relevant INFO screens the term "ARC" is indicated between arc waypoints involved in nonprecision approaches, the term "PT" does not appear in either of these sources for those transitions requiring a procedure turn.

Another and most confusing operation reported by S. Coyle (personal communication, September 12, 1998) relates to use of the NRST (nearest) function, differing as it does according to subtle nonintuitive distinctions in input. The anomaly was discovered during a test of the use of this function in the event of a need to make an emergency landing. The test began by loading the flight plan CYOW-YOW-CYSH, the approach at CYSH to be loaded (approach enabled) while en route to YOW. Prior to reaching YOW, the decision was made to divert to and land at CYND and, accordingly, NRST was pressed followed by -D->, with the expectation of subsequently getting the approach for CYND. However the unit's response was to go to CYND followed by YOW and the rest of the CYSH approach waypoints, thus indicating that it would not allow the approach to CYND to be loaded into the new flight plan. All waypoints then had to be deleted and the input repeated. Making effort to understand the machine's logic, Coyle made a second attempt. This time, the original flight plan was loaded, and NRST and -D-> were pressed *after* YOW, the intention again being to divert to CYND rather than to CYSH. Now the box produced the comment "UNLOADING APPROACH" and relegated CYND to the last destination, obviously planning to take the aircraft *over* CYSH rather than letting it immediately divert to CYND. Discussions with the manufacturer revealed an option in the configuration which would delete the flight plan if the Direct To point were not included in the original flight plan. As this option would indeed solve the confusion, it was recommended that the manufacturer incorporate it as the default configuration.

The account above describes an experience that occurred in the safe environment of computer simulation. The pilot untrained or unpractised in use of a GPS receiver is likely to discover its capriciousness only through frustratingly confounding trial and error, after which he/she (if still alive) must store it solidly in memory with the equally solid hope that it will not be forgotten during a stressful situation. Even once trained in the operation of a given receiver, the pilot faced with using that receiver in a different aircraft must check to ensure that the software versions are the same. A case in point involves one receiver's use of a signal to indicate the availability of RAIM, with only the disappearance of that signal indicating RAIM's unavailability. As the *absence* of a signal is a counterintuitive way of issuing a warning to humans, a corrective version of the software was developed. However, now, it is incumbent on the pilot to know the difference between the two versions.

The operational problems already described are compounded by the existence of database irregularities. Although RAIM protects against position errors, the pilot is at the mercy of the manufacturer of a GPS receiver with respect to errors, omissions, and other database irregularities that originate in the idiosyncratic software used by the manufacturer to reformat the databases obtained from suppliers. Apart from errors, some waypoints and even the entire navaid record may be missing and, contrariwise, the database may contain waypoints *not* on the existing approach plate. In the latter case, the pilot must make inferences regarding the sites of such waypoints so as to be on top of the aircraft's autosequencing. Needless to say, database anomalies of *any* kind create operational difficulties that dispose the pilot to adopt a head-down position. Moreover, although it is the pilot's responsibility to update the database every 28 days, the updated version may be no less trouble-free than the preceding one. Additionally, of course, the pilot must verify co-ordinates before undertaking an approach, but this is an indirect process involving reference to many types of information. A more detailed review of database problems can be found in Coyle (1997).

Another adverse influence on receiver operation lies in the fact that virtually no provisions are yet in place to standardise the design of GPS receiver architecture, labels, controls, displays, operational logic, database codification procedures, or operating manuals. As a consequence of this fact, manufacturers have been virtually free to follow their own fancies with respect to these important features, a state of affairs that has unfortunate implications for the way pilots perform with their units in the cockpit. The relevant issues were first raised by O'Hare and St. George (1994) and then highlighted in Nendick and St. George's (1995) survey of New Zealand GA pilots, carried out with a view to identifying human factors problems with GPS receiver design and operation.

The problem referred to above will occur when, having been trained and/or become well practised through private learning in use of one model, a pilot must for one reason or another fly an aircraft in which a different model is installed. Eysenck and Keane (1990) present a concise treatment of various theories of both working and long-term memory processes, from which inferences can be made about the cognitive demands that would be placed on such a pilot. There is no doubt, for example, that learning to use the second model would be difficult because of negative transfer from the learning and memory processes relevant to the first. Moreover, even if learning and remembering with respect to the second model appeared to take place, retrieval from long-term memory could nevertheless falter or fail at some future point in time. This outcome—due perhaps to faulty encoding from working memory into long-term memory as a result of interference from experience with the first model, or to the not infrequently found state of underlearning—is most apt to occur when the pilot is experiencing stress, fatigue, and/or high workload.

Displays Two major features of receiver displays will be treated in this section, namely, signals (such as annunciator lights) and alphanumeric characters. The ergonomics of both of these display elements involve important sensory/perceptual processes. For efficient information processing by an observer, signals, for example, must be detected (i.e., sensed as figures against their backgrounds), discriminated (i.e., from other stimulus patterns and 'noise' in the environment), and recognised (i.e., acquire meaning through reference to long-term memory). However Coyle (personal communication, December 1996) reports that, during field tests of various GPS models, certain annunciator lights were apparently missed by pilot subjects and, therefore, may not have been detectable; unfortunately, although there are specific standards for annunciator lights, these are often ignored. Without information available regarding the adequacy of levels of brightness and brightness contrast, the size of the lights, their displacement from range of view, or their colour, it is impossible to say which one or more of these signal properties might have influenced their detectability. Glare also might have been a factor.

Apart from the factors mentioned above, after-the-fact activation of other signals (e.g., a "set altimeter" signal occurring after the altimeter had been set) or other interface events, may have distracted the attention of the pilots referred to above, or impaired the discrimination process. Another possibility is that the missed signals were detected and discriminated from other stimulus events, but not recognised in terms of their meanings which either could not be dredged up from long-term memory, or were never understood in the first place.

Alphanumeric characters are described in Sanders and McCormick (1992) as requiring visibility, legibility, and readability, properties that are essentially the same as those referred to above as detectability, discriminability, and recognisability, respectively. More explicitly, visibility refers to the ability of a character to be distinguished from its background. Legibility refers to the discriminability of a character and is dependent primarily on such typographic features as font, stroke width, and width/height ratio. Readability refers to the recognisability of groups of characters as meaningful words and of groups of words as meaningful messages; features such as the spacing of characters and the spacing between lines are important determinants.

Undoubtedly the limited numbers of pixels and scan lines on a GPS receiver screen, along with the vast amount of potential display information, make these requirements extremely difficult to satisfy. Nevertheless manufacturers seem not to have availed themselves of relevant guidelines which are to be found in almost any good elementary ergonomics text, one by Sanders and McCormick (1992) being a good choice. Even the simple measure they advise for light characters on a dark screen—that of reducing stroke width so as to offset irradiation effects which make strokes appear thicker than they actually are—would help to reduce screen density and, thereby, enhance legibility and readability. Using lower case letters where possible would also help in this respect. A wealth of other information relevant to maximising the effectiveness of screen displays awaits the interested reader in Stokes et al. (1990).

Some attention has already been given to maps in respect of their use with handheld receivers. However further comment is now made to cover their use with receivers for IFR navigation on approach to landing. If an electronic map is to be used during this phase of flight, what the pilot requires from it is both local guidance and situation awareness. The fact is, though, as Olmos, Liang, and Wickens (1997) point out, a vehicle-centred (track-up) map is best at providing local guidance (where I am), a world-centred (north-up) one at providing situation awareness (where I need to be). However, the task of integrating the necessary information from both a track-up and north-up view places excessive cognitive workload on the pilot who is, thereby, left open to error and confusion at the most critical phase of flight. On the basis of results of a simulator investigation of this problem, Olmos et al. propose that, whenever a map is rotated to the north-up position, a wedge with an angle equal to the number of degrees of rotation from the track-up position should appear on the rotated map between that position and the new one. Such a wedge, they explain, provides a "visual momentum" (p. 63) that enhances the pilot's ability to integrate the information from both map perspectives; the relevant mechanism, whose aim is to create correspondence between the two displays, accords with the concept first advanced by Woods (1984).

Other ergonomics features of maps also are significant. As alphanumeric characters are usually black on a lighter ground, the problem of irradiation does not arise. However, unless a decluttering function is available, the over-abundance of information often appearing on these maps can have a harmful effect on the visibility, legibility, and readability of characters and, hence, on information processing. Additionally, it can be noted that various colours (or grey-scale values) are used for coding areas, landmarks, and other details of interest to pilots. As these appear to differ from model to model, the complex issue of which colours and their properties will maximise discrimination in the cockpit environment appears to need investigation, with results being presented to regulatory authorities as a basis for standardisation.

Effects of ergonomics deficiencies In sum, many ergonomics deficiencies inhere in today's panel-mounted GPS receivers. Physical properties of controls result in keystrike errors, nonintuitive labels and operating procedures place too much load on both working and long-term memory, database irregularities give rise to error and confusion, and display characteristics result in sensory/perceptual and cognitive difficulties. In each case, information processing in the pilot/GPS loop of Figure 9.1 is delayed or in some way rendered inefficient, information processing in the pilot-aircraft loop suffers accordingly, head-down time to work through the confusion adds to the effect, and situation awareness is consequently decreased or lost.

The safety implications of the process just described are underscored in accounts of numerous incidents and accidents. The Transportation Safety Board of Canada (1994a; 1994b) reports two occurrences that are germane. In the first of these, a pilot had flown a half hour over his fuel-on-board time while trying to operate his GPS receiver. As a result, he landed 800 feet short of a runway in Timmins, Ontario Canada, while attempting to make an emergency landing, demolishing his newly constructed Pulsar but, fortunately, suffering no injury himself. In the second case, a near disaster was avoided because of the astuteness of ATC's observations. Crew of a Dornier departing Winnipeg had climbed so far beyond the altitude specified by ATC that their aircraft was only 500 ft vertically distant and 1.25 nm horizontally distant from an inbound B727 before ATC advised immediate descent. The Safety Board's report explained that the Dornier pilot had given over his attention to the co-pilot who was struggling to reprogram their "broken down" GPS unit. More recent is NTSB's (1997b) account of a collision between a Beech BE-77 and a Cessna Skylane 182R. After departing Felts Field, the pilot of the latter became preoccupied with programming his GPS unit and impacted the Beech, the student pilot of which was practising ground reference manoeuvres at the time. The Beech sustained major damage, the Cessna minor; miraculously, neither pilot was injured.

The small number of examples outlined above typify the kinds of experiences, albeit not all as dramatic, that are frequently reported by pilots using GPS receivers. Their safety implications are readily traceable to design features that violate ergonomics principles.

Pilot Attitudes and Behaviour

Although many aviation incidents and accidents can be attributed to ergonomically adverse design features of GPS receivers, the roles of pilot attitudes and behaviours are not to be ignored. As Parasuraman, Molloy, and Singh (1993) have commented, the accuracy and power of GPS may foster an attitude of complacency among pilots equipped with this technology. Similarly, over-confidence and over-reliance may be co-partners of an unshakeable faith in the machine. Although the overlap among these three attitudes is recognised, some small distinctions can nevertheless be made, and so they are treated separately below.

Complacency

For the purposes of this chapter, complacency is defined as the pilot attitude inferred from negligent behaviours such as failing to complete standard cross checks or to carry out flight-planning and monitoring tasks. A complacent pilot is one who relaxes, undoubtedly in the belief that, if the GPS "magic box" is on, he/she can do so. An NTSB (1995b) report describes a tragic CFIT accident, in which a Cessna 402C en route from Nome to Koyuk, Alaska, was totally destroyed, and all aboard—the pilot and four passengers—were fatally injured. This pilot—using another pilot's GPS receiver and without having filed an IFR flight plan—flew in cloudy, overcast, heavy snow-fall conditions through mountainous terrain, towards a destination for which no IFR routes from his departure point were established, nor GPS IFR routes authorised. The owner of the receiver later conveyed that he and the accident pilot programmed waypoints differently. The accident pilot's complacency lies in the fact that he apparently so believed in GPS receivers—even a borrowed differently programmed receiver—that he did not undertake to file a flight plan, broke rules, and ignored weather conditions.

Over-Confidence

This term is attributed to pilots who fly in conditions in which they would not consider flying without a GPS receiver. To be more explicit, such a pilot is one who, equipped with a GPS unit but lacking an instrument rating, might nevertheless go above cloud, assuming that he/she will get down through an advantageous break at destination, while at the same time being unmindful of the fact that position information could be lost because of power failure or interruption of reception. This profile is consistent with Nendick and St. George's (1996) finding that the small proportion of their GPS-equipped subjects who reported flying above cloud without an instrument rating, were also less likely to monitor their units and, therefore, less likely to realise when the information was no longer valid.

D. Freedman of the University of Newcastle recounts an incident (personal communication, April 1998) that underscores the nature of over-confident piloting behaviour. According to Freedman, the pilot in question conveyed to him that he was flying a VFR-only Pitts Special from Sydney, Australia to a central destination in New South Wales. Despite the fact that broken cloud had become more solid, he felt confident of getting to his destination because he was tracking on his handheld GPS. When cloud reached 8/8, he was forced to climb to 8000 ft to get above it. Apparently unperturbed by communication with another aircraft confirming the severity of conditions, he tracked toward an area that, according to his map, was clear of high ground. Then, running short of fuel, he put the aircraft into a stable spin at 8000 ft, entered cloud at 7000 ft and stayed IMC (instrument meteorological conditions) in the spin until he emerged from cloud at 3000 ft. He then recovered from the spin and, with a story-ending little short of stupendous, landed without further incident.

Not so blessed was another pilot, similarly without an instrument rating, who was involved in an accident on October 26, 1997. On that day, while on a private flight from Winchester, Virginia to West Palm Beach, Florida, he flew through IFR conditions to his death when the Cessna 172E he was flying impacted trees on a ridgeline about 3100 ft msl. In an abbreviated NTSB (1998b) report of this accident, his wife—whom he had telephoned before departure, and who had expressed her concern about weather conditions—is quoted as saying "He was anxious to get going. He felt he could get above the clouds. His GPS was working and he said as long as he kept that instrument steady [attitude indicator] he'd be all right. He really felt he was going to get above the clouds" (p. 1).

Comments with a similar ring are all too common. R. St. George, Field Safety Advisor with the New Zealand Civil Aviation Authority (personal communication, April 1998) tells of an occasion when he and fellow pilots had landed for an overnight stay, interrupting their flight northward because of reports of sea fog at 200 ft. Another northbound pilot, who had landed his aircraft for refuelling, offered assistance, saying "You can follow us...we have GPS."

Over-Reliance

A pilot who is over-reliant on GPS may be thought of as one who depends on the box to perform the entire navigation task to the point where his/her navigation skills (e.g., map-reading, flight planning, carrying out reversionary procedures, etc.) begin to suffer. Under these circumstances, self-confidence declines so that, if the GPS unit fails during flight, the pilot becomes stressed and finds it difficult to revert to basic navigation procedures.

Pilot reports obtained in a survey examining the usability of GPS avionics (Joseph, Jahns, Nendick, & St. George, 1998, October) were to the effect that, as they gain GPS experience, they make fewer operational errors, improve their tracking skills, and become more confident in the operational intricacies of receivers but, concomitantly, become more reliant on GPS (as well as more complacent with it). An NTSB (1994) report of an accident with bizarre overtones seems to exemplify this tendency. Certainly it reflects questionable flying competency which may well represent a loss of skill and/or confidence due to over-reliance on GPS, with disregard for procedures and complacency being attendant factors. It concerns a pilot flying a Cessna Caravan C-208 from Kotzebue to Cape Sabine, Alaska. The pilot made several excursive attempts to land through overcast windy conditions at a location where no instrument procedure was available while the one passenger aboard held a handheld GPS receiver for the pilot's use. On one of these likely terrifying trials, the aircraft crossed the runway at a 45^0 angle, striking a wing.

The episode referred to above appears to have been a hair-raising struggle against wind by the pilot during which sea and land alternated underneath the aircraft—the behaviour of which was later described by the passenger as being all over the sky. The passenger was engaged for further assistance during the final attempt to land, for now he was instructed by the pilot to hold the yoke strongly to the left while the pilot, exhausted from effort, used the engine and prop to bring the aircraft down to a hard landing and, with full reverse prop and brake, to an eventual stop. Here, it would seem, was a pilot who was apparently so reliant on the GPS receiver that he had become unpractised in and/or diffident about the skill of handling an aircraft in windy overcast weather, and perhaps also had lost sight of the importance of maintaining such a skill.

More recently (NTSB, 1997c), a pilot, who although instrument rated admitted that he had not remained current, had landed because of bad weather at Sky Ranch airport in Nevada after "scud running" with a handheld GPS receiver from an initial departure site in California. Without calling for a weather briefing, he took off again in his Cessna 172K when rain stopped, bound for Pahrump airport about 33 miles northwest of Sky Ranch. During the flight, he reported over CTAF/UNICOM that he had the airport in sight and was descending on approach. Alas, though, the wreckage of his aircraft was found about 3900 ft msl on the south side of an east-west ridge *about 9 miles southeast of the Sky Ranch airport*. This obviously disoriented pilot paid with his life, apparently for not attending to his instrument skills while, instead, relying on a handheld GPS receiver to get him through IMC weather.

Comment

Of the numerous incidents/accidents reviewed by the authors as potential illustrations of one or more of the pilot attitudes discussed above, almost all involved use of a handheld receiver, a fact that likely reflects GA pilots' unawareness of the limitations of these small portable, albeit affordable, units. It is unfortunate, indeed, that some pilots place undue confidence in these devices which, lacking RAIM as they do, cannot detect faulty satellites and may, therefore, be associated with errors of position up to 80 nm (SatNav Program Office, 1995, September 15). It is also unfortunate that pilots may abandon the basic visual technique of VFR flying in favour of an instrument that could lead them hopelessly astray because of a simple easily made keystrike error. What perhaps gives greatest pause, however, is that some pilots, unknowledgeable about the nontrivial nature of approach design, undertake to design their own approaches with VFR units. As the SatNav Program Office (1995, September 15, p. 5) puts it, "Homemade approach designers are very good candidates for CFIT accidents." The Australian Civil Aviation Safety Authority, I. Mallett (personal communication, June 1997) tells of just such a pilot who, conducting a home-made approach, tracked his GPS CDI bar directly into a mountain in France.

Apropos of the pilot attitudes of complacency, over-confidence, and over-reliance, it was acknowledged earlier that distinctions among them were somewhat arbitrarily defined. It will now be evident from the official and anecdotal accounts intended to illustrate these attitudes that, indeed, they seem to blend or merge in almost every occurrence. Yet, the large pool of reports from which the writers selected the small number of cited examples appeared to fall without too much coercion into the categories delineated above. Perhaps, then, it is not unrealistic to speculate that these categories may, in fact, represent a progression of dispositions whereby the negligent preparatory behaviours stemming from complacency are followed, first, by an over-confidence that enables the pilot to fly undaunted in threatening conditions and, thence, by an over-reliance on the machine that results in impairment of basic pilot skills. However, this issue can perhaps be put aside for a while, for the sake of two more obvious and more important facts, namely, that pilot attitudes about the capabilities of GPS receivers, especially those that are handheld and meant only for VFR flying, are associated with dangerous performance in the sky, and that the concern of the aviation community is therefore warranted.

Remedial Approaches

In preceding sections, the writers have presented material attempting to provide an answer to the question posed in the Introduction: Why should a lost-in-space warning about satellite navigation be issued? Their attempts will have failed if by now it is not transparent that the answer can be reduced to a simple bifurcate form: (a) because ergonomics deficiencies in the design of GPS receivers interfere with human information processing in a way that brings about confusion and head-down time which, in turn, can lead to unsafe pilot behaviour, and (b) because pilot attitudes such as complacency, over-confidence, and over-reliance have developed, and these, too, are associated with unsafe GA pilot behaviour. Given this unacceptable state of affairs, the burning question that now arises is: What can be done about it? The answer here appears to lie mainly within each of four interdependent remedial approaches, namely, government effort and regulation, research, training, and ergonomics input. These are discussed in turn below.

Government Effort and Regulation

The arrival of the GPS receiver on the GA scene occurred in the absence of any preliminary research to determine how the task of operating it could be integrated into the existing task of piloting an aircraft. It was, in fact, *imposed* on the existing task, thereby creating the double-loop pilot information-processing system referred to in an earlier section and illustrated diagrammatically in Figure 9.1. Moreover, as the design and development processes undertaken by competing manufacturers appear not to have had the benefit of ergonomics expertise, the units that quickly flooded the market have missed the mark with respect to ergonomics efficiency. A serious consequence of this shortcoming is that the information processing workload in using the receiver is so much greater than it needs to be that it risks disrupting the pilot's overall task of flying the aircraft in an efficient and safe manner.

Although unprepared for the early flash flood of models on the market, governments are now aware not only of the human factors problems of GPS receivers, but also of the fact that these problems have been associated with numerous incidents and accidents of the kinds described earlier. Moreover, they appear to be making serious effort to meet the needs of the existing situation. Certainly the U.S. TSO C129, along with provisions in certain paragraphs of the FARs, have exerted much needed control. Additionally, despite its as yet nonregulatory status, a deep-probing and comprehensive set of bench tests—the

FAA checklist (Huntley, Turner, Donovan, & Madigan, 1995)—has been designed to screen out GPS models not meeting desirable human factors and operational criteria. Further, with now at hand a set of requirements for standardisation of features and functions of a next-generation GPS receiver, developed jointly by the FAA and the Aircraft Owners and Pilots Association (AOPA), the FAA has funded development and simulator testing of a prototype of a minimum-standard receiver (S. Coyle, personal communication, September 1998). Evaluation of this unit should reveal the need for any design modifications that can be applied to production models.

In Canada, the Aircraft Certification Directorate (ACD) of Transport Canada Aviation (TCA) works co-operatively with SOIT in examining and field-testing prototypes and new arrivals on the market, as well as in carefully documenting immense amounts of relevant human factors information, all with the aim of bringing about eventual improvement of certification standards. As part of this effort, Coyle (1997) has delved into such problems as published procedures, altitude restrictions, waypoint identifications, vertical navigation pseudo glide-path detail, and others similarly pertinent to the integrity of on-board navigation data. His report points up the role of database suppliers and the need for quality control after data are supplied to them by Canadian national authorities.

Following on this work, Transport Canada's Database Integrity Working Group has used results of its penetrating analyses of database problems to raise the level of awareness of concerned parties: in particular, a presentation was made to the Charting Harmonization for Avionics and Databases (CHAD) panel formed by the Flight Management Systems Group of the Air Transport Association (S. Coyle, personal communication, September 1998). According to CHAD's Chairman, J. Terpstra—Senior Corporate Vice President, Flight Information Technology and Aviation Affairs, Jeppesen Sanderson, Inc—his committee is now working toward improving compatibility between charts, databases, and their relevant avionics, while identifying for training those items that, for various reasons, cannot be compatible (J. Terpstra personal communication, September 1998). Terpstra adds that the group is also considering ways in which government aeronautical source providers might better tailor their information for the digital database. Complementing this effort, explains Terpstra, is that of the RTCA SC-181 Working Group 2, Industry Requirements for Aeronautical Data, which has set out specifications for governments regarding the preparation of approach procedures, SIDs, STARs, airways, and other information, so as to make the relevant data work in ground and airborne databases.

The thrust of still another area of endeavour is training. Although Canada does not require GA pilots to undergo an approved training programme qualifying them for navigation of stand-alone GPS approaches, nor to be licensed for such (Bowie, 1995), TCA has prepared a Flight Instructor Guide for GPS Training, a manual focussing comprehensively on such topics as preflight preparation; preparation for departure; en route, holding, approach, and missed approach procedures; diverting to an alternate; handling system malfunctions; and post flight procedures. Presently there is no compunction for instructors to use this guide, nor for pilots to avail themselves of its contents. Nevertheless pilots are constantly updated on the pitfalls of GPS receiver usage through such publications as NAV CANADA's SatNav which is available on the internet at:

http://www.navcanada.ca/publications/satnav/index.htm

Other countries are also trying to do what they can to deal with potentially untoward effects of GPS technology. For example, Australia already has an approved GPS training programme, following which flight testing and licence endorsement are required for flying GPS IFR, although not for GPS VFR. However, according to Air Services Australia (1996), this country's Civil Aviation Safety Authority (CASA) is now considering various recommendations made by the industry GNSS Implementation Team (GIT), one of which would require training similar to that for radio navigation aids for VFR flight above cloud. Additionally, GIT called for inclusion of certain advisory material in a Civil Aviation Advisory Publication (CAAP), as well as an industry education programme for GPS VFR navigation through safety seminars, pilot aid cards, magazine articles, and amendment of the CASA GPS booklet. GIT recommended also that CASA include GPS training requirements in the Day VFR ground and flying training syllabus.

New Zealand's Civil Aviation Authority (CAA) regards GPS only as an aid to, not as a primary means for, VFR navigation, the rationale being that certification is made problematic because of the many different types of equipment (CAA, 1996). However a formal training syllabus for IFR GPS has been established and is set out, along with the CAA Flight Evaluation Schedule for GPS IFR Approval, in the Aviation Information Circular-General (AIC-Gen) for February 1998 (CAA, 1998).

It is beyond the scope of this chapter to present a treatise on what is happening in all countries confronted with problems arising in general aviation since the introduction of GPS technology. Rather, the intention is to give some idea of the directions governments world-wide may be taking to deal with these problems. However, as much as it is the responsibility of governments to investigate, regulate, and inform, an invaluable aid to their efforts is surely independent research activity of the kind discussed in the next section.

Research

The considerable amount of research already undertaken concerning the human factors related to the design of GPS receivers has been useful not only in drawing attention to relevant issues but also in generating a host of new research questions. For example, Nendick and St. George's (1996) finding that pilots perceived their workload to be considerably reduced by use of a GPS receiver raises, as a start, the question of whether or not the receiver actually *does* reduce workload. This and related questions await empirical test, perhaps using a simulator and obtaining electrophysiological and other well defined workload and performance measurements so as to produce data suggesting ways in which the design and operation of receivers can be modified to reduce workload to an absolute minimum.

Another study carried out by these investigators (Nendick & St. George, 1995), indicating pilot preferences with respect to helpful screen information, could be followed up usefully by one designed to identify the totality of a receiver's information domain, and to establish not only the nature and order of selections from that domain for screen display, but also the screen-display form that will make these elements most easily readable by the pilot. Above all, the new study needs to determine the maximum amount of information that will be useful to the pilot, and to identify and eliminate frills in existing databanks, with the aim of defining a lean and mean information system that could be readily integrated with the pilot's task of safely operating the aircraft.

As output from their usability study, Wreggit and Marsh II (1998) were able to make many recommendations regarding function allocation and terminology, button identification, feedback, and other features of receivers. However, in the hands of different manufacturers, many of these recommendations could translate into designs that differ substantially from model to model. This possibility might be offset by the results of new research to determine precisely how the design features Wreggit and Marsh II mention can be modified so as to maximise their ergonomics efficiency, its output being a set of specifications that can be easily implemented by manufacturers, or used by governments to set standards for such implementation. Work of this kind may well be within the scope of the U.S. Civil Aeromedical Institute's (CAMI) General Aviation Human Factors Research Program. As Wreggit and Marsh II convey, this programme focuses on how newly developed devices can be integrated into the GA cockpit according to human factors principles, and carries out systematic testing procedures in the Institute's Usability Testing Laboratory with the aim of producing relevant implementation guidelines.

Various investigators (Langholz & Nendick, 1997; Nendick, Langholz, & Houghton, 1997; Wickens, Liang, Prevett, & Olmos, 1996) have recently examined the issue of the dimensionality of maps, while Olmos et al. (1997) have proposed Woods' (1984) concept of visual momentum as a means of facilitating integration of information from track-up and north-up views. What is needed now, it would seem, is research bent on determining (a) the extent to which a moving map improves the GPS-equipped GA pilot's situation awareness, (b) specifications for maximally effective colour coding, and (c) the nature and form of information to be displayed. An encouraging development in this respect is that RTCA's SC-181 Working Group 4, Minimum Operational Performance Standards for the Depiction of Navigational Information on Electronic Maps, is well into its task of developing standards for moving map displays (J. Terpstra, personal communication, September 1998). This work, which is addressing the standardisation of symbols, colours, fonts, and other visual properties of such displays, will be applicable to GA GPS receivers, as well as to systems in the larger aircraft.

On balance, the usefulness of existing research notwithstanding, what is needed now with respect to GPS receivers is problem-oriented human factors research producing hard-core data that can be turned over to government regulating authorities for their use in developing standards or, in the meantime, easily used by manufacturers.

Training

Obviously, modification in the design of GPS receivers is not going to occur overnight and, according to all indications, will proceed only gradually over time. Meanwhile, the GA pilot is left to bull his/her way through flying situations demanding the highest level of cognitive and perceptual/motor skill with a piece of technology that, while bringing the complexity of a heavy jet FMS into his/her cockpit, has not been designed to accommodate relevant human information-processing capabilities. What, then, is the GA pilot to do? In the absence of a quick solution to the design deficiency problem, training and practice appear to be the only ways in which a GA pilot can be prepared for the sky using a GPS receiver.

A GPS training course, when available, is undoubtedly the best option for the GA pilot. Such a course should be both broad and deep in scope, including theoretical, system component, and operational material in its content, as well as navigation and performance requirements. Moreover, the human factors of performance should be covered in detail, with strong emphasis not only on those related to design deficiency, but also on the pilot attitudes of complacency, over-confidence, and over-reliance and their corresponding imprudent practices. A good instructor will assist students in adopting suitable mnemonics to assist in committing to memory the complex detail relevant to each model studied. Ideally, the course would also include a practicum, perhaps using a simulator or computer simulation, and be followed at the end by an examination which, if passed, would result in competence certification for the given model(s) covered by the training.

The GA pilot, however, does not have the advantage of the airline pilot whose training curriculum has been designed and administered with the aid of sophisticated equipment by experts in their fields. Nevertheless, options are not lacking. Already, manufacturers of some GPS receivers have computer simulations for their products, and the advent of courses on the world-wide web is perhaps not in the realm of wild speculation. Supplementary to these approaches, the pilot might gain certain information from one of the texts now appearing on the market (e.g., Clarke, 1996); however, these may not be sufficiently user-oriented to provide the pilot with useful operational insights. Other than—if not to say in conjunction with—these lines of endeavour, the GA pilot must simply take time to sit down with a receiver and explore its intricacies and nuances on his/her own, being sure to practise and rehearse those operations that will be used prior to each flight.

Ergonomics Input

Training, whether formally or privately undertaken, is undoubtedly an important and necessary step toward alleviating the problems now being experienced by GA pilots and resulting in loss of situation awareness. It should not be thought of, however, as a substitute for design that is ergonomically poor and, hence, associated with safety implications. Numerous aviation psychologists and ergonomists have drawn attention to the need for human factors input in the design, development, and implementation of automation avionics (Billings, 1997; Heron,

1997; Heron et al., 1997b; Joseph et al., 1998; O'Hare & Roscoe, 1990; Wiener, 1988). Heron makes explicit the point that use of ergonomics expertise should begin at the conceptual stage of development and continue throughout drawing, mockup, prototype, field evaluation, modification, and even manufacturing stages. An essential part of this input by the ergonomist would be a thoroughgoing task analysis designed specifically to provide indications as to how the new technology can be introduced into the existing task without increasing workload.

Similar cries for a human-centred design approach have been heard for the last fifty years, that is, since Fitts and Jones (1947) explained results of their analyses of large numbers of adverse pilot experiences in terms of design features of the interface rather than of so-called "pilot error". Yet technology consistently appears to win the race, imposing its products on the interface without consideration of how they affect the existing task of the pilot. In this respect, it is perhaps noteworthy that the prestigious Avionics Conference and Exhibition, held annually in London, U.K., and attracting large numbers of manufacturers and technological professionals from around the world, does not include a session on human factors.

Conclusion

This chapter has presented, first, the perspectives of both GNSS technology and ergonomics so as to provide a context for the reader in understanding subsequent sections on the human factors problems being experienced by GA pilots using GPS handheld and panel-mounted receivers. These problems, it was demonstrated, fall into two categories—those that are related to design, and those that are related to pilot attitudes and behaviour—for both of which numerous accidents and incidents were cited. Remedial approaches were then outlined.

The picture emerging from the brief treatment of these approaches, while indicating that progress is indeed being made, is not altogether encouraging. Although governments are making headway in areas of analysis and dissemination of information, training, and regulation, there is still much to be done. Although ergonomics/human factors research is being undertaken by universities and other agencies, more is needed, particularly the problem-oriented kind that will produce data that can be readily used by governments and/or manufacturers. GA pilots desperately need training and guided practice, but have few options in this respect. Manufacturers have not applied high-level ergonomics to the design of their GPS receivers and will, no doubt, continue in this way until standardised specifications are required of them by regulating authorities. Apropos of this last unfortunate state of affairs, Joseph et al. (1998) have proposed a most promising approach, namely, that of a collaborative effort among certification personnel, manufacturers, and human factors specialists who would work *together*, systematically, on the information gathering and research required to lay out the requisites for standardisation.

In the meantime, however, while there is no gainsaying that current efforts augur well for improvement in the design of future GPS models, it will likely be some time before a model that is maximally ergonomically efficient appears on the market. Until then, GA pilots can only heed all warnings concerning both design difficulties and pilot attitudes, and strenuously apply themselves in taking relevant precaution. Fortunately, information regarding the human factors problems associated with use of GPS receivers is becoming more available both in print and electronically. The hope with respect to this chapter is that it will contribute, even in some small way, to the enlightenment of those who use these units and of the aviation community of which they are a part—and, thereby, to greater safety in the sky.

Acknowledgements

The work involved in preparing this chapter was supported in part by University of Newcastle RMC Grant 45/299/643 to the junior author. The authors are grateful for important contributions by the following: R. Bowie, SatNav Program Manager, NAV CANADA; S. Coyle, Engineering Test Pilot, Transport Canada Civil Aviation, Aircraft Certification Directorate; D. Freedman, University of Newcastle, Australia; I. Mallett, Civil Aviation Safety Authority, Australia; D. O'Hare, Senior Lecturer, Psychology Department, University of Otago, Dunedin, NZ; R. Potter, Lecturer, University of Newcastle, Australia; R. St. George, Field Safety Advisor, CAA, NZ; J. Terpstra, Senior Corporate Vice President, Flight Information Technology and Aviation Affairs, Jeppesen Sanderson, Inc.

References

Air Services Australia. (1996). *GNSS implementation team 5: Minutes*. Canberra: GNSS Program Office.

Billings, C. E. (1997). *Aviation automation: The search for a human-centered approach*. Mahwah, NJ: Lawrence Erlbaum.

Bowie, R. (1995). New stars to steer by: The SatNav revolution. *Leading Edge, 6,* 1-7.

Civil Aviation Authority. (1996). *Conditions of installation of Global Positioning System (GPS) equipment in New Zealand registered aircraft for use in VFR operations* (AIC-GEN A4/96). Wellington, NZ: Aeronautical Information Service.

Civil Aviation Authority. (1998). *GNSS IFR operations* (AIC-GEN A2/98). Wellington, NZ: Aeronautical Information Service.

Clarke, B. (1996). *Aviators guide to GPS*. (2nd ed.). New York: McGraw Hill.

Coyle, S. (1997). Aircraft on-board navigation data integrity: A serious problem and solutions. *Bluecoat Digest*, http://bluecoat.eurocontrol.fr/reports/.

Evarts, E. V., Shinoda, Y., & Wise, S. (1984). Cell types and information-processing circuits in the cerebral cortex. In *Neurophysiological approaches to higher brain function* (pp. 49-64). New York: Wiley.

Eysenck, M. & Keane, M. T. (1993). *Cognitive psychology.* London: Lawrence Erlbaum.

Federal Aviation Administration. (1992). *Airborne supplemental navigation equipment using the global positioning system (GPS)* (TSO C129 A1). Washington, DC: Author.

Fitts, P. M. & Jones, R. E. (1947). *Analysis of factors contributing to 460 "pilot error" experiences in operating aircraft controls* (Report TSEAA-694-12). Air Material Command, Wright-Patterson Air Force Base.

Galotti, V.P. (1997). *The Future Air Navigation System (FANS).* Aldershot, UK: Ashgate.

GNSS Working Group. (1993, June 23). *GNSS update* (Issue 2). Ottawa: Transport Canada Aviation.

GNSS Working Group. (1993, December 10). *GNSS update* (Issue 3). Ottawa: Transport Canada Aviation.

GNSS Working Group. (1995, September 15). *GNSS update* (Issue 6). (TP11916E). Ottawa: Transport Canada Aviation.

Heijl, M. (1997, April). Aviation community working on development of infrastructure needed to support free flight. *ICAO Journal, 52*(3), 7-9.

Heron, R. M. (1997, February). Ergonomics and interface design: How long will it take us to learn? *Aeronews,* 2-3.

Heron, R. M., Krolak, W., & Coyle, S. (1997a). A human factors approach to use of GPS receivers. *Proceedings of the 9th Canadian Aviation Safety Seminar.* Calgary, Alberta, Canada. (See also *Bluecoat Digest,* http://bluecoat.eurocontrol.fr/reports/).

Heron, R. M., Krolak, W., & Coyle, S. (1997b). An ergonomics assessment of GPS receivers. *Proceedings of the 1997 Avionics Conference and Exhibition* (pp. 10.2.1-10.2.11). Surrey, UK: Era Technology.

Horne, T. A. (1996, October). Countdown to 2010. *AOPA Pilot,* 73-85.

Huntley, M. S., Turner, J. W., Donovan, C. S., & Madigan, E. (1995). *FAA aircraft certification human factors and operations checklist for standalone GPS receivers (TSO C129 Class A)* (NTIS No. DOT-VNTSC-FAA-95-12). Washington, DC: Department of Transportation.

Johns, J.C. (1997). Enhanced capability of GPS and its augmentation systems meets navigation needs of 21st century. *ICAO Journal, 52,* 97-100.

Joseph, K. M., Jahns, D., Nendick, M. D., & St. George, R. (1998, October). *An international usability survey of GPS avionics equipment.* Paper presented at the IEEE/AIAA 17th Digital Systems Avionics Conference, Seattle, WA.

Kantowitz, B. H., & Sorkin, R. D. (1983). *Human factors: Understanding people-system relationships.* New York: Wiley.

Langholz, C., & Nendick, M. (1997). Interpretability of global positioning system (GPS) alphanumeric and moving map displays: An exploratory study. *Proceedings of the 3rd Australian Symposium on Satellite Navigation Technology* (pp. 140-149). Brisbane, Australia: Space Centre for Satellite Navigation, Queensland University of Technology.

Lawson-Smith, G. (1997). *GNSS planning & implementation: Airservices Australia strategic direction*. Paper presented at the 7th Global Navigation Satellite System (GNSS) Implementation Team (GIT/T) Meeting, Canberra, Australia.

National Transportation Safety Board. (1994a). *NTSB report ANC94LA126*. Washington, DC: Author. Retrieved from the World Wide Web: http://www.ntsb.gov/Aviation/aviation.htm.

National Transportation Safety Board. (1995a). *NTSB report MIA95LA028*. Washington, DC: Author. Retrieved from the World Wide Web: http://www.ntsb.gov/Aviation/aviation.htm.

National Transportation Safety Board. (1995b). *NTSB report ANC95FA016*. Washington, DC: Author. Retrieved from the World Wide Web: http://www.ntsb.gov/Aviation/aviation.htm.

National Transportation Safety Board. (1997a). *NTSB report NYC97LA120*. Washington, DC: Author. Retrieved from the World Wide Web: http://www.ntsb.gov/Aviation/aviation.htm.

National Transportation Safety Board. (1997b). *NTSB report SEA97LA038A*. Washington, DC: Author. Retrieved from the World Wide Web: http://www.ntsb.gov/Aviation/aviation.htm.

National Transportation Safety Board. (1997c). *NTSB report LAX97FA331*. Washington, DC: Author. Retrieved from the World Wide Web: http://www.ntsb.gov/Aviation/aviation.htm.

National Transportation Safety Board. (1998a). *NTSB Report ANC98A044* Washington DC: Author. Retrieved May 11, 1998 from the World Wide Web: http://www.ntsb.gov/Aviation/aviation.htm.

National Transportation Safety Board. (1998b). *NTSB report NYC98FA020*. Washington, DC: Author. Retrieved from the World Wide Web: http://www.ntsb.gov/Aviation/aviation.htm.

Nendick, M., Langholz, C., & Houghton, R. (1997). GPS moving map interpretation: North-up or track-up? *Proceedings of the 9th World Congress of the International Association of Institutes of Navigation* (pp. 218-226). Amsterdam: International Association of Institutes of Navigation.

Nendick, M., & St. George, R. (1995). Human factors aspects of global positioning systems (GPS) equipment: A study with New Zealand pilots. *Proceedings of the 8th International Symposium on Aviation Psychology* (pp. 152-157). Columbus, OH: The Ohio State University.

Nendick, M., & St. George, R. (1996). GPS: Developing a human factors training course for pilots. In B. Hayward & R. Lowe (Eds.), *Applied aviation psychology: Achievement, change and challenge* (pp. 152-157). Aldershot, UK: Avebury Aviation.

O'Hare, D., & Roscoe, S. (1990). *Flightdeck performance: The human factor*. Iowa: Iowa State University Press.

O'Hare, D., & St. George, R. (1994). GPS: (Pre)cautionary tales. *Airways, 6* (2), 12-15.

Olmos, O., Liang, C., & Wickens, C. D. (1997). Electronic map evaluation in simulated visual meteorological conditions. *The International Journal of Aviation Psychology, 7*(1), 37-66.

Parasuraman, R., Molloy, R., & Singh, I. L. (1993). Performance consequences of automation-induced 'complacency'. *The International Journal of Aviation Psychology, 3*(1), 1-23.

Safety and Security Directorate. (1996). *Issues related to the introduction of GPS non-precision approaches* (TP12745E). Ottawa: Transport Canada.

Sanders, M. S., & McCormick, E. J. (1992). *Human factors in engineering and design* (7th ed.). Singapore: McGraw-Hill.

SatNav Program Office. (1995, September 15). *GNSS update* (Issue 6, p. 5) (TP 11916E). Ottawa: Transport Canada Aviation.

SatNav Program Office. (1997, July). *SatNav update* (Issue No. 1). Ottawa: NAV CANADA.

SatNav Program Office. (1998, June). *SatNav update* (Issue No. 2). Ottawa: NAV CANADA.

St. George, R., & Nendick, M. (1997). GPS='got position sussed': Some challenges for engineering and cognitive psychology in the general aviation environment. In D. Harris (Ed.), *Engineering psychology and cognitive ergonomics* (Vol 1, pp. 81-92). Aldershot: Ashgate.

Stokes, A., Wickens, C. D., & Kite, K. (1990). *Display technology: Human factors concepts.* Warrendale, PA: SAE.

Transportation Safety Board of Canada. (1994a). *Aviation occurrence report No. A9400003.* Ottawa: Author.

Transportation Safety Board of Canada. (1994b). *Aviation occurrence report No. A94C0009.* Ottawa: Author.

Transportation Safety Board of Canada. (1995). *Aviation occurrence report No. A95H0008.* Ottawa: Author.

Wickens, C. D., & Flach, J. M. (1988). Information processing. In E. L. Wiener & D. C. Nagel (Eds.), *Human factors in aviation* (pp. 111-155). New York: Academic Press.

Wickens, C. D., Liang, C., Prevett, T., & Olmos, O. Electronic maps for terminal area navigation: Effects of frame of reference and dimensionality. *The International Journal of Aviation Psychology, 6*(3), 241-271.

Wiener, E. L. (1988). Cockpit automation. In E. L. Wiener & D. C. Nagel (Eds.), *Human factors in aviation* (pp. 433-461). New York: Academic Press.

Williams, K. W. (1998). *GPS design considerations displaying nearest airport information* (NTIS No. DOT/FAA/AM-98/12). Oklahoma City, OK: FAA Civil Aeromedical Institute.

Woods, D. D. (1984). Visual momentum: A concept to improve the cognitive coupling of person and computer. *International Journal of Man-Machine Studies, 21,* 229-224.

Wreggit, S. S., & Marsh II, D.K. (1996). *Cockpit Integration Of GPS: Initial assessment - menu formats and procedures* (NTIS No. DOT/FAA/AM-98/9). Oklahoma City, OK: FAA Civil Aeromedical Institute, Human Factors Research Laboratory.

10 Innovative Trends in General Aviation: Promises and Problems

Dennis Beringer

Introduction

After a number of years of economic depression in the General Aviation (GA) sector, using sales of new aircraft as an index, a resurgence appears to be underway. Cessna has begun manufacturing light single-engine aircraft again, sales of GA aircraft are on the rise, and new technologies are offering capabilities to the pilot heretofore unrealised in this area. The face of General Aviation is changing, and how it evolves is dependent upon both the technology driving some of the change and the processes by which these technologies become incorporated. The intent of this chapter is to identify some of the major technologies that can contribute to the GA evolution/revolution and the factors that may limit the rate at which change can occur.

GA as a Good Safety Investment

General Aviation is a potentially fertile ground for safety intervention as compared with the commercial transport sector. It is important to keep in mind that GA spans a very wide range of aircraft types, from Piper Cubs to the Learjets, including aircraft that cruise anywhere from 65 knots to Mach .9, with service ceilings ranging from 12,000 feet to 35,000 feet. If we compare the number of deaths attributable to GA to those in scheduled carriers, there is an order of magnitude difference. In the U.S. in 1995, for example, 732 individuals died in GA accidents whereas only 52 died in air carrier accidents (National Transportation Safety Board, Aviation Statistics, 1995 preliminary data). A word of caution is in order, however, that these figures must be put into perspective. Safety is relative; over 42,000 people died in automobile accidents in the U.S. during the same period. Thus, the gain from investing in GA safety is comparatively good only if one does not leave the aviation domain.[1] One need also keep in mind that statistics for specific commercial carriers by country are variable; although only 1 flight in 10 million for U.S. carriers results in a controlled-flight-into-terrain (CFIT) accident, the world-wide statistics are less assuring.

The way in which "investment" in safety is handled, however, varies according to how the industry is regulated. Actual accident "statistics," as are used to determine what we generally refer to as "objective" probabilities of mortality resulting from a single flight, seem to be used rarely to apportion public money for the purpose of statistically increasing the public well-being. Rather, "perceived risk" seems to drive the desire for regulatory interventions (Cross, 1996; Starr, 1969). Thus, the regulatory response to aircraft incidents and accidents is generally more extreme by both the public and regulatory agencies than would be warranted by the quantitative data. This phenomenon will be discussed further when we look at those forces that either facilitate or retard the implementation of new procedures or technologies in aviation.

Potential Areas for Intervention

It is useful to examine the GA performance record to define the areas where some application of technological advancement might produce benefits. Where do problems exist in GA operations today? Several situations are consistently associated with incidents or accidents, including: controlled flight into terrain, continued flight into deteriorating meteorological conditions, disorientation/mis-orientation, emergency partial-panel operations, low-altitude stalls, failure to maintain separation from other aircraft, and ground-air/air-ground communications errors. The major information-processing/responding activities involved include sensing/perceiving, remembering, deciding, and manipulating controls/systems (Figure 10.1). Any method by which we can simplify these tasks for the pilot should allow us to enhance performance and increase the safety of GA operations. This suggests that focus should be directed in the areas of displays and alerting systems (sensing/perceiving and remembering), pilot decision support systems (remembering and deciding), and control systems (manipulating).

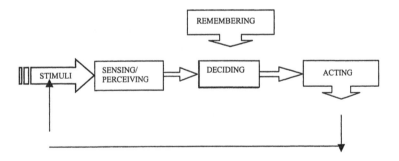

Figure 10.1   A simplified schematic of pilot information processing and responding

Where Accidents Occur and Why

It is also instructive to examine which points in a flight are most likely to benefit from these interventions. This helps us to focus on areas of maximum potential impact. Figure 10.2 (Boeing, 1985) illustrates the disproportionate frequency of accident occurrence by flight phase for air carrier aircraft. One can see that Take-off and Climb account for only about 2 percent of the time spent in flight whereas close to 22 percent of accidents occur during this time. Similarly, Final Approach and Landing account for only about 4 percent of the time spent during flight, but fully 40 percent (an order of magnitude greater in both comparisons) of the accidents occur in this realm of flight. A similar distribution of accidents versus exposure can be found for corporate flight operations (also Figure 10.2), a significant component of General Aviation (data from O'Hare & Roscoe, 1990). Thus, the statistics are somewhat similar across the types of flying and aircraft involved in the different operations.

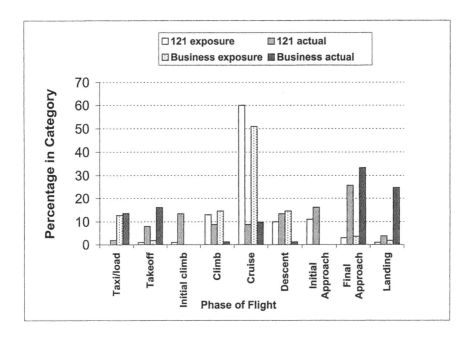

Figure 10.2   Distribution of accidents by phase of flight as compared with exposure time

What is it that causes this significant departure from rates that are commensurate with time of exposure? Simply put, aircraft accidents are ultimately defined as having occurred when the aircraft unintentionally strikes the ground.

Thus, it is more likely that this will occur when there is less separation between the ground and the aircraft, during the beginning and ending stages of the flight, allowing less time for the pilot to respond prior to possible ground contact. However, there are several other factors that are correlated with the passage of time during the flight. First, flights generally begin and end in "terminal" areas, or places where there is a high density of air traffic. Thus, pilot workload is higher during these phases of flight and errors are more likely to occur, whether they be in the cockpit or in the ground-based control of the aircraft. Second, aircraft configuration is changing during these times (gear, flaps, trim, etc.), and the aircraft is generally in transition from one configuration to another. It is during these transitions that problems can surface with either equipment or pilot functioning, sometimes involving design-induced errors resulting from shortcomings in the design of equipment, systems, or procedures. Finally, those accidents that occur during the terminal phases of flight may also involve pilot fatigue, particularly after long flights or after a series of short flights.

The fatigue issue may be particularly relevant for the flight instruction community, where one instructor may have to book a number of students throughout the course of a day if it is a particularly good one for flying. Not only did I observe this phenomenon frequently during my own instruction (I often flew with an instructor, from 10:30 PM to midnight, shooting instrument approaches to Hollywood-Burbank Airport after the instructor had a full day of VFR instruction with other students), but I have also seen it in my current geographic location, Oklahoma, where really good flying days don't come that often due to the frequent winds and discouraging weather in the winter and spring, causing student time to "pile up" on those good Saturdays and Sundays when winds and visibility are favourable.

Technological Opportunities

When all is said and done, however, there are numerous potential interventions that may be applied to the more problematic realms of flight to enhance pilot performance and increase the safety of GA operations. These include technological innovations in displays, control systems, pilot support systems, and communication of data both to and from the pilot.

Integrated Displays

Given that over 90% of our information about the world around us comes to us through vision, our highest-bandwidth sense, it is no surprise that a significant emphasis area in technological development has been visual data displays. Up through the L-1011 and the Concord, the general philosophy of panel design was to add dedicated instrumentation for each new system or subsystem that was added. The result of this process, however, is an ever-increasing appetite for panel space. This trend was first reversed, in a small but significant way, by the development of the horizontal situation indicator (HSI) (proposed by Williams in

1949 and manufactured a few years later, see Williams, 1949). This instrument combines the indications of the directional gyro (heading) and the course deviation indicator (CDI; driven by radio navigation aids) within a common reference frame. Not only does this increase the accuracy of both intercepting and tracking courselines, but it also can reduce the incidence of procedural errors by 75% (Beringer & Harris, 1996). Here we have the best of both worlds: increased performance and decreased panel space requirements.

What has opened up this area for more global and profound development is the arrival of more capable and affordable computing power (faster processors, graphics accelerators, cheaper memory, CDs), cheaper, lighter-weight, and larger displays (flat-panel LCDs), and multi-tasking software (e.g., windows) to allow parallel operation of several processes. A laptop Pentium-based computer has more capability than some earlier-era main-frame computers, and developments in the field are continuing at a rapid pace (e.g., the recent unveiling of a process to use copper in place of aluminium to produce faster, cheaper integrated circuits). A portable global positioning system (GPS) unit is already available that fits in the palm of one's hand, has a map display, integrated antenna, power, and an extensive database of obstacles, cultural features, and navigation aids (Figure 10.3).

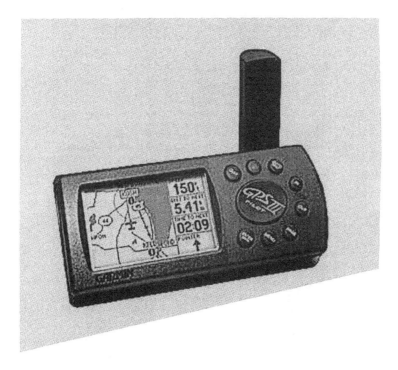

Figure 10.3   Example of compact integrated control-display GPS units available for General Aviation (courtesy of Garmin)

It is critical to keep in mind, however, that technology should serve good design rather than technology driving design (the age-old question can be entertained here: Does need drive technology or does technology drive need? Do we really need cellular personal telephones, for example, or did the need develop in response to the technology becoming available? These and related questions will be addressed under the heading of Need).

There are presently three areas where integrated displays can enhance performance and reduce panel-space needs: primary flight displays (PFDs), multi-function displays (MFDs), and datalink displays.

Primary Flight Displays (PFDs) There are two lines of emphasis in PFDs at the moment. The first can be seen as a transitional one, taking more-or-less conventional instrumentation and representing it on electronic flight instrument systems (EFIS). One such example can be seen in the more conventional aspects of Honeywell's Primus Epic system depicted in Figure 10.4. While some of the data displays are somewhat more integrated than their electro-mechanical counterparts, the presentation is mostly conventional. This emphasis began with small, single-instrument CRTs replacing electro-mechanical instruments on a one-for-one basis, allowing more flexibility and reconfigurability for any given instrument. This was followed by larger display surfaces becoming available wherein multiple instruments or a larger integrated presentation could be depicted.

Figure 10.4   Primus Epic integrated display system (courtesy of Honeywell)

These technical developments opened the way for a second emphasis: integrated pictorial guidance displays, including forward-looking perspective and other formats. Figure 10.5 depicts a highway-in-the-sky format proposed by NASA and being evaluated for use in the Advanced General Aviation Transport Experiments (AGATE) aircraft proposed for the 21st century. Large-format pictorial guidance displays do have definite benefits in that they can be used to present both orientation and guidance information in an integrated form. There are many issues that influence the efficacy of these displays, and research detailing these factors and their effects can be found in a number of recent articles (e.g., Dorighi, Ellis, & Grunwald, 1993; Reising, Liggett, Solz, & Hartsock, 1995).

It is apparent that different flight tasks stand to benefit to different degrees as a function of the display formats used (perspective/forward view, plan view, profile view). However, the key here is that GA now stands to benefit from a new-found freedom to explore new and innovative formats of data display. One cockpit concept that has been developed to represent the possibilities arising from the AGATE program is represented in Figure 10.6. Note that the concept has two integrated flight displays (PFD, MFD) in the panel, and a head-up display (HUD) depicting guidance information and using a pathway-in-the-sky representation. This move toward wide-aspect displays of graphical information and use of graphical analogues of desired flight path was anticipated in earlier military programs (Furness, 1986), and there are striking similarities between the Furness cockpit drawings and Figure 10.6.

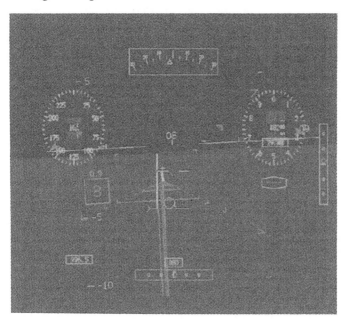

Figure 10.5   **Modified NASA/Parrish pathway-in-the-sky display with round-dial (in place of original vertical-tape) indicators**

Figure 10.6   Artist's conception of AGATE cockpit (courtesy AGATE program office, NASA)

One of the interesting options for presenting primary flight-guidance information in an integrated fashion is the head-up display. This type of display was developed for military operations involving ordinance delivery so that pilots could keep their vision directed out of the cockpit during low-level operations while monitoring flight-control data within their forward-directed field of view. The HUD has also been found to be a valuable tool for low-visibility operations where one anticipates a late transition during an approach from instrument-guided flight to visually referenced flight. An example of this display type can be seen in Figure 10.7 (Honeywell/Marconi-GEC HUD). These displays generally use a partially reflective combining glass to reflect projected data into the pilot's field of view. The view of the outside world is achieved by looking through and around the combining glass.

There are numerous issues associated with these types of displays, which include: collimation (apparent focal distance of the data images), conformality (does HUD symbology representing features from the outside world map directly upon the true real-world features when viewed through the HUD), cognitive capture (how compelling is the imagery relative to the outside world), and the usual set of display concerns involving brightness, contrast, clutter, display element motion, etc. The topic is an involved one and is beyond adequate discussion here. The reader is referred to Roscoe (1987), Newman (1987), and Wickens (1997) for further discussions of the issues. Let it suffice to say that this type of display can allow operations to be performed that would otherwise be difficult with conventional head-down instrumentation. Recent efforts at NASA have led to the development of yet another application for the HUD: presentation of low-visibility taxi information in a perspective format with synthetic guidance cues and "billboards" providing alphanumeric data (McCann, Andre, Begault, Foyle, & Wenzel, 1997).

Figure 10.7   GEC-Marconi head-up display (courtesy of Honeywell)

It should be noted that some of the same tasks that the HUD supports can be performed on head-down displays that are optimally located and contain similarly configured symbology. In fact, the NASA low-visibility taxi display is coupled with a head-down display using a hybrid perspective view to present a map of the airport environs. The cost of HUDs is presently prohibitive, however, for the majority of GA owners and operators. These devices are not likely to find extensive use beyond selected high-end GA aircraft until a low-cost version is available. Efforts are presently underway to produce low-cost HUDs that can provide the same functionality as those presently seen in military, commercial, and high-end corporate aircraft. Some limitations of this type of device should be kept in mind. The displays are generally monochrome due to brightness and contrast problems with the variable outside-world background; collimated (appearing at optical infinity) HUDs generally have a comparatively small volume through which one can move the head and still see the display; and these devices require some considerable real estate for installation, be it overhead or buried within the instrument panel.

Flight controls are inextricably tied to primary flight displays and the nature of the one generally determines the nature of the other. At present we see primarily conventional control systems (direct 1:1 relationship between control yoke/stick position and aerodynamic control surface position) and the adaptation of displays to function with this type of control system. However, there are better ways of controlling the progress of the flight other than moment-to-moment attitude control, and the options now becoming available will be discussed when we arrive at the section on flight controls.

Multi-function displays In the AGATE effort coordinated by NASA and participated in by many manufacturers and some government laboratories, multi-function displays are being envisioned that include data for the major functions of navigation (momentary determination of position in space and direction of travel) and terrain avoidance, weather avoidance, traffic avoidance, and communications. This becomes even more important when one examines the concept of "free flight" and the consequences for the National Air Space and Air Traffic Control (or, alternately, Air Traffic Management). One of the major premises of this approach is the shift to more reliance on the pilot for maintaining separation from weather and from other traffic (to be discussed under Communication).

Ultimately, the criterion defining the successful completion of the flight is that the ground is not contacted prematurely and unintentionally. Thus, terrain data are likely to be most useful to the pilot in accomplishing this task, particularly when flying in the vicinity of significant terrain features. Part 121 operators who are members of the Airline Transport Association have recently agreed to voluntarily adopt one version of this type of display, the Enhanced Ground Proximity Warning System (EGPWS; an Allied Signal product). This system, although too expensive for GA, uses GPS (Global Positioning System) data to determine position; an onboard database then allows the display of terrain in the vicinity of the aircraft. The system also provides warnings of near approaches to the terrain using radar altimetry and comparative altitude measurements. Similar functionality on a much-reduced scale is now becoming available for the GA community in the form of a software package that costs less than $300 and requires input from a GPS device. Although it does not warn of proximity to terrain in the same way that the EGPWS does, it has options for both a forward-looking view of the terrain ahead of the aircraft, seen in perspective view as shaded and textured objects and the display of an electronic sectional chart with aircraft position and heading/track overlayed. Simplified plan-view depiction of courselines, waypoints, and airports is also possible to provide navigational orientation with reduced display clutter.

The multi-function display, in its optimal form, is an integrated display that effectively combines data from many sources to allow the pilot to develop a coherent and complete picture of the flight situation, examine system "health" data, engage in flight "replanning," and other requisite activities. There are already software packages, beyond the terrain-viewing tool described above, that allow the presentation of some subsets of data, through the use of dedicated navigation devices or laptop computers, not only for representing shaded perspective-view terrain data, but also downloaded weather data (base reflectivity, etc.), real-time "sensed" weather (airborne radar and lightning sensors), some navigation and flight-planning activities, and even some limited representation of traffic data in initial experiments for GA aircraft. The important difference, however, is that the true multi-function display must effectively combine or integrate the data from all the separate sources within a single framework, rather than in separate displays or in a single display without a meaningful combination of the separate pieces. For example, it is presently possible to obtain separate data concerning precipitation and lightning strikes. However, one can obtain a better overall picture of weather activity by overlaying these data to see where cell activity is either increasing or dying out.

It is yet another issue to provide a meaningful framework within which to place many different types of data and allow coordinated access to them (Wickens, Liang, Prevett, & Olmos, 1996). The issues attendant here involve some of the principal concerns inherent in any human-computer interface. The most important function to be performed in the evolution of these systems is the meaningful integration of the separate functions for effective presentation and easy access. This will require the intelligent use of access structures (menus, function actuators, etc.), integrated displays (using graphics and display layering), and data prioritisation to present the most relevant data for the task at hand while reducing clutter and eliminating mode confusion. Several strategies are being considered, dependent upon the size of the display surface that can be used. These strategies range from part-time sharing of the display surface (one function at a time) for the smaller units, to multiple windows for the larger displays, and from set, standardised presentations to those allowing tailoring of the displays by the user. Some significant work has already been done in this area, and some examples of different display "pages" are presented in Figure 10.8 that represent what might be done for traffic, weather, and terrain as well as one example of how access "keys" might be available. The means of accessing the data are also crucial, and some of those strategies are mentioned under Communications: Interface Devices.

Note that an additional function, that of datalink, has not been specifically illustrated. It is presently conceived of as textual in nature, allowing ATC/M to uplink clearances and instructions to the aircraft. It is unlikely that a workable interface would require the pilot to enter responses in their entirety from "scratch," so present strategies are centred about using a repertoire of stored responses that can be easily called up by the pilot. These options, whether to be included in the multi-functional display or in some other form, are discussed in Communications, along with interface devices.

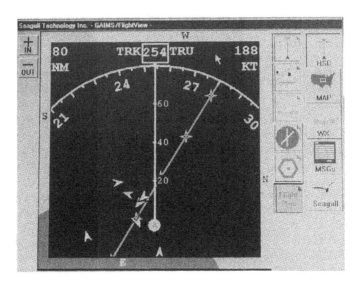

Figure 10.8   **Example of an integrated multi-functional display (courtesy of Seagull Technologies)**

"Virtual" display devices A possible successor to the HUD could be the head-mounted display (HMD), presently achieved by using see-through optics and projected imagery much the same as that used in the HUD. This device can be either monoptic or dioptic, the latter allowing true stereoptic presentation. The major advantages of the HMD over the HUD are that (1) head motion is no longer constrained by a narrow viewing "box" as the display is affixed to the head of the viewer and follows head motions and (2) as a result of this "head tracking," the pilot can look anywhere in the environment, even down through the floor of the aircraft, and see a representation of things that may be contained in the HMD database. This has a clear benefit to locating temporarily obscured terrain features, obstructions, and airport facilities by using the same "look-around" behavior as pilots presently use during Visual Flight Rules (VFR) flight. The present drawback is in the need of the user to wear display devices on the head, usually placing increased loads on the head and neck muscles through the increased weight and distribution of that weight (lengthened moment arms from the head's Center of mass). There are current efforts to reduce the optical system requirements for the displays through the use of direct retinal projection. Success in this area may be necessary before an affordable and acceptable device can be fielded for general use.

One can present several types of data on this form of display including those for system status, flight performance, command guidance, and any of a number of conformal displays that are intended to directly overlay the external "contact" world view. Each has specific requirements to achieve an effective presentation and allow the pilot to use the data optimally within the limits imposed by the device.

The simplest form of data presentation is one containing system status displays. These data are generally depicted in a fixed location in the HMD field, and thus are fixed with reference to the viewer's head. This form of presentation allows the pilot to view critical data and monitor system operation without going "head-down" in the cockpit, thus providing a better opportunity for continued visual surveillance of the surrounding airspace. This category of display is likely the easiest to present because all data "fields" (this includes graphical presentations) and objects are generally fixed within the greater display frame and only indices in each data field are being modified. Thus, no coordinate transformations are necessary to produce the display and the computing power required to support the HMD is minimal.

Flight performance data may be similar in its requirements until one comes to continuous, instantaneous indicators of specific real-time items, including aircraft attitude. Recent attention has been given in the literature to the effects of presenting attitude data in a fixed position in the HMD field of view whilst the head is rotated away from the forward-view vector. This situation causes motions in pitch and roll to be nonconformal with the outside world when the head is rotated to any position other than that of the aircraft heading vector, with accompanying distortions in the perception of aircraft response. Recent research has outlined the potential performance disruptions, and it should suffice to say that such arrangements are certainly less than optimal. Use of this type of presentation is unlikely in General Aviation, and the present cost of a HMD is not likely to justify its use for only status or attitude data.

One of the more compelling presentation formats that can use the HMD to great advantage is that of conformal data representing terrain, obstructions, weather, or runways/airports shown against the outside world. These data, however, are ultimately referenced to the outside world, and thus must change as the pilot's head moves. This requires sensing of head translation and rotation so that the presented data can be appropriately justified in the pilot's field of view. Systems presently used for this purpose are electromagnetic and have specific limitations, a major one being cost. There is, along with installation and calibration requirements, the need for considerable computer processing power to receive the head-sensor data, manipulate the database, and present the appropriate view of those data in a conformal format. It is likely that this technology will ultimately become affordable and usable as the price of computing and displays drops relative to processing and display capabilities. There are also different technologies being explored for the sensing of head position (Liggett, Kustra, & Reising, 1997) that may allow minimal functioning of such systems at greatly reduced cost and with fewer calibration requirements.

One projected use, being researched currently, is the presentation of skeletal synthetic cues in the field of view to assist the pilot in finding and flying approaches to airports in conditions of degraded visibility. This is illustrated in Figure 10.9 where Frame A shows the external view of the world under acceptable

VFR conditions. Frame B shows the same scene with smoke and haze obscuring features and with the HMD skeletal data overlay locating the airport. Both the benefits of use and the need for highly accurate data supporting the display are evident. It is anticipated that the database, possibly updated using compact optical disks (vis-à-vis Jeppesen's products), would be stored onboard. Positional reference relative to the database would be through the Global Positioning System (GPS) or a comparable source of locational information.

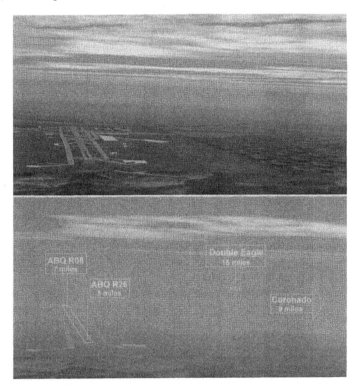

Figure 10.9   (A) View of the "contact" world with good visibility and without overlaid display and (B) same scene shown with aerial haze and overlaid computer-generated imagery

Flight Controls

Flight controls, as alluded to in the earlier section on primary flight displays, are inextricably intertwined with primary guidance displays. There is a series of questions that ultimately take us to our goal of effective flight control over any system: (1) What things do we wish ultimately to control? (alternately, what degree of control authority do we require?) (2) What are the display indices that we can use to determine (a) what desired performance is and (b) what actual performance is? and (3) Is it possible and/or desirable to delegate control of some of these flight-control parameters to automated systems? The question of which aspects of flight to automate and which to retain as "manual" or pilot initiated/guided has plagued aviation for many years. There are proponents on both sides who are equally vocal concerning the appropriateness of their approach and, in the commercial air carrier world, this division may be seen to be embodied by the differences between Airbus and Boeing. The issues that must be considered for a general-aviation application include available technology, human factors and function allocation, cost, and reliability. Each of these shall be considered in turn.

Available technology First, let us examine what we can do, to be followed by what we should do regarding continuous flight control. Today, even General Aviation aircraft have the option of autopilot control. Complex (constant-speed propeller, retractable landing gear) single-engine aircraft now are routinely available with two-axis autopilots, able to maintain heading (yaw) and pitch attitude. Additionally, these autopilots can be obtained with the options to directly set and control altitude (hold current, level at target), ascent and descent rates, courseline/VOR tracking, and instrument-landing-system (ILS) approaches. These systems are, in comparative terms, both available and affordable in the present General Aviation context, using servo motors to activate the aerodynamic control surfaces attached to conventional mechanical control linkages.

What one generally does not see in the majority of GA aircraft is any form of powerplant control or management associated with the autopilot. However, some new developments in power control, specifically the single-lever power control (full-authority digital engine control or FADEC) for complex aircraft, promises to allow integration of engine control into the autopilot, producing the equivalent of a flight management system for low-end GA aircraft. The FADEC is also a way to simplify powerplant management for aircraft with constant-speed props even without an autopilot link, eliminating the need to separately set propeller (RPM) and throttle (manifold pressure) controls. Ultimately, digital control of the engine can allow one to set desired thrust levels, as one does with turbojet engines. It should also be noted that small turbojet engines are now being developed that are intended to power small, single-engine GA aircraft carrying 4 to 6 passengers. This achievement will even further simplify the control of aircraft systems and, hopefully, increase their reliability. One aircraft of this design that is currently undergoing certification is the Visionaire Vantage shown in Figure 10.10.

Human factors considerations The autopilot system, however, usually requires that control be exercised through a control-display unit or associated "status-setting" controls. Thus, one generally doesn't exercise continuous control of the flight parameters but, rather, sets a value (e.g., heading) and allows the autopilot to achieve and hold that value. There are a number of potential problems associated with this type of flying, as detailed by Wiener (1988), that are associated with potential loss of manual flying skills, perceived loss of control, and uncertainty concerning what the automation is doing. This was recently underscored in a series of flight simulator studies of General Aviation pilots' responses to autopilot malfunctions (Beringer & Harris, 1997).

Figure 10.10   The Visionaire Vantage; (top) aircraft in first flight and (bottom) cockpit (courtesy of Visionaire)

Findings indicated that pilots were generally able to cope effectively with simple and straightforward malfunctions, but the more complex or extremely subtle ones were not easily dealt with and, in 13 of 24 exposures to runaway pitch-trim down, led to flight-terminating conditions (aircraft structural overspeed or imminent contact with the ground). Additionally, responses to the malfunctions required considerably longer times, in many cases, than the response intervals specified for the certification testing of autopilots. Mode confusions were also observed, as have been seen in flight management systems on scheduled air carrier aircraft. There are several reasons for these results, including a lack of an accurate conceptual model of the autopilot system, lack of experience with specific malfunctions, and general complacency (see Funk, Lyall, & Riley, 1995, for a more comprehensive discussion of human factors problems in automation).

This brings us to the point of what we should be doing to facilitate flight control. There are three requirements that help to delineate our desired goal. First, it is necessary to identify the aspects of flight we need/wish to control. Present discussions on future control systems centre on the control of the performance of the aircraft (rates of climb/descent, rate of turn, etc.) and not its attitude in space (pitch, roll, yaw). Although the ultimate interface would be goal-oriented control in the form of our destination (push the button with the name of your destination airport), anticipated near-future ATC practices and procedures, particularly in terminal areas, will probably require that we control performance (heading, airspeed, altitude) to conform with clearances designed to resolve tactical traffic problems. Thus we will probably seek schemes by which we can directly communicate desired performance to the aircraft systems. Second, we will need to be able to match, either by manual control or through automated systems, actual performance with desired performance. Either option will require displays of both sets of parameters so that the pilot can either manually direct the aircraft performance until target performance is reached or verify, through the displays, that the automated system has indeed caused the aircraft to reach the goals. Finally, we must resolve to what degree we want the pilot involved in the moment-to-moment operation and control of the aircraft, considering not only the potential errors in performance but also the value of pilot involvement in the control loop.

The more involved the pilot is in the operation of the aircraft, the more likely it is that out-of-tolerance conditions or malfunctions will be detected early and corrected without mishap (supported by recent work of Young, 1997). One way to use the power of the autopilot and yet keep the pilot involved is to realise a performance-controlled system (Roscoe & Bergman, 1980). The 1970s instantiation of this concept was a side-arm controller that allowed direct control of rate of turn and rate of ascent/descent. Thus, there was a 1:1 mapping of performance to control positions: So many degrees deflection of the control to the right would establish a standard-rate turn to the right. Releasing the control back to "neutral" would return the aircraft to straight-and-level flight. It was not necessary to be concerned with the usual initial deflection and subsequent neutralising of controls when reaching the desired attitude that corresponded with the desired performance.

However, one does not achieve this simplification in aircraft control without sacrificing something: Commanding performance directly removes direct control of attitude, making it impossible to perform barrel rolls or loops. Then again, in aircraft used primarily for general transportation, this would appear to present little problem. This scheme can be realised in current-day aircraft by directly connecting a side-arm or other manual controller directly to the autopilot so that control displacements can be interpreted as commands for changes in desired performance. The result is an analogue control that is spatially compatible with directional response of the aircraft but that commands performance rather than rate of change of attitude (for a more extensive explanation of order of control and the mathematics involved, see Roscoe, Eisele, & Bergman, 1980).

The present-day instantiation of this concept is the fuzzy-logic controller (Duerksen, 1996). This approach incorporates not only the basic performance controls described above but also allows one to define performance limitations in the underlying computer programming that defines system response. Thus, one can say that certain performance is not possible, given the values of some flight performance variables (e.g., one can hardly command a high climb rate if airspeed is very low and the throttle is retarded), and attempt to prevent the aircraft from entering potentially dangerous configurations. This idea of limits is embodied in the present debate in the air transport arena between the philosophies embodied by the Airbus approach and the Boeing approach. The two key items to be considered are (1) decoupled controls and (2) hard versus soft limits. Airbus uses decoupled controls—controls that use intervening electronic devices to interpret control positions and activate control surfaces. They are decoupled to the point where the captain and first officer can deflect their controls in opposite directions and the control system will average the input (excepted when one pilot uses the override option to disconnect the other controller). These controls are not force back-driven and, thus, do not serve as (tactile) displays in and of themselves. Although Boeing, in its most recent products, also uses decoupled controls in the strictest sense of the definition, the controls respond as though they were linked directly and provide force feedback to control inputs.

The second item, hard vs. soft performance limits, is another area in which the major airframe manufacturers appear to disagree. A true hard-limit approach (as adopted by Airbus) defines performance and manoeuvre limits that can never be exceeded. Thus, if a pilot wished to pitch the aircraft up 20 degrees and enter a 45-degree bank, and the control logic determined that the aircraft, in its present configuration, should not exceed 10 degrees pitch up and 30 degrees of bank, the aircraft's attitude would only be allowed to reach those values determined by the hard limits. This can be said to be roughly analogous to the often-heard joke about the crews of advanced automated aircraft: the pilot and the dog. The pilot is there to feed the dog and the dog is there to bite the hand of the pilot who tries to touch anything.

A soft-limit approach (Boeing's), however, usually contains the same limits, but it does not prohibit the pilot from exceeding them. Rather, it usually provides some kind of feedback, often tactile, to indicate that a limit has been reached. It is then up to the pilot to decide whether or not to exceed that limit. The argument for this latter philosophy is that the highly trained pilot should be able to evaluate potentially hazardous situations and determine if exceeding limits may be necessary to avoid catastrophic termination of the flight. The counter argument is that one could, if done carefully, define limits that were not exceedingly restrictive but would prevent damage from being done to the aircraft or keep it from entering marginal and hazardous operational modes. The resolution of this dispute is ultimately tied to the task requirements and expectations about what behaviours and tasks might be required given the worst-case scenario. A number of pilots, as would be expected, appear to favour soft limits.

Flight control artifices One should also consider the actual devices to be used to affect continuous manual flight control. The two most prevalent forms have historically been two-axis joysticks and yokes (as seen in Figure 10.10, bottom; derivative of the original "control wheel") with independent rudder pedals. There are several variants on these two forms, some more advantageous than others (e.g., the side-mounted "half yokes" in the Cirrus SR-20 that free up panel space and the centre console), but most in General Aviation aircraft are still direct linkages to control surfaces.

With consideration of fly-by-wire performance-controlled systems now possible, other forms of the physical interface are possible. Indeed, with a system that generated side force directly, not unlike that in the AFTI F-16, one could directly command lateral drift rates on final approach to compensate for crosswinds without the need for the normal compensatory controlling behaviours. This also would allow the landing gear to remain aligned with the runway without adopting B-52 style rotatable main gear or Cessna 195-style gear, something that was never popular. The experimental control in Figure 10.11 (Beringer, Gruetzmacher, and Swanson, 1984) allows input of such a command through the yaw axis of the four controller axes available (pitch, roll, yaw, and longitudinal slide). The upper photo shows the control centred; the lower photo shows pitch, roll, and yaw axes deflected. These three axes of motion are centred within the wrist with a minimum of the usual bio-mechanical coupling found in multi-axis controls (derived from a two-axis Hughes design for the Apollo manned space-flight program; Bauerschmidt & Besco, 1962).

Not only are displacement controls useful to (not to mention preferred by) pilots, but one can also envision further developments that provide multiple axes of control and an economy in the use of cockpit space. Figure 10.12 shows the Spaceball™, a six-axis force-transducing controller. It does have some slight displacement during activation but responds primarily to forces applied in the three rotational and three translational axes. The Spaceball™ bears a striking resemblance to an earlier experimental device developed by Wenger, (personal communication, 1987) which he termed the "grip ball." It, however, was a spring-centred displacement device with ridges to facilitate easier rotation and identification of control orientation.

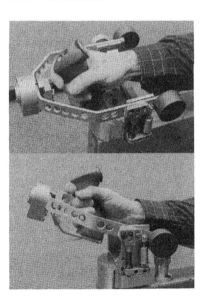

Figure 10.11   4 -axis side-arm controller

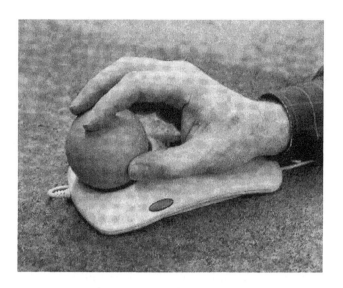

Figure 10.12   Spaceball™ 6-axis controller

One problem with early performance controllers was that one had to hold them in position for prolonged climbs or descents. The current fix, rather than providing a control locking mechanism as considered previously, is to achieve the desired performance of the aircraft and then engage the autopilot to maintain that performance. The one drawback to this strategy is that if the position of the controller is not back-driven by system performance, the controller no longer serves as a display for system performance. (For additional discussion of the problems associated with multi-axis controls, see Lippay, King, Kruk & Morgan, 1985; Kruk, Runnings, King, Lippay, & McKinnon, 1986; and Beringer, Gruetzmacher, & Swanson, 1984.)

Communications

There are a number of lines of communication that flow in and about the cockpit and each of these has its own unique requirements. Most GA aircraft operators find that the majority of voice communications are with Air Traffic Control (ATC), with smaller proportions devoted to position reports at uncontrolled airports, requests for information from various ground sources, and communications with other aircraft.

ATC/ATM Traditionally, communications with ATC have been by voice, transmitted by FM radio, and subject to all of the problems of a single-channel communications medium. Although there is usually a number of assignable frequencies for enroute traffic, the problem is greatest at or near airports where significant numbers of aircraft may be queuing up to land on the same runway and, thus, sharing the same frequency. The problems usually associated with this communication medium include difficulty getting a word in edgewise at busy airports, understanding and remembering the content of a message, and receiving a correct response to a read-back (verification). One proposed solution to some of these problems is some form of digital datalink (Hrebec, Infield, Rhodes, & Fiedler, 1994). Such a system (similar to some used by commercial air carriers) allows reception of alphanumeric data in digital form for visual display in the cockpit. The transmission can be bi-directional, allowing the pilot to formulate and transmit responses to ground stations and/or other aircraft. One could ultimately transmit any form of digitally encoded data (alphanumeric, graphical, auditory) and have it specifically encoded for the recipient based upon some identifier like the transponder code.

There are some limiting factors that may influence how GA applications of this technology are employed. Although the capability to store and display messages that are no longer time dependent, as are speech communications, helps to eliminate forgetting, miscomprehending, and failing to receive, the manner in which the pilot interacts with the system is important. We already know that it is unrealistic, even dangerous, to ask a pilot to manipulate a full alphanumeric keyboard whilst flying the aircraft, even more so in GA aircraft where there simply are no places to put such a device. Thus, some other means may be necessary to enter and transmit responses. Two notions are presently being evaluated as to their potential. The first is to store a series of the most frequently used responses, and allow them to be selected and transmitted by single "key" presses. This greatly reduces the task loading on the pilot. Second, novel interface strategies may be required (e.g., voice or touch panel) to minimise the impact on pilot workload in transitioning from all-voice analogue communications to some hybrid or fully digital systems. Current options will be discussed under Interface Devices. One should keep in mind that loss of "party-line" information, i.e., not hearing transmissions from other aircraft, is seen by many as a decided disadvantage. It will be incumbent upon system architects to assure that the data gleaned from these "overheard" conversations are available in other forms, be they CDTI, graphic weather, or instantly available pilot reports.

Weather and traffic Two potential sources of conflict that may threaten the safe conduct of the flight and involve the communication of data are significant weather activity and traffic. Data concerning these items have previously been obtained, in the lower-end GA aircraft, from briefings and ATC advisories or by direct observation. It is now possible to uplink weather data to the cockpit. These data include radar imagery (NEXRAD, etc.), lightning, precipitation, temperature, wind, etc. These data are most likely to be presented on the multi-function display mentioned previously. There are significant issues concerning the format of the displayed data, including colour use, resolution, and prioritisation (Boyer, Campbell, May, Merwin, & Wickens, 1995), but there is insufficient space here to discuss these issues to any level of satisfaction. Suffice it to say that the data are available and are being used, and that future systems will likely make these data available in the cockpit routinely as is beginning to happen in air carrier aircraft.

Traffic data have recently been presented in varying ways, and TCAS (Traffic Collision Avoidance System) devices of one type or another have been used to present data on other traffic in the immediate area, often using data from the aircraft's transponder (Mode S). Recent developments using ADS-B (Automatic Dependent Surveillance) present even more interesting options for traffic data in the cockpit. The future of CDTI (Cockpit Display of Traffic Information) is also tied to the multi-function display, allowing the integration of traffic data with other relevant data. Again, there are the problems of coding of data, use of colours, and numerous other issues that need to be addressed to ensure that the pilot can successfully use the device to maintain separation (Beringer, Allen, Kozak, & Young, 1993; Ellis, McGreevy, & Hitchcock, 1987; O'Brien and Wickens, 1997).

Interface devices Having raised the issue of interface devices, we must now examine what types may have the most utility for a GA environment. The operating environment is generally characterised by low-to-moderate levels of vibration, a wide range of noise levels, and limited space (both panel and cockpit in many cases). There are several realisable interface devices, each of which has advantages and disadvantages according to the tasks to be performed. The most likely candidates include the touch panel, the glide pad, thumb stick, and voice recognition. Rather than recap all of the pros and cons, a brief summary follows.

The touch panel presents the option for direct manual access to spatially presented data, allowing one to develop applications like "touch maps" (Beringer, 1980; Beringer, Hendrich, & Lee, 1989). Data concerning specific features (airports, navigation facilities, waypoints) can be directly accessed by touch without an intervening interface or the need to learn the use of same. It is a direct and natural form of response for spatially related data and is classified as a position-control device. It requires adaptations to enter alphanumeric data (on-screen keypads or menus). The limiting factors that may affect its use are several. First, anti-reflective display coatings do not fare well with repeated finger contact, picking up oils and contaminants and causing reflective spots to develop. Second, despite high-resolution devices being available, resolution is limited to the placing accuracy of the human finger and the coarseness of the "entry stylus," so-to-speak. Third, if displays are restricted to a vertical orientation, vertical accelerations due to turbulence may affect accuracy, and repeated arm extensions to touch the vertical surface can cause arm fatigue. Optimal implementations should use a slanted display/input surface (30 to 45-degrees to horizontal) short of full-arm extension.

A related input device, the glide pad, can be seen as a mini off-display touch panel. This type of device has been used in Boeing's 777 as a cursor control device. It has the same benefits of relative spatial movement as a mouse, but does not allow the same kind of direct instantaneous spatial access to graphically presented data as does the on-display touch panel. However, it is much smaller, can be located at the end of an armrest at the resting arm's length, and does not have any effect on display transmissivity, as can be the case with touch panels. It also depends upon some kind of on-screen display of menus or keypads for alphanumeric data entry. Being a cursor-control device, it does provide for relative spatial access to spatially presented data, but only by reference to cursor location on the display.

The thumb stick can be mounted on the control yoke or stick, allowing the pilot to keep one hand on the flight controls whilst moving the cursor across the display. Thus, the thumb stick is probably least affected by minor turbulence or acceleration. This device, however, is generally implemented as a rate-control device (degree of deflection or force exerted in a direction determines rate of cursor movement), rather than as a position-control device, as seen in the touch panel and glide pad. It requires the least space of the considered interface devices and allows continued use of the activating hand for flight control, but it is the most dependent of the devices on a visual cursor and is prone to overshoot errors because it is a rate-controlled device. It is also, in general, the slowest of the interface devices listed here (longest data accession times). It has already been used extensively in the cockpits of combat aircraft.

An interesting possibility for the transmission of ATC/ATM related data and potential selection of system functions is speech activation. Much has been accomplished since earlier efforts using exhaustively trained speaker-dependent systems. It is now possible to have reasonably accurate speech recognition for a selected set of functions, and accuracy can be improved with some training of the system. Some of the usual concerns are present in the GA cockpit and some new devices may pose interesting challenges for implementation of speech recognition. First, there is the issue of signal-to-noise ratio given that a good number of GA aircraft have anything but "quiet" cockpits. This ambient noise does two things to human speech. First, it makes it a little more difficult to sort out speech sound from engine sounds. Second, the pattern of speech is likely to be affected as the pilot may speak a little more forcefully in the noisy environment. Thus, a "trained" system needs to be trained in the same environmental conditions as it is to be used (engine running, pilot task loaded). This may provide a good opportunity for using moderate-fidelity flight simulators in the set-up of cockpit speech recognition systems.

Additional concerns are pilot acceptance and system reliability. Acceptance of speech activation has been variable, particularly if the operator had to talk extensively. Some of the experiments with voice "typing" indicated that operators became fatigued much more easily when using voice-entry methods than when performing keyboard-based entry tasks. Thus, one requirement may be that the pilot not be required to speak any more than is now required. Second, for more elaborate entries, it may be advisable to develop a verbal "shorthand" that uses verbal abbreviations for longer often-used expressions. These must, however, be designed to minimise the length of training required. The reliability of recognition of the speech also plays a major role in pilot acceptance. Clearly, error rates must be low for the pilot to be willing to use this means of interfacing. Voice systems have been used in other applications, including one version of the B-52 and numerous computer system interfaces and related applications of those systems.

The increasing power of personal computers and the increasing capability of software will greatly enhance the viability of this form of interface.

Auditory "displays" have also been considered in two application areas, one for conveying unidimensional information (alerts, alarms) and one for conveying multidimensional information (spatial location of objects; for example, the location of other airborne traffic). The first category has seen significant work conducted in changing the way in which auditory alarms are designed so that more differentiable and detectable signals can be created for the cockpit, with urgency of the signal coded in ways other than "louder is more urgent." Recent testing of GA pilots (Beringer, Harris, & Joseph, 1998) has revealed localised hearing decrements around 3kHz, which are consistent with observed difficulties that older pilots have in detecting warning tones near that frequency. Although efforts are underway to improve the efficacy of non-verbal auditory warnings (Edworthy, 1994; Hellier, Edworthy & Dennis, 1993), some manufacturers of GA equipment have already begun to incorporate voice messages into their auditory warning schemes.

Voice messaging has, of course, some definite advantages over non-verbal warnings. Most notable is the elimination of any need to memorise what specific sounds mean in relation to malfunctions or cautions. The words specifically identify the problems. For example, Allied Signal now has voice annunciations available for (and which state) "autopilot," "trim in motion," "check trim," "altitude," and "check altitude." It should be kept in mind, however, that there are potential drawbacks to using voice messaging. It is always possible, unless the voice is kept "machine-like" in nature, that system utterances can be confused with other voice communications in the cockpit (copilot, ATC). There is also the consideration of the primary language of the pilot, and potential misinterpretations if the pilot's primary language and the voice-warning language are not the same. The potential benefits of a well-designed system, however, are substantial.

An auditory display for conveying multidimensional (spatial) information was proposed and considered for the "super cockpit" (Furness, 1986) and research was performed to examine the possibility of using such a device. Recent research has examined this topic again (McKinley, D'Angelo, Haas, Perrot, Nelson, Hettinger, & Brickman, 1995), and it is possible that some type of auditory spatial cueing could be useful in the detection of visual targets. However, implementation will depend upon the development of low-cost reliable head-position sensors (Liggett, et al., 1997) if the auditory information is to be presented through headphones and head movement is unconstrained. One must also consider the possible effects of active noise reduction systems, either in the cockpit overall or localised in the headphones (Jordan, Harris, Goernert, & Roberts, 1996). These systems may prove to be very valuable in reducing fatigue and hearing loss due to noise exposure if they are effectively designed and implemented, but they may also pose the potential problem of masking some parts of the spectrum containing sounds pilots need to hear.

Interface logic Well-designed physical interfaces, although necessary, are not sufficient to guarantee effective and efficient performance of data manipulation tasks. It has been recognised for some time that the logic underlying an interface is equally crucial to the operation of the system. One can observe the effects of logic changes in the iterations of Windows ™ for personal computers, particularly as seen in the changes between version 3.X (many displayed icons) and later '95 and NT versions (nested automated pop-up menus). The importance of menu structure received attention as early as 1965 (USAF; the Master Monitor Concept) and recent research (Riley, DeMers, & Misiak, 1998) emphasised the importance of congruence between the user's mental model of the system operation and the true functioning of the system:

> *"... we want to make the system work like the pilot thinks, so we don't have to train the pilot to think like the system works."*

While one may dispute the grammar, the intent is clear. The work of Riley, et al., has produced a user interface metaphor called "Cockpit Control Language" that uses targets (physical parameters that define the flight path) and actions (things that can be done relative to the targets). These actions and targets can then be strung together into logical statements representing, in this case, an instrument clearance. The syntax of the statement closely approximates that of an instrument clearance. These statements then serve as the basis of operation for the autopilot/flight management system. An initial version of one such interface is depicted in Figure 10.13.

Although this logic can be implemented through a number of interface techniques, one can imagine the advantages to using speech input. Ultimately, however, one could use direct digital datalink with options to: accept as is; reject; modify and send for approval. This approach might well represent the "negotiated clearance" type of Free Flight, a transfer of responsibility for separation from hazards from the air traffic controller to the pilot (O'Brien & Wickens, 1997), that could ultimately become commonplace. One already sees this, to some degree, in requests for altitude changes and vectors around significant weather, particularly with aircraft having appropriate weather-sensing equipment.

Pilot Support Systems

So far, we have seen developments that may enhance the securing, presentation, and subsequent transmission of data, both to and from the cockpit and within the cockpit. One cannot stop there, however; one must proceed to the next step, which is the decision of what to do with the data once it has been received. It is in this role, the "filtering/processing" and "recommending" tasks, that a support system might render some assistance to the pilot. Some work has already been done in this area, but the majority of the early work was oriented towards combat aircraft and tasks (a program called Pilot Associate; see Hammer & Small, 1995). There has recently been more consideration of the tasks in General Aviation, with particular attention to assistance during emergency or non-routine events. The places where a support system can be useful are several, including monitoring of pilot performance of required procedures, recommendation of procedural responses to emergencies, recommendation of diagnostic procedures, assistance in flight re-planning to account for variations in weather, traffic, and aircraft system status, and any other activity where the pilot must weigh data from several sources and make decisions based upon that weighing.

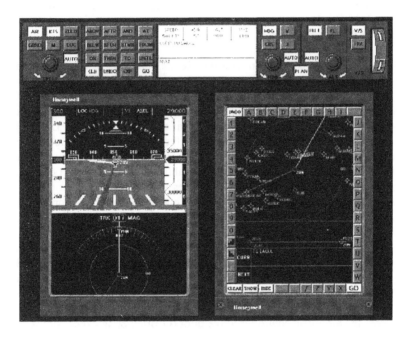

Figure 10.13   Honeywell's initial interface using Cockpit Control Language (Patent Pending) (courtesy of Honeywell, Inc.)

There are some basic principles that should be observed while implementing pilot support systems. Most notably, the system must notify the pilot as to what actions are desirable prior to actually taking any action. Otherwise, we have the possibility of a semi-autonomous flight management system that evaluates options, makes decisions, and executes the required actions, all without any intervention or participation by the pilot. This was exactly the concern expressed by a number of operators of automated flight systems in previous surveys and examinations of automation (Funk, Lyall, & Riley, 1995; Wiener, 1988).

Limiting Factors

The research and development community often comes up with new and interesting ways of accomplishing aviation tasks, both for pilots and for controllers. The implementation of any of the innovations discussed in this chapter is limited by a number of real-world concerns that are attendant to any fielding of an operational system. These limiting factors include need (perceived or actual), cost, certifiability, liability, availability, ability to interface with the existing structures in the aviation environment, and user acceptance.

Need

One of the interesting considerations that plagues implementation of new technology is the issue of need. Need has ripple effects into other categories and so precedes those in our discussion. Need is not often determined from a rational and quantitative assessment of the requirements of the task and the ability of the proposed change/addition to increase system safety (as defined by reliability, efficiency, workload requirements, and/or accident statistics). Need is sometimes defined as a function of *perceived* hazard, as mentioned earlier (Starr, 1969), or as a function of capability.

Perceived need greater than actual need Starr (1969) and Cross (1996) both identify the case where public policy is determined as a function of a perception of the magnitude of the hazard carrying more weight than the statistically demonstrated magnitude of the hazard. They express the concern that, in a climate of shrinking budgets and limited funding, the allocation of public funds to the solution of a perceived public hazard may, in fact, actually produce fewer safety benefits across the exposed population than if those same resources were allocated to a hazard with a higher statistical frequency of occurrence. This particular type of problem may not be one that is easily resolved, however. Public agencies that try to prioritise their responses to hazards using statistical bases for the allocation of resources may in fact be perceived as being unresponsive by the general public. Thus, it may be necessary to accept a more subjectively based prioritisation in some cases.

Perceived need less than actual need On the other hand, we sometimes find that a modification of a procedure or change in equipment can enhance the safety of flight, but the perceived need for the change is not great. This is often the case where the perceived cost or inconvenience required for change is greater than the perceived benefit of that change. One example is a recent study (Beringer & Harris, 1996) in which the superiority of the integrated HSI (horizontal situation indicator) was indicated over performance attained with the separate directional gyro compass and VOR (very-high-frequency omni range) courseline indicator. Although the former produced a 4:1 reduction in procedural errors over the non-integrated instruments, there was no great rush by the GA community to equip by those who were not so equipped. Part of this was due to the comparative criticality of the supported navigation tasks and the other was related to cost (acquisition and installation at about $5,000).

    There is also the issue of *technology-driven need*, as mentioned earlier. Technology-driven need can arise through several circumstances. One such circumstance is when a piece of aviation technology becomes so affordable relative to the features it provides that it is unthinkable to proceed without it. Two recent examples of this situation include GPS receivers and PC-based flight simulators. Technology has made each of these possible and, recently, affordable. This is even more the case where the new "applications" can be added on to existing systems (e.g., a new software package that does flight planning for someone who already owns a laptop computer). The more that the technology diversifies and uses common underlying tools to generate displays, secure information, and manage databases, the more likely this type of "convenient" acquisition will be commonplace in general aviation.

Cost

Cost is a major driver of the implementation of technological innovation. The saying one often hears among manufacturers is, "There are three things that determine how well our product markets: cost, cost, and cost." This factor is particularly salient in any attempt to revive or revitalise any segment of aviation at times when "the bottom line" is so often the final criterion for decisions. It is particularly salient for the general aviation community when one considers that yesterday's $20,000 airplane is today's $180,000 airplane (the new Cessna 172 is supposedly selling for $175,000), and that many certified and installed systems now cost exceptional amounts of money. This has fuelled a trend toward the use of "hand-held" or add-on avionics, particularly in the area of GPS units that tend to cost less than panel-mounted units. This cost differential often results from the requirement that equipment permanently installed in the aircraft be certified for use (by the FAA Aircraft Certification Service). The add-on "portable" devices are not permanently installed and, as such, do not receive the same certification as the permanent installations (further discussion will be found under Certifiability). This cost factor also comes into play in the homebuilt market, where aircraft can be classified in the Experimental category and avoid some of the issues and costs involved in the certification of production aircraft. The economics of scale also figure prominently here, particularly when comparisons are made between the costs of automotive parts and aircraft parts. There have been recent efforts to investigate the use of automotive parts for aviation, which ultimately leads to the issue of certification.

Certifiability

Any equipment that is permanently installed in an aircraft must be certified by the FAA. Manufacturers indicate that a substantial proportion of the cost of producing an aircraft is in the certification process. This is the cost of complying with, in the U.S., the standards set forth by the Code of Federal Regulations, Part 14, which contains the FARs (Federal Aviation Regulations). This process involves documentation to demonstrate that all applicable criteria have been met and flight tested to demonstrate that the aircraft meets all safety and performance standards for its category.

Although the intent of most regulatory processes is to provide safe systems in which the flying public can have confidence, there are occasional wrinkles that produce anomalous results. Consider, for instance, the regulations governing the installation of electronic displays. The FARs for Part 23 aircraft presently call for two independent sources of attitude information for IFR (instrument flight rules) operation. This is generally met by having an attitude indicator and a turn-and-bank indicator (slip/skid and turn rate). However, if one wished to replace the electro-mechanical attitude indicator with an electronic one, the rules do not presently allow this to be done. Rather, one would be required to also have a second attitude indicator, as well as the turn-and-bank indicator, or, alternately, three attitude indicators, despite the fact that the electronic device is presently every bit as reliable as the older electro-mechanical one (and, in many cases, more reliable). This is a legacy of rules that were written before many of today's systems existed; these regulations are in the process of being examined for possible revision. The problem, as one can imagine, is that change in bureaucracy takes time, and technological developments are happening much faster than rule-making can keep up with. As a result, accidents are often catalysts for change.

There is also the case where the existing regulations are not sufficiently specific to require that any particular human-factors convention be used for the design of the system under consideration, despite the fact that known conventions have been demonstrated to be superior in a number of respects. One example of this was an arrangement of powerplant indicators I saw recently that was proposed by an applicant for certification. There were the usual round-dial indicators (six in number), arranged in two rows by three columns, and containing one double indicator (two needles). Above these, on the proposed panel, was a digital display area, separated from the analogue indicators, that contained digital data mirroring the analogue indicators, arranged, this time, in three rows by two columns (Figure 10.14a). The initial response was that integrated instruments, like the counter-pointer altimeter and many other analogue/digital combination instruments, would be preferred, would produce superior performance, and would conserve panel space (Figure 10.14c). The response was apparently negative, so the second, or "fall-back" recommendation was to arrange the digital readouts to match, spatially, the layout of the analogue indicators (both in the 2 by 3 arrangement, and spatially corresponding 1:1) (Figure 10.14b). This display arrangement problem is similar to the age-old one of arranging the knobs on a stove to correspond with the burner elements. There was also, apparently, an appeal to the "We've done it this way before without any problems" defence.

Whether the applicant will relent and use these suggestions is a matter of the choice of the applicant, under current rules, unless the particular configuration can be shown to be "unsafe." Thus, although by all human-factors engineering criteria the arrangement of displays violates known principles of proper design, it has to be "demonstrated" that the particular proposed design is unsafe or likely to pose an operational hazard. It was then suggested that perhaps the liability issue would be the way to encourage the applicant to alter the design. That is to say, if an accident occurred in which the powerplant indicator arrangement could somehow be implicated as a source of confusion in determining or setting levels, it was likely that an experienced human-factors engineer serving as an expert witness would testify to the deficiency of the design, adding weight to the theory that the manufacturer was at fault. Although we generally cringe at the product liability litigation that often plagues the field, this appeared to be one case where the spectre of litigation could be used to prevent later legal action and promote good design.

Reliability is the often-used buzz word in discussions of the certification of systems. Reliability requirements may make certain products unavailable or unusable in the cockpit because the cost of producing a single unit that meets the criterion may make that unit completely unaffordable, either monetarily or as a function of power, interface and/or space requirements in the cockpit. The present rules under which many of us must work at the moment in the U.S. require that flight-critical systems (those required to continue flight) be reliable to the level of p(failure)<.000000001 (or, 10^{-9}). It is interesting to note that the human operator is generally no more reliable than p(failure)<.0001 (10^{-4}, or even 10^{-3} in some cases and tasks; see Swain & Guttman, 1980; Adams, 1982). One might then question the wisdom or necessity of having systems that are far more reliable than the human operator, who is, in fact, just one more component in the overall aircraft system. Although as of this writing these requirements are also being reviewed for possible modification, means of working around this limitation are being examined, including use of multiple display surfaces and processors that are interchangeable in flight. Thus, having identical display surfaces for the primary and multi-function displays would allow the primary flight guidance symbology to be depicted upon either display surface, increasing the reliability of the overall display system through display redundancy and allowing display components with lower individual component reliabilities to be used.

One also has the interesting problem of the process of certification itself and how it is conducted. Presently the U.S. uses a system of distributed regional aircraft certification offices (ACOs) that operate in a relatively autonomous fashion. It is not surprising, given the usual differences in staffing and staff experience, that there is some variation between the various offices in the way in which the certification criteria are interpreted. Thus, one sometimes hears of a seeking out of these variations in order to maximise the probability of that manufacturer's system receiving certification. These differences have also produced occasional circumstances where aircraft that have been certified at one ACO and then, through sale of the company to another holder, have been brought in with minor modifications to a different ACO, are required to go through an entirely new and often more demanding certification process. Significant efforts are presently under way at the FAA's Small Airplane Directorate to provide guidance in unambiguous interpretation of the certification rules so that a greater measure of uniformity can be achieved in the certification process. These changes should further promote and ease the entry of new technologies into the GA cockpit.

Liability

Change in the cockpit is also governed by perceived liability on the part of the manufacturer. Cessna said it had left the business of manufacturing single-engine (reciprocating) aircraft because of the litigations that were occurring at the time. Now that a 15-year U.S. limit has been placed on these product liability litigations involving aircraft, Cessna has re-entered the business, beginning with the new Skyhawk (C-172). The current state of law is such that only new problems not discovered within the first 15 years after manufacture can be litigated after the 15-year period has elapsed. The problem remains, however, that one can comply with all existing state and federal regulations in the manufacture of a system and still be subject to product liability litigation. This has surfaced recently in litigations involving halogen lamps and home fires, where some consumers are claiming inherent design deficiencies (although existing regulations were met) and manufacturers are claiming the devices were used in unintended ways.

However, there is still the matter of making changes to the technology and the question of how the consumers and, more importantly, the courts will interpret such changes. Simply put, there is a concern among some manufacturers that any change that is made in an aircraft system will be interpreted as an indication that something was wrong with the previous incarnation of that system. Let us use a non-aviation example to illustrate. Some years ago, some graduate students and I performed an examination of hand-guided powered lawn mowers to determine how likely various models were to produce vibration-induced white finger (a condition involving restricted blood flow to the hands). Some models transmitted fair amounts of vibration to the hands, while others had reasonably effective motor mounts and vibration isolation so that minimal vibration was conducted to the operator. We inquired of one of the manufacturers of the vibration-conducting mowers as to why they had not chosen to use the more effective dampening means used by another manufacturer, thinking that it might involve proprietary design or licensing fees. The answer we received was surprising. They told us that to do so would imply that something was wrong with their previous models and that the company would then be vulnerable to litigations on the older models. Never mind that the present model was conducting unacceptable levels of vibration and potentially producing chronic ailments for the consumer.

Availability

An interesting limiting factor is the ability of industry to produce and distribute devices or aircraft. If we take the case of the Enhanced Ground Proximity Warning System (EGPWS; Allied Signal), the questioning by the press corps during the public announcement of intent to voluntarily equip the commercial fleet was quite telling. When asked if the component systems were available to equip the remaining aircraft in the fleet, representatives indicated that the company could manufacture 600 units per year (it would require eight years to produce enough to equip the 4700 aircraft involved), but that production could likely be brought up to 1000 units per year. When asked if there was any plan to license other manufacturers to produce the units, representatives responded, albeit tactfully, that this was unnecessary.

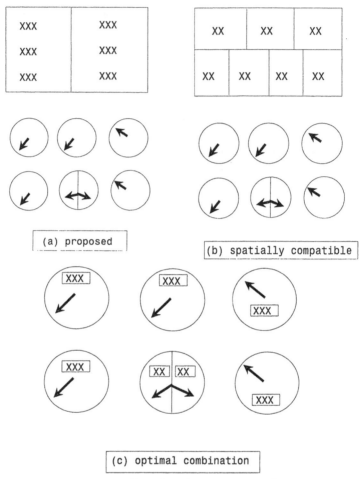

Figure 10.14   Non-corresponding digital and analogue display layouts

There are other availability limits as well, particularly if the technologies supporting the innovations are new. One can recall the lack of availability of certain Pentium chips early in the introduction of that processor line. Demand was great and production somewhat underestimated the response. One can also encounter the situation where there are working prototypes of devices or systems, but final production and distribution of marketable devices awaits either the ability of industry to mass produce the device for it to be economically viable, or for the technology (e.g., blue lasers) to mature so that devices can be made in other than ones or twos or as bench prototypes.

User Acceptance

The ultimate final approval comes from the user. The interesting thing about human operators is that they often do not reject outright the devices and designs that have human engineering flaws, instead choosing to work around the problems. Why? This often comes back to the investment required to either change the system or the investment already in the existing device or system. If one has a $295 GPS receiver that is reasonably functional and provides for sufficient piloting tasks, such that it is easier to use it than to rely on the more conventional navigation means, one is much more likely to learn to live with its annoyances rather than seek to acquire a more expensive device that may have had benefit of human engineering. It should be kept in mind that usability testing is often not as formalised a process with the smaller manufacturers as it is in the industrial giants, and the smaller avionics houses usually do not have the luxury of in-house human factors support. This is both good news and bad news. The bad news, of course, is that more safety or usability issues are likely to slip by in the design of some of these products. The good news is that this would appear to be a fertile ground for those who have the credentials to provide human-factors support, providing the opportunity both to improve the safety and efficiency of the product and to be employed in an increasingly competitive field.

Summary and Prospectus

What have been outlined herein are the various technological, social, legal, political, and behavioural influences that may come into play in defining the future of General Aviation. It is reasonably certain that we will see some forms of each of the following in the General Aviation aircraft of the 21st century:

- Wide-aspect flat-panel displays
- Graphical analogue path guidance displays
- Enhanced manual control systems that command system performance directly
- Bi-directional digital data transfer
- Pilot decision support systems

The successful implementation of these types of systems and devices will depend upon the following being accomplished:

Affordability of systems to General Aviation through:

- Product liability reduced to reasonable limits for General Aviation
- Certification simplified and uniformly applied
- Increased application of human factors criteria during system design and certification

- Production methods achieved that produce reliable products (survive exposure to aviation environment) at reasonable cost

We have had neither space nor time to consider here what the training for the GA pilot may look like in the not-too-distant future. Training will also be affected by technology and by the new things that can be accomplished through intelligent applications of the new technologies. Some changes are already under way, regarding the use of new technologies for flight training, to accommodate the changing nature of aviation and to pave the way for even more interesting applications. What will be crucial to the success of these efforts is the consistent and intelligent application of human factors design criteria so that we can continue to have comparatively safe and efficient operation of our aviation systems across the wide range represented by General Aviation.

Note

1   These figures are presented only to illustrate the magnitude of the hazard across the general population, i.e., how many fatalities were attributable to a source, and not the relative safety of the activities. One would have to include exposure data (fatalities per mile travelled, etc.) to obtain statistics for such comparisons.

References

Adams, J. A. (1982). Issues in human reliability. *Human Factors, 24,* 1-10.

Bauerschmidt, D. K., & Besco, R. O. (1962*). Human engineering criteria for manned space flight: Minimum manual systems* (Report AMRL-TDR-62-87, 23-24). Wright-Patterson AFB, OH: Aerospace Medical Research Laboratories, Tech. Doc.

Beringer, D. B., Allen, R. C., Kozak, K. A., & Young, G. E. (1993). Responses of pilots and nonpilots to color-coded altitude information in a cockpit display of traffic information. *Proceedings of the 37th Annual Meeting of the Human Factors and Ergonomics Society* (pp. 84-87). Santa Monica, CA: Human Factors and Ergonomics Society.

Beringer, D. B. (1980). The design and evaluation of complex systems: Application to a man-machine interface for aerial navigation. *Navigation, 27*(3), 200-206.

Beringer, D. B., Gruetzmacher, G. R., & Swanson, N. (1984). Behavioral stereotypes and control design strategies for a six-degree-of-freedom "walking" vehicle. *Proceedings of the 28th Annual Meeting of the Human Factors Society* (pp. 492-496). Santa Monica, CA: Human Factors Society.

Beringer, D. B., & Harris, H. C., Jr. (1996). *A comparison of the effects of navigational display formats and memory aids on pilot performance* (Technical Report DOT/FAA/AM-96/16). Springfield, VA: National Technical Information Service.

Beringer, D. B. and Harris, H. C., Jr. (1997). *Automation in General Aviation: Two studies of pilot responses to autopilot malfunctions* (Technical Report DOT/FAA/AM-97/24). Springfield, VA: National Technical Information Service.

Beringer, D. B., Harris, H. C., Jr., & Joseph, K. M. (in press). Hearing thresholds among pilots and nonpilot: Implications for auditory warning design. *Proceedings of the 42nd Annual Meeting of the Human Factors and Ergonomics Society.* Santa Monica, CA: Human Factors and Ergonomics Society.

Beringer, D. B., Hendrich, R. C., & Lee, K. M. (1989). *Parameters influencing the airborne application of touch devices for data entry and retrieval: Phase 3A; keypad and touch panel performance comparisons using flight gloves or protective (CBR) gloves* (Technical Report No. HFRL-89-1/MDHC-89-1). Human Factors Research Laboratory, Department of Psychology, New Mexico State University.

Boeing Commercial Airplane Company (1985). Statistical summary of commercial jet aircraft accidents, worldwide operations, 1959-1984.

Boyer, B., Campbell, M., May, P., Merwin, D., & Wickens, C. D. (1995). Three-dimensional displays for terrain and weather awareness in the National Airspace System. *Proceedings of the 39th Annual Meeting of the Human Factors and Ergonomics Society* (pp. 6-10). Santa Monica, CA: Human Factors and Ergonomics Society.

Cross, F. B. (1996). The risk of reliance on perceived risk. Risk: Health, Safety & Environment, Vol. 3, Winter (web site: www.flpc.edu/RISK/vol3/winter/cross.htm).

Dorighi, N. S., Ellis, S. R., & Grunwald, A. J. (1993). Perspective format for a primary flight display (ADI) and its effect on pilot spatial awareness. *Proceedings of the 37th Annual Meeting of the Human Factors and Ergonomics Society* (pp. 88-92). Santa Monica, CA: Human Factors and Ergonomics Society.

Duerksen, N. (1996). *Fuzzy logic decoupled lateral control for general aviation airplanes. NASA Contractor Report 201735.* Springfield, VA: National Technical Information Service.

Edworthy, J. (1994). The design and implementation of non-verbal auditory warnings. *Applied Ergonomics, 25*(4), 202-210.

Ellis, S. R., McGreevy, M. W., & Hitchcock, R. J. (1987). Perspective traffic display format and airline pilot traffic avoidance. *Human Factors, 29,* 371-382.

Funk, K., Lyall, E., & Riley, V. (1995). *Perceived human factors problems of flightdeck automation. A comparative analysis of flightdecks with varying levels of automation,* Federal Aviation Administration Grant 93-G-039, Phase 1 Final Report.

Furness, T. A., III (1986). The super cockpit and its human factors challenges. *Proceedings of the 30th Annual Meeting of the Human Factors Society* (pp. 48-52). Santa Monica, CA: Human Factors Society.

Hammer, J. M., & Small, R. L. (1995). An intelligent interface in an associate system. In W. B. Rouse (Ed.), *Human/technology interaction in complex systems* (Vol. 7, 1-44). Greenwich, CT: JAI Press.

Hellier, E. J., Edworthy, J., & Dennis, I. (1993). Improving auditory warning design: Quantifying and predicting the effects of different warning parameters on perceived urgency. *Human Factors, 35(4),* 693-706.

Hrebec, D. G., Infield, S. E., Rhodes, S. and Fiedler, F. E. (1994). A simulator study of the effect of datalink on crew error. *Proceedings of the 38th Annual Meeting of the Human Factors and Ergonomics Society* (pp. 66-70). Santa Monica, CA: Human Factors and Ergonomics Society.

Jordan, J., Harris, W. C., Goernert, P. N., & Roberts, J. (1996). The effect of active noise reduction technology on noise-induced pilot fatigue and associated cognitive performance decrements. *Proceedings of the 40th Annual Meeting of the Human Factors and Ergonomics Society* (pp. 67-71). Santa Monica, CA: Human Factors and Ergonomics Society.

Kruk, R. V., Runnings, D. W., King, M., Lippay, A. L., & McKinnon, G. M. (1986). Development and evaluation of a proportional displacement sidearm controller for helicopters. *Proceedings of the 30th Annual Meeting of the Human Factors Society* (pp. 865-869). Santa Monica, CA: The Human Factors Society.

Liggett, K. K, Kustra, T. W., & Reising, J. M. (1997). A comparison of touch, manual, and head controllers for cursor movement. *Proceedings of the 41st Annual Meeting of the Human Factors and Ergonomics Society* (pp. 80-84). Santa Monica, CA: Human Factors and Ergonomics Society.

Lippay, A. L., King, M., Kruk, R. V., & Morgan, M. (1985) Helicopter flight control with one hand. *Canadian Aeronautics and Space Journal, 31*(4), 335-345.

McCann, R. S., Andre, A. D., Begault, D., Foyle, D. C., & Wenzel, E. (1997). Enhancing taxi performance under low visibility: Are moving maps enough? *Proceedings of the 41st Annual Meeting of the Human Factors and Ergonomics Society* (pp. 37-42). Santa Monica, CA: Human Factors and Ergonomics Society.

McKinley, R. L., D'Angelo, W. R., Haas, M. W., Perrot, D. R., Nelson, W. T., Hettinger, L. J., & Brickman, B. J. (1995). An initial study of the effects of 3-dimensional auditory cueing on visual target detection. *Proceedings of the 39th Annual Meeting of the Human Factors and Ergonomics Society* (pp. 119-123). Santa Monica, CA: Human Factors and Ergonomics Society.

National Safety Council (1996). *Accident Facts,* 1996 Edition. Itasca, IL: National Safety Council.

Newman, R.L. (1987). Responses to Roscoe, "The trouble with HUDs and HMDs." *Human Factors Society Bulletin, 30*(10), 3-5. Santa Monica, CA: The Human Factors Society.

O'Brien, J. V., & Wickens, C. D. (1997). Free flight cockpit displays of traffic and weather: Effects of dimensionality and data base integration. *Proceedings of the 41st Annual Meeting of the Human Factors and Ergonomics Society* (pp. 18-22). Santa Monica, CA: Human Factors and Ergonomics Society.

O'Hare, D., & Roscoe, S. (1990). *Flightdeck performance: The human factor* (pp. 181-207). Ames, Iowa: Iowa State University Press.

Reising, J. M., Liggett, K. K., Solz, T. J., & Hartsock, D. C. (1995). A comparison of two head-up display formats used to fly curved instrument approaches. *Proceedings of the 39th Annual Meeting of the Human Factors and Ergonomics Society* (pp. 1-5). Santa Monica, CA: Human Factors and Ergonomics Society.

Riley, V., DeMers, B., Misiak, C., & Schmalz, B. (1998). The cockpit control language: A pilot-centered avionics interface. *Proceedings of the 1998 International Conference on Human-Computer Interaction in Aeronautics* May 27-29 (pp. 35-39). Montreal, Canada.

Roscoe, S. N. (Ed.). (1980). *Aviation psychology*. Ames, Iowa: Iowa State University Press.

Roscoe, S. N. (1987). The trouble with HUDs and HMDs. *Human Factors Society Bulletin, 30*(7), 1-2. Santa Monica, CA: The Human Factors Society.

Roscoe, S. N. and Bergman, C. A. (1980). Flight performance control. In S. N. Roscoe (Ed.), *Aviation psychology* (pp. 39-47). Ames, Iowa: Iowa State University Press.

Roscoe, S. N., Eisele, J. E., & Bergman, C. A. (1980). Information and control requirements. In S. N. Roscoe (Ed.), *Aviation psychology* (pp. 33-38). Ames, Iowa: Iowa State University Press.

Starr, C. (1969). Social benefit versus technological risk: What is our society willing to pay for safety? *Science, 165,* December, 1232-1238.

Swain, A., & Guttman, H. E. (1980). *Handbook of human reliability analysis with emphasis on nuclear power plant operations.* Washington, DC: US Nuclear Regulatory Commission.

Wickens, C.D. (1997). Attentional issues in head-up displays. *Engineering psychology and cognitive ergonomics: integration of theory and application.* London: Avebury Technical Publishing Company.

Wickens, D.C., Liang, C.C., Prevett, T., & Olmos, O. (1996). Electronic maps for terminal area navigation: effects of frame of reference and dimensionality. *The International Journal of Aviation Psychology, 6*(3), 241-271.

Wiener, E. L. (1988). Cockpit Automation. In E. L. Wiener & D. C. Nagel (Eds.), *Human factors in aviation* (pp. 433-461). New York: Academic Press.

Williams, A. C. (1949). Suggestions concerning desirable display characteristics for aircraft instruments. Interim Report SDC 71-16-4, July 1949; reprinted in S. N. Roscoe (Ed.), *Aviation Research Monographs, 1*(1), July 1971, University of Illinois Institute of Aviation.

Young, G. (1997). *The transfer of inferred vs. signaled failure detection performance in monitors and controllers of a complex dynamic task.* Unpublished doctoral dissertation, New Mexico State University, Department of Psychology, Las Cruces, New Mexico.

Part 6
Safety and Accident Investigation

11 Safety is more than Accident Prevention: Risk Factors for Crashes and Injuries in General Aviation

David O'Hare

Introduction

Transportation accidents are a major public health problem in most societies, with motor vehicle accidents the leading cause of accidental death in most cases. Motor vehicle crashes in the United States alone are responsible for "between 44,000 and 52,000 deaths and between 4 million and 5 million injuries each year" (Baker, O'Neill, Ginsberg, & Li, 1992). There are approximately 2000 road fatalities in Australia each year, and even in a small country such as New Zealand the annual road death toll is between 500 and 600. Whilst the annual number of fatalities associated with general aviation accidents is very much less, at around 750 for the U.S. and 28 for the U.K., the rates of death and injury for participants in aviation appear to be much greater. For example, approximately 72 out of every 100,000 general aviation pilots in the U.S. are killed each year, compared to around 25 per 100,000 male drivers (Baker & Lamb, 1989).

The fatality rates for certain categories of general aviation, notably that of agricultural and aerial work, may be higher than for any other occupational group. For example, Hall (1991) reports the rate for Australian agricultural pilots at 460 per 100,000 participants. Cryer and Fleming (1987) reported rates of 1,100 per 100,000 participants for fixed-wing agricultural pilots in New Zealand, giving a forty-year lifetime risk of death of 1 in 3! Baker and Lamb (1989) have summarised the situation in general aviation as follows: "...crashes are both common and severe. Approximately one aircraft in four will crash during a 20-year lifespan" (Baker & Lamb, 1989, p. 535).

Much of the literature in aviation human factors has been concerned with the identification and amelioration of factors thought to be associated with the risk of aviation crashes. Sometimes the identification of risk factors has been done empirically, as in Jensen and Benel's well known study of the NTSB database (Jensen & Benel, 1977). On other occasions, large scale educational efforts have been based on little or no empirical foundations, as in the case of the development of programmes designed to overcome the so-called 'hazardous attitudes' (Berlin, Gruber, Jensen, Holmes, Lau, Mills, & O'Kane, 1982). Most of these efforts have in common a focus on the prevention of accidents. However, as Lane (1973) has pointed out: "Safety is not an end in itself: the objective is reduction of losses, whether in personal injury or in damage to property" (p. 15). Improvements in safety can therefore be achieved by modifying the energy release in the actual crash as well as by interventions designed to reduce losses once a crash impact has taken place. The three phases (pre-crash, in-crash, and post-crash) must all be considered in a well-balanced safety programme:

> *It is to be emphasised that loss reduction is not simply accident prevention...some failures will occur and, in that context, there is no particular reason to prefer a pre-crash to an in-crash or a post-crash measure simply because it is intended to operate earlier in the accident sequence. (Lane, 1973, p. 16).*

Researchers have used the pre-crash, in-crash, post-crash framework to organise proposals for improved safety in driving (e.g., Haddon, 1980) and in aviation (e.g., Richter, Gordon, Halamish & Gribetz, 1981). In the latter case, the authors suggest that their various proposals, ranging from aircraft cooling to reduce heat stress and dehydration (pre-crash), energy absorbing cockpits (in-crash) and availability of atropine and syringes (post-crash), should be evaluated on the basis of "effectiveness and cost rather than earliness or lateness in the sequence of injury causation" (Richter et al., 1981, p. 56).

The most appropriate basis for the development of safety improvement measures comes from empirical investigation of the factors which increase the risk of injury and loss. Such factors can operate at any of the three stages. For example, if male pilots were, like male drivers, found to be significantly over-involved in crashes then gender would be a risk factor at the pre-crash stage. If older pilots were more likely to receive fatal injuries in a crash, then age would be a risk factor at the in-crash or post-crash phase. Epidemiologic methods, used to study a wide variety of problems in public health and medicine, can be applied to the identification of risk factors at any of the three stages. In practice, as Booze (1977) noted: "Very little attention has been devoted to classical epidemiologic methods and variables as an approach to a better understanding of the aetiology of aircraft accidents" (p. 1081).

In a recent review of epidemiologic studies of pilot-related factors in aircraft crashes, Li (1994) found a total of 24 controlled epidemiologic investigations, but very few of the more rigorously designed case-control or cohort studies (only three of each type were located). Li (1994) points out that: "As a result, pilots as a high-risk occupational group are greatly under-served by public health researchers" (Li, 1994, p. 951). The purpose of the present chapter is to review the findings from epidemiologic studies into the risk factors which determine the risk of involvement in an aviation crash (pre-crash) and the risk factors which determine the likelihood of injury once a crash has occurred (in-crash and post-crash). In each case, the potential risk factors are organised under the following headings: pilot, aircraft, environment, and operational factors.

Risk Factors for Involvement in Aviation Accidents

Pilot Factors

The classic study of Booze (1977) was one of the first large-scale controlled epidemiologic investigations of involvement in aviation accidents. Booze examined the records held by the FAA on 4,491 accident involved general aviation pilots and compared them with the records of a second group of 9,414 active pilots drawn from the complete files of all 762,604 active pilots at that time. The main factors of interest were age, occupation, total flight time, and recent flight time. Brief reference was also made to gender, state of occurrence, month and time of day, time into flight, and purpose of flight.

Age Booze (1977) reported clear increases in accident risk as a function of age with the older age groups four times more likely to be involved in accidents than the younger age groups. This is a somewhat surprising finding, particularly in view of the clear findings in the automobile driving literature that the younger age groups are at much greater risk of vehicle accident involvement (Evans, 1991). Li (1994) has pointed out that there is a significant flaw in Booze's study in that he compared accident involved general aviation pilots with a sample drawn from the total pilot population—including air transport pilots. This could have biased the results, although in this case one might expect the comparison sample to have included slightly older, more experienced airline pilots, and thus to have made it more likely that the bias would be towards overestimating the involvement of younger pilots. However, in support of Booze's findings, the U.K. Civil Aviation Authority have pointed to an over-involvement of older pilots (age 37+) in a series of general aviation accidents in 1987 (Civil Aviation Authority, 1988).

Other studies have found an increased risk of involvement for younger pilots (e.g., Lubner, Markowitz, & Isherwood, 1991). Guide and Gibson (1991) found much higher accident involvement amongst younger holders of CPLs or ATPLs than amongst older CPLs and ATPLs. In contrast, for PPL holders there was a steadily increasing accident involvement with age. These differences seemed to largely disappear once flight hours were taken into account. Bruckart (1992) has pointed out that as the average age of the pilot population has increased, so accident rates have fallen. Mohler (1969) found no direct relationship between crash risk and age amongst general aviation pilots. The same conclusion was reached in a recent study of New Zealand pilots (O'Hare, Chalmers, & Bagnall, 1996).

Occupation The highest rates of accident involvement were found by Booze (1977) to be professional pilots, followed by lawyers, farmers, sales representatives and physicians. The lowest rates were found amongst engineers, teachers, "housewives", students, and members of the armed forces. When adjusted for flight exposure, professional pilots had the lowest rates, whilst students and "housewives" topped the list. Lawyers and physicians remained relatively high, and teachers and armed forces members relatively low. Booze attributed the over-involvement of the former groups to common personality traits: "All are likely to be aggressive, busy, independent, and self-sufficient" (Booze, 1977, p. 1086). However, rather than sharing a raft of personality characteristics, it is more likely that lawyers and physicians have access to more complex, higher performance aircraft which they use as a means of carrying out their business, thus exposing themselves to an increased variety of operational risks.

Total flight time Booze (1977) found a steadily increasing risk of accident involvement as a function of total flight time. Pilots with 1000-2000 total hours were three times as likely to be involved in a crash as pilots with less than 50 hours. Given that the former group had at least twenty times the exposure to accident risk, the results of Booze can be taken to suggest that there is in fact a protective effect of cumulative flight experience. This is consistent with other evidence that risk is markedly lower for pilots holding the highest levels of qualification (the Airline Transport Pilot Licence) who "have fewer accidents than private pilots in proportion to both their numbers and their hours of exposure to risk" (Salvatore, Stearns, Huntley, & Mengert, 1986, p. 1460). Hall (1991) found that the probability of having a crash in a given year declined with increasing levels of total experience.

A number of investigators have suggested that there is a special period of vulnerability to accident involvement at around the 100 hour mark (Olsen & Rasmussen, 1989) or between the 100 and 300 hour period (Jensen, 1995). O'Hare, Chalmers, and Bagnall, (1996) compared the flight experience of pilots who had been involved in an accident with data on flight experience provided by a nationwide survey of all active pilots in the country. Every active pilot in New Zealand was sent a survey form which asked them to transcribe flight times from their log books for each of the previous six years. The survey form contained additional questions concerning demographics, types of aircraft flown etc.

The comparison between the accident involved pilots and the national sample is shown in Figure 11.1. It can be seen that at the lowest levels of total flight time (< 100 hours) the proportion of pilots in the accident involved group (12%) is nearly twice as high as the proportion of pilots in the total population (7%). For pilots with more than 100, but less than 800 total flight hours, the situation is reversed with these pilots slightly under-represented in the accident involved group. To test the hypothesis of special vulnerability to accident involvement in the 100 to 300 hour period, a chi-square calculation was performed on the proportion of accident involved pilots in the 100-300 hour range (21.2%) versus the proportion of pilots in the total population with total flight experience in the same range (25.8%). The chi-square (1, n=890) = 1.16, p > 0.05 confirmed that there is no special vulnerability to accident involvement in this period. If anything, there is a slight tendency towards under-involvement (O'Hare & Chalmers, 1999).

For pilots with more than 800 but less than 2000 total flight hours, the proportion in the accident involved group (18.5%) was slightly greater than the proportion in the total population (15.9%). For pilots with more than 2000 hours total flight time the proportions were similar for both the accident involved pilots and the population as a whole.

Recent flight time Booze (1977) found that 33% of all crashes were experienced by the 11% of pilots with more than 100 hours of flight time in the previous twelve months: "Recent exposure appears to be the best discriminator of accident risk" (Booze, 1977, p. 1090). Yacavone, Borowsky, Bason, and Alkov (1992) found that the highest mishap rates for U.S. Naval fighter/attack pilots were for groups with the greatest amount of recent experience (> 40 hours in previous 30 days). O'Hare, Chalmers, and Bagnall (1996) compared the flight hours reported in the previous 90 days by those pilots who had been involved in an accident with the recent flight hours reported by pilots in the population as a whole. The results are shown in Figure 11.2.

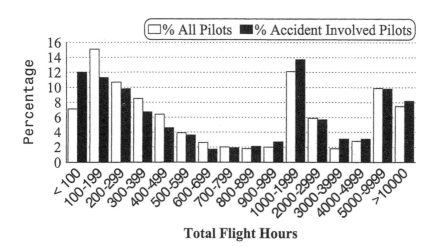

Figure 11.1   **Total flight hours as a risk factor for accident involvement**
Source: O'Hare, Chalmers, and Bagnall (1996)

Figure 11.2 shows that the majority of pilots fly relatively infrequently. Nearly two-thirds (61.6%) of all pilots have recorded less than 30 hours flight time in the previous 90 days. Given that the median annual reported flight time for New Zealand private pilots was only 22 hours, it is clear that a substantial proportion of pilots are flying very irregularly. This seems to be true worldwide, as Hunter (1995) found that the median annual flight time for U.S. private pilots was

30 hours. However, this lack of recent flight experience does not in itself constitute a risk factor for accident involvement, as a much lower proportion (48.7%) of accident involved pilots had less than 30 hours flight time in the previous 90 days, and thus pilots in this lower recent experience range are under-represented in accidents. On the other hand, pilots with between 30 and 89 hours in the previous 90 days made up only 13.1% of the total population, but comprised 27.2% of the accident involved group. For recent experience levels above 90 hours in the previous 90 days there was no difference between the proportion found amongst accident involved pilots and the proportion found in the total population.

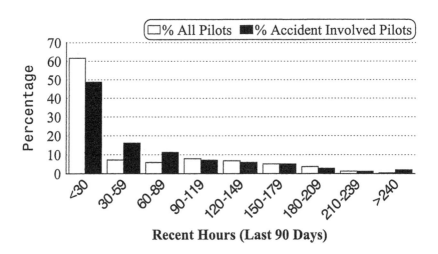

Figure 11.2   Recent (last 90 days) flight hours as a risk factor for accident involvement
Source: O'Hare, Chalmers, and Bagnall (1996)

As Li (1994) has pointed out, there are two factors underlying the relationship between flight time and accident risk. One is simply exposure—the more hours flown the greater the exposure to the hazards involved in aviation and therefore the greater the risk. The second factor is the presumed greater expertise that comes with higher levels of flight time, both total and recent. Li (1994) suggests that when exposure is held constant the evidence shows a positive effect of recent experience on accident risk. The data shown in Figure 11.2 suggests a more complex relationship with the highest levels of risk associated with intermediate levels of recent experience. The fact that pilots with higher levels of exposure had lower levels of accident involvement suggests that there may be some other factor or factors associated with intermediate recent experience levels that leads to greater accident liability.

One possibility is age—pilots with this level of recent experience may tend to be young, commercial pilots in their first jobs. They may therefore be exposed to flying in a more challenging range of conditions and environments than they had previously experienced. A second factor may be the nature of the operational environment. The companies which employ such pilots are likely to be small operations involved in tourist, charter and other aerial work outside the highly structured and proceduralised environment of an airline operation. They are more likely to operate smaller single-engine and low-performance twin-engine aircraft. It is likely that further inquiry into the nature of such factors and their effects on pilots may be more productive in defining accident risk than a continued focus on flight time *per se*.

Recently some attempts have been made to better understand the various paths and career profiles of aviators in the United States to determine the true nature of the risks faced by pilots and enable better targeted safety interventions. Hunter (Chapter 3) presents some of the findings from these surveys of GA pilots. O'Hare et al. (1996) present three risk profiles based on their findings of accident involvement by New Zealand pilots.

• *The novice* with less than 100 hours and prone to (mostly minor) accident involvement.

• *The recreational pilot* with between 100 and 800 total flight hours and an annual accumulation of less than 120 flight hours. This group is under-represented in accidents. Pilots in this group may be more cautious in their choice of activities bearing in mind their overall experience and currency levels.

• *The potential high flyer* with over 800 total flight hours and an annual flight experience of 120 to 360 hours. This group is over-represented in accidents. These pilots may have come to feel quite confident and may be undertaking activities (e.g., aerial work) with greater risk levels than those typically flown by more experienced pilots. They are keen to rapidly accrue flight time to progress further up the career ladder.

Many further refinements and extensions of these profiles could be developed, but these would be beyond the reach of currently available data to support.

Gender Vail and Ekman (1986) examined the NTSB accident records of all general aviation accidents between 1972 and 1981. The proportion of male pilots who were involved in an accident in any given year (0.52%) was more than double the proportion of female pilots involved in an accident in any given year (0.23%). At higher levels of total flight time (above 1000 hours), the difference was even greater with males being more than five times as likely to be involved in a crash. Accidents involving male pilots were also twice as likely to result in a fatal injury. In their case-control study, Lubner et al. (1991) found males to be only slightly more likely (1.3 times) to be accident involved, but almost twice as likely to have been the subject of an enforcement action by the FAA.

The situation is complicated by differences in flight time and occupational positions between male and female pilots. O'Hare and Chalmers (1999) report that in their nationwide survey of New Zealand pilots males reported significantly higher recent (last 90 days) and lifetime total flight hours than female pilots. For recent times the mean values were 38.6 hrs for males and 19.4 hrs for females; for lifetime total hours the mean values were 2231 hrs and 491 hrs respectively. These differences largely reflect the uneven distribution of female pilots within the aviation system. For example, none of the sample of female pilots held ATPLs. This was not surprising, as there were only a handful of female ATPL holders nationwide.

Male pilots comprised 94.6% of the pilot population of New Zealand, comprised 96.1% of the pilots involved in accidents, and reported 98% of the hours flown in our survey (O'Hare, Chalmers, and Bagnall, 1986). These figures illustrate the difficulty of drawing conclusions about gender and accident involvement. Males appear to be both over-represented (on a population basis) and under-represented (on an hours flown basis) in accidents. In fact, using the flight hours reported in the survey we calculated the overall accident rate for male pilots to be 13.6 per 100,000 hours flown compared to 26.5 per 100,000 hours for female pilots. However, as noted above, this comparison is invalidated by the lack of high-time female pilots holding ATPLs and flying commercial transport aircraft where the accident rates are extremely low.

On the basis of the flight hours reported, we found that the overall difference between male and female annual flight hours was smallest in the category of fixed-wing aircraft under 2,730Kg MTOW (Males = 77.9 hours; Females = 62.8 hours). The difference was least in the subset of pilots who flew predominantly for pleasure (rather than for aerial work, transport, or training) where the mean annual hours were 30.9 (M) and 29.8 (F). The accident rates for male and female pilots per 100,000 flight hours for small fixed-wing aircraft (< 2,730Kg MTOW) flown for pleasure were 30.1 and 28.4 respectively. Contrary to overseas research (Vail & Ekman, 1986), when type of operation and aircraft category are equated, the difference between male and female accident rates appears to be negligible. McFadden (1996) has recently found no difference between the pilot-error accident rates of male and female airline pilots. Gender does not appear to be a risk factor for accidents when exposure is controlled for.

Licence category Lubner et al. (1991) found that commercial pilot licence (CPL) holders were at twice the risk of being involved in an accident or receiving a FAA enforcement action as other types of licence holders.

Alcohol No U.S. airline has ever experienced an alcohol-related accident (Modell & Mountz, 1990). However, the crash of a Japan Air Lines DC-8 freighter at Anchorage, Alaska in 1977 was due to the effects of alcohol on the American captain of the aircraft (National Transportation Safety Board, 1979). Li (1994) has pointed out that reported alcohol involvement rates in U.S. general aviation have apparently decreased from around 40% in the 1960s to 10% in the late 1980s. In the 1984 the NTSB reported positive alcohol tests from 6.4% of fatally injured pilots of scheduled commuter aircraft and 7.4% from fatally injured pilots of air taxi operations, and an overall GA positive alcohol rate of 10.5%. (National Transportation Safety Board, 1984). However, between 1983 and 1988 only a very small number of fatally injured pilots of air taxi flights were found to test positive for alcohol with no involvement for pilots of commuter aircraft (Li, 1994). The most recent evidence (Li, Hooten, Baker, & Butts, 1998) also shows a positive alcohol test rate of 7% for pilots and 15% for other occupants fatally injured in air crashes in North Carolina between 1985 and 1994. Although the numbers are very small, it is interesting to note that the four victims with the highest blood alcohol levels were male GA pilots aged 20-29 who crashed at night.

The role of alcohol in aviation accidents appears to be even lower in other countries. For example, alcohol was not a factor in any of the 166 fatal GA accidents in the U.K between 1985 and 1994 (Civil Aviation Authority, 1997). Because of the zero rate of alcohol-related accidents amongst U.S. airlines the value of random alcohol testing of U.S. airline employees has now been challenged (McFadden, 1997).

Previous history Li and Baker (1994) examined the role of commuter pilots' previous crash and FAA violation history as risk factors for crash involvement. The study involved comparing the records of 846 pilots-in-command who were involved in a commuter or air taxi crash during 1983 and 1988 (the 'cases'), with the records of 1,555 'control' pilots who were also active full-time pilots with a scheduled commuter carrier during the same period. The 'cases' had nearly twice the rate of previous crashes in the three years prior to their accident (11%) than the controls (6.7%). The cases were also more likely to have been cited by the FAA in the same period (9.5%) than controls (6.3%). Both these differences were statistically significant. In subsequent regression modelling, both previous crash history and previous violation history were shown to be significant risk factors for crashes of scheduled commuter aircraft. These findings were replicated in a subsequent historical cohort study (Li & Baker, 1995).

It appears that previous automobile driving history may also be relevant in predicting aviation crash risk. McFadden (1997) evaluated the usefulness of the FAA's policy of monitoring driving-while-intoxicated (DWI) convictions as a basis for assessing a pilot's fitness to fly. Data were obtained on all airline crashes (major air carrier and commuter) during the period 1986 to 1992. Whilst the probability of a pilot being involved in a crash was considerably lower if that pilot was flying for a major airline, the probability is about doubled if that pilot has any DWI conviction. The same was true for pilots in commuter operations. There were 1,372 pilots, or 1.96% of the airline pilot population with a DWI conviction.

A second analysis suggested that there is a linear increase in risk of accident involvement with increasing number of DWI convictions.

Aircraft Factors

Size and performance At the broadest level, aircraft can be categorised into fixed-wing and rotary-wing varieties. Fixed-wing aircraft cover a very wide range from the largest B747-400 to the most humble Cessna 150. One way of grouping aircraft with relatively similar performance characteristics together is by using the maximum certificated takeoff weight (MTOW). For example:

- Over 13,610 Kg (e.g., B767, A320 etc.)
- Between 5,670 and 13, 610 Kg (e.g., Saab 340, Jetstream 41 etc.)
- Between 2,730 and 5,670 Kg (e.g., Cessna Caravan)
- Under 2,730 Kg (e.g., Cessna 172, Piper PA-28 etc.)

Accident rate statistics have consistently shown an advantage to larger, higher performance aircraft. In the U.S. for example, the accident rate for scheduled airliners (almost exclusively the highest weight band above) is about 25 times better than for all of general aviation, covering the remaining three bands. Comparing rotary-wing with fixed-wing aircraft overall shows that rotary-wing aircraft have an accident rate (per 100,000 flight hours) almost 50% higher than fixed-wing aircraft in general aviation (NTSB, 1987).

In addition to the rotary-wing and four bands of fixed-wing aircraft, general aviation also encompasses a variety of sport aircraft, including experimental or 'homebuilt' aircraft, microlights and gliders. There are some additional classifications such as gyrocoptors and amphibians which are relatively rare. The study carried out in New Zealand for the Civil Aviation Authority (O'Hare, et al., 1996) showed very substantial differences in accident rates for aircraft in these different categories (see Figure 11.3).

There was a 175-fold difference between the accident rate of the riskiest category (microlight aircraft) and the safest (fixed-wing aircraft above 13,610 Kg MTOW). The accident rate for rotary-wing aircraft was considerably higher (33 per 100,000 flight hours) than for any of the categories of fixed-wing aircraft. The accident rate for microlight aircraft was two and a half times that of the next riskiest category (experimental/homebuilt aircraft).

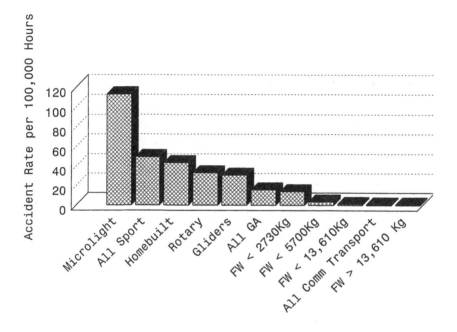

Figure 11.3   Comparative accident rates for different categories of aircraft
Source: O'Hare, Chalmers and Bagnall, 1996

Looked at another way, microlight aircraft were involved in 11% of all reported aircraft accidents in New Zealand but recorded only 1.3% of all flight hours. Similarly, experimental/homebuilt aircraft were involved in 4.4% of accidents whilst flying 1.4% of total flight hours. Rotary-wing aircraft (25.7% of accidents, 10.5% of flight hours) and gliders (9.1% of accidents, 4% of flight hours) were also over-represented in the accident statistics. On the other side of the coin, large transport aircraft (> 13,610 Kg MTOW) flew 24.2% of the flight hours recorded in New Zealand but were involved in only 1.1% of the accidents. These data can be summarised as follows:

Above average risk of accident involvement
- Microlight
- Experimental/Homebuilt
- Rotary-wing
- Glider

Average risk of accident involvement
- Fixed-wing aircraft < 2,730 Kg MTOW

Below average risk of accident involvement
- Fixed-wing aircraft > 2, 730 Kg MTOW

Within a particular category of aircraft (e.g., fixed-wing aircraft below 2,730 Kg MTOW) there can be substantial variations in accident rates between one particular aircraft type and another. For example, in the U.K. between 1985 and 1994, the rate of fatal accidents per 100,000 flight hours varied by a factor of almost fifteen between the best (0.41 for the Cessna 172) and the worst (6.1 for the Cessna 182). Interestingly, the fatal accident rate for the often maligned Robinson R22 helicopter at 3.3 was about midway between these two (Civil Aviation Authority, 1997).

In their analysis of commuter air crashes Baker, Lamb, Li, and Dodd (1993) note the over-involvement of certain types in specific categories of crashes. For example, every recorded case of a gear-up landing involved just one aircraft type— the Metroliner. Similarly, all cases of pilot-related fuel-starvation accidents involved either Cessna 402s or Piper Navajos.

Powerplant The accident rate statistics (e.g., NTSB, 1987) show clear advantages of having more than one engine, and of having turbine rather than reciprocating/piston engines. For fixed-wing non-turbine aircraft, the overall accident rate drops 42%, from around 11 per 100,000 flight hours for single engine aircraft to around 6 for multi-engine aircraft. However, the **fatal** accident rate actually increases by approximately 2%. Significant falls in both overall and fatal accident rates were seen for turbine power—turboprops had an overall accident rate 64% lower than reciprocating engined aircraft, and turbojets were 85% safer.

Environmental Factors

State of occurrence Booze (1977) noted that Alaska had four times as many accidents as would have been expected given the number of pilots in the state population. Other 'frontier' states in the U.S. such as Wyoming, Nevada and New Mexico had double the expected rate, whilst the northeastern states such as New Jersey, Connecticut and Maryland had half the expected number of accidents. Baker and Lamb (1989) noted that the mountain states experienced fatal accident rates twice those of other states, with Colorado's 2.6 times the national average.

Urban (1983) compared the crash rates of pilots from different size communities in Colorado and found that smaller communities had double the crash rate of the larger metropolitan areas. Since no exposure data was collected, it cannot be determined if this simply reflects differences in flying activities between pilots from small rural areas and those from large cities. Urban advanced two hypotheses to account for his data: (1) that there is a greater emphasis on risk-taking and aggressiveness as values in smaller communities and (2) there is less opportunity for social comparison to develop judgement abilities and compare safety information in smaller communities.

The differences in accident involvement between U.S. states has been replicated by Lubner et al. (1991). Using a case-control design (see below), pilots from Alaska were found to be more than three times as likely as pilots from elsewhere to have been involved in an accident and more than three times as likely to have been the subject of a FAA enforcement action. In contrast, pilots from the northeastern states were about half as likely as other pilots to have been involved in either an accident or an enforcement action. As Lubner et al. note "it is intriguing that the geographic variable is a robust predictor " (Lubner et al., 1991, p. 240) although the psychological mechanisms which underpin this effect have not been determined. It would be interesting to examine data from other countries, such as Australia, Canada, and New Zealand to see if there are discernible regional differences and whether these reflect similar 'sociological' distinctions between different communities.

Time (Day/Month) The numbers of accidents tends to be greater during the summer months, presumably reflecting the greater amount of flying which takes place during this period. In New Zealand, for example, there are more than twice as many accidents during the summer quarter (37.5%) as during the winter quarter (16%) (Aviation Co-Operating Underwriters Ltd, 1997). Baker et al. (1992) note that the majority of aviation fatalities occur between Friday and Sunday and that almost one-third occur during the Northern-hemisphere summer months of July and August.

The difficulty with variables such as time of flight, purpose of flight etc. is in the lack of exposure data. Whilst these facts are sometimes noted when a crash has occurred, there is no comparable information for flights that have not crashed. In reporting, for example, that "85% of all accidents occurred between 9am and 9pm as one might expect" (Booze, 1977, p. 1090), the author simply appeals to one's intuition that this seems about the same as the proportion of all flights that would have taken place between these times.

Operational Factors

Purpose of flight Booze (1977) noted that "40% of all accident flights were originated for business purposes, and the remainder for pleasure" (Booze, 1977, p. 1090). As noted above, the lack of comparison data (i.e., the percentage of all flights which are carried out for business purposes) makes this statement uninformative with respect to accident risk. An attempt to gather comparative exposure data was made by O'Hare et al. (1996). As part of a large scale study of aircraft accidents in New Zealand over a seven year period, all active pilots were sent a questionnaire which asked them to supply the hours written in their personal log books under various headings. These data were used to calculate the basic accident rates of different categories of aircraft and types of operation.

Unfortunately, it was not possible to obtain a detailed breakdown of hours flown in different operational categories (e.g., training, aerial work etc.). However, respondents did indicate the operational category in which the majority of their hours was flown. Bearing in mind the very approximate basis of this comparison, it was found that for both rotary-wing and fixed-wing aircraft, the highest level of accident risk was associated with private operations. For example, in rotary wing aircraft this category accounted for only 7.5% of reported flight hours, but accounted for 44.7% of all accidents. Because of the rather crude method of estimating flight hours in this category, the 7.5% is most certainly an underestimate of the true flight hours. Nevertheless, the extreme difference between the proportions of accidents versus hours flown, together with similar findings in respect of small, fixed-wing aircraft, suggests that there almost certainly is some over-representation of this operational category in the accident statistics. In contrast, flights undertaken for training and for commercial transport purposes were under-represented in the accident statistics.

Phase of flight There are marked differences between the risk profiles found in general aviation and in scheduled air transport. The greatest risk during air carrier operations is found during the descent and approach phases of flight. Over 55% of air carrier crashes between 1959 and 1983 occurred during these phases of flight which account for only 24% of air carrier flight time (Boeing, 1985 cited in Nagel, 1988). In general aviation by contrast, only about 15% of crashes occur during descent and approach (NTSB, 1987). The takeoff and initial climb phases are over-represented in crashes involving both general aviation and air carrier aircraft, accounting for 2% of air carrier flight time but 21.4% of crashes. Crashes in these phases account for almost the same proportion of general aviation crashes (23.6%). Finally, crashes during the cruise phase are much more characteristic of general aviation (16.3%) than air carrier operations (8%). In the latter case, there is evidence that with 60% of average flight time spent in cruise, there is significantly less risk associated with the cruise phase of flight. Beringer (Chapter 10) discusses these issues further and the distribution of crashes by flight phase for air carrier (U.S. Part 121 operations) can be seen in Figure 10.2.

Summary

It is perhaps somewhat surprising that the aviation data do not reflect the overwhelming over-representation of young males as is found in the road vehicle statistics. As Evans (1991) puts it: "The over-involvement of young, and male, road users is one of the largest and most consistently observed phenomena in traffic throughout the world" (Evans, 1991, p. 41). In many Western nations, access to motor vehicles is relatively easy with fairly minimal entry requirements. Automobile use is also bound up with certain aspects of male youth culture. In contrast, access to aviation is more restricted and the entry requirements are harder to surmount. These differences may explain, in part at least, the lack of a clear youth effect in aviation crash risk.

There is some evidence to suggest however, that youthfulness in combination with other factors may be a risk factor. Booze (1977) identified the combination of 'low age, high exposure' as representing a four-fold increase in risk when looked at over total flight times, and almost a three-fold increase in risk when considering recent flight experience during the preceding 6 months. From the evidence reviewed above, the person at greatest risk of accident involvement would be a younger, full-time commercial pilot, currently flying between 30 and 90 hours every 90 days. If that person has a history of previous accidents or violations (including automobile violations for DWI) then the risk increases again. The risk is also significantly higher in certain geographical areas, most notably Alaska. As mentioned previously, it is not presently possible to determine if the increased risk is generated by personal characteristics or whether such pilots are likely to be flying more hazardous operations (single pilot, less well-equipped aircraft, more challenging environments) or are located in less well-established organisations with fewer defences against economic and operational pressures.

Personal characteristics certainly play a part. O'Hare (1990) found that younger pilots were more likely to embark on a VFR flight in marginal weather conditions. These tended to be the younger pilots with relatively high recent and total flight hours. Behaviour in this simulated flight seems to have mirrored the findings from epidemiologic investigation rather closely. These pilots also rated themselves as more likely to take risks and had higher scores on a measure of 'personal invulnerability' to the most likely causes of accidents. Such pilots are aware of the overall risks, but are prepared to take more risky decisions and are more likely to distance themselves from the usual causes of aircraft accidents. Wichman and Ball (1983) found that increasing experience was associated with a stronger tendency towards self-serving biases in judgement.

This view has implications for pilot training. Providing information about risks and hazards will not change the behaviour of these pilots. Changes need to be made to their willingness to undertake additional risk by providing incentives for less risky behaviour and disincentives for more risky behaviour. This might involve changes in some or all of the following areas: regulation and licensing, monitoring and supervision, social norms, and organisational practices. Future research should move beyond investigations of bivariate relationships (e.g., age and accident risk) to the development of more detailed models of accident risk incorporating the personal, social, and organisational aspects discussed.

Risk Factors for Injury in Aviation Accidents

The focus of attention in this section is on those factors which determine the injury outcome of a crash. As I pointed out in the introduction to this chapter, significant gains in aviation safety may be made by reducing the losses incurred during the in-crash and post-crash phases of an accident. Whilst there have been some detailed studies of passenger injuries sustained in air transport crashes, including the injuries in the Kegworth B737-400 crash in the U.K (Air Accidents Investigation Branch, 1989), the Continental Airlines DC-9 crash at Denver (Lillehei & Robinson, 1994), and the Avianca B707 crash on Long Island, New York, (Barancik, Kramer, Thode, Kahn, Greensher, & Schechter, 1992), there have been few well-conducted epidemiologic investigations of air crash injury across the board.

In addition to the individual case studies noted above, the main sources of information on aircrash injuries are the studies conducted at the Johns Hopkins University (e.g., Krebs, Li, & Baker, 1995; Li & Baker, 1993) and a recent nationwide study of air crash injuries in New Zealand (O'Hare, Chalmers, & Bagnall, 1996; O'Hare, Chalmers & Scuffham, in preparation). Information on risk factors for injury outcome is organised under the same headings (pilot, aircraft, environment, and operational factors) as the information on risk factors for air crashes discussed previously.

One of the main forms of analytical studies in epidemiology is the case-control study. In this approach, a group of individuals with some characteristic in common (often this is the presence of some disease, such as lung cancer) are compared with another group of individuals who do not share this characteristic (in this case, people who do not have lung cancer). Obviously, the selection of the comparison or control individuals is critical in this kind of study. The desired information must be available for both the 'case' group and the 'control' group. It was this kind of study which initially led to the identification of cigarette smoking as a key risk factor in the development of lung cancer (Friedman, 1980). There have been very few case-control studies carried out in aviation (Li, 1994). The information discussed below comes from the case-control studies of crashes involving commuter and air taxi pilots in the U.S. (Krebs, Li, & Baker, 1995; Li & Baker, 1993) and a case-control study of air crash injury in all sectors of New Zealand aviation (O'Hare, et al., in preparation).

Two kinds of explicit comparisons have been made between 'cases' and 'controls'. In the first case, these were designed to identify the risk factors for fatal injury, where 'cases' refer to pilots-in-command who were fatally injured in a crash and 'controls' refer to pilots-in-command who survived the crash. In the second case, to identify the risk factors for injury of any kind, a comparison was made between pilots-in-command who were injured in some form (including fatally) and pilots-in-command who were uninjured. Simple comparisons can be made between the proportion of pilots with a particular characteristic (e.g., male, not wearing a seat belt, flying a helicopter etc.) who survived versus the proportion without those characteristics (e.g., were not male, were wearing a seat belt, were not flying a helicopter etc.) who survived. A more sophisticated technique known as logistic regression can then be used to calculate the odds ratios for each characteristic or variable. The odds ratio, which can be statistically adjusted to take into account the influence of other variables, provides an overall estimate of the size of the risk associated with a particular characteristic or variable.

Pilot Factors

Age In the Avianca crash (Berencik et al., 1992) survival rates decreased with age from an 88% survival rate amongst those aged 15 and under to a 0% survival rate amongst those aged 60 and over. Krebs et al. (1995) reported that pilots who were fatally injured in helicopter commuter and air taxi crashes were slightly older (average age of 41 years) than pilots who survived (average age 38 years), but this difference was not statistically significant. Of the 104 people killed in New Zealand air crashes between 1988 and 1992, 17% were aged 50 or over and 53% were between 30 and 49 (O'Hare et al., 1996). A case-control study using only the pilots-in-command (O'Hare, et al., in preparation) found no difference between the age distribution of pilots who were fatally injured and those who were not. Again, the proportion aged 50 and over was 18% and 17% respectively. Similar findings were reported for the overall risk of injury.

Occupation Full-time pilots were more likely to be fatally injured than pilots who had other occupations (O'Hare, et al., in preparation). The proportion of pilots who were fatally injured ('cases') who were full-time pilots was 63% whereas only 42% of controls fell into this category. When other variables such as post-crash fire, and aircraft category were controlled for, the odds of a full-time pilot receiving fatal injuries were found to be 2.4 times higher than for non full-time pilots.

Total flight hours and recent flight hours There is no evidence from any of the studies that injury outcome varies with any aspect of flight experience.

Gender Similarly, there is no evidence that fatality rates or non-fatal injury rates differ between males and females.

Restraint use Li and Baker (1993) reported that pilots in fixed-wing commuter and air taxi crashes who were not using 'shoulder restraints' were three times more likely to be killed than those who were. It is not clear whether 'shoulder restraints' refers only to a full shoulder harness, or whether it includes the 'diagonal strap' commonly used in light single-engine aircraft. In the tables of Odds Ratios reported by Li and Baker (1993) and by Krebs et al. (1995), the term 'shoulder harness' is used with Odds Ratios of 9.2 for helicopter crash survival and 3.7 for fixed-wing crash survival. Both studies reported very low rates of non seat-belt wearing. This was also the case in the New Zealand study (O'Hare et al., in preparation) with only 2.5% of accident-involved pilots not wearing a belt. Non-belt wearing was a risk factor for fatal injury (Odds Ratio = 4.36).

Helmet wearing Protective helmets are worn by pilots in a variety of circumstances, most commonly when undertaking agricultural and aerial work. Pilots of sport aircraft, particularly microlights and homebuilt/experimental aircraft may also wear helmets. In the New Zealand study (O'Hare et al., in preparation), the overall rate of helmet wearing amongst pilots involved in a crash was almost 50%. An unexpected finding was that whilst 16% of pilots who were wearing a helmet were killed, only 10% of those not wearing a helmet were fatally injured. The Odds Ratio of 0.6 indicates that **not** wearing a helmet is a protective factor against the risk of fatal injury! Similar results were obtained for overall risk of injury (Odds Ratio = 0.48).

In interpreting such a finding there are two points to be borne in mind. The first is that pilots who are wearing a helmet may be doing so because they are engaged in much more hazardous operations than those who are not using a helmet. The finding of decreased risk for non helmet-wearing may simply reflect these operational differences. The second possibility is that wearing a protective device such as a helmet may encourage pilots to undertake more risky manoeuvres in the belief that they are well protected from any adverse consequences. This is consistent with the 'risk homeostasis' theory (Wilde, 1982) which suggests that people act to maintain a constant level of risk in their environment, so that the provision of improved safety devices (e.g., better road surfaces, ABS braking systems) may actually lead to lower safety levels.

Aircraft Factors

Size and performance Using fixed-wing aeroplanes as the comparison, pilots of experimental/homebuilt aircraft are more likely to suffer fatal injury. After adjusting for post-crash fire and the location of the crash, experimental/homebuilt aircraft were nearly three times as risky (Odds Ratio = 2.83). In looking at overall risk of injury, microlight aircraft were three times as risky as fixed-wing aircraft (Odds Ratio = 3.07) and helicopters nearly twice as risky (Odds Ratio = 1.86). Li and Baker (1993) also reported an Odds Ratio of around 2 for helicopters in comparison to fixed-wing aircraft. There seems to be a tendency for glider/sailplane pilots to be at less risk of fatal injury and at less risk of overall injury, although the latter was not statistically significant (O'Hare, et al., in preparation).

Pilots-in-command of twin-engine aircraft are more likely to be fatally injured in a crash. Although only 4% of 'controls' were in command of a twin, 12% of 'cases' were. After adjusting for post-crash fire, crash location, and aircraft category, pilots of twins are three times more likely to be killed (Odds Ratio = 3.04). This confirms the findings noted previously, that twin-engine aircraft have a higher fatal accident rate than single-engine aircraft. The findings for overall risk of injury were not statistically significant (O'Hare et al., in preparation).

Environmental Factors

Location of crash In their study of pilot survival in fixed-wing commuter and air taxi crashes (Li & Baker, 1993) fatal injuries were much more likely to occur when the crash took place 'off airport'. Crashes which occurred 'on airport' were only one-fifth as likely to be fatal as these 'off airport' crashes. In our study of

New Zealand crashes (O'Hare et al., in preparation) we also found crashes 'off airport' to be much more likely to result in fatal injuries (Odds Ratio = 9.42).

We were able to obtain more detailed information on the circumstances of each crash from the files of the Civil Aviation Authority (CAA) of New Zealand. and the Transport Accident Investigation Commission (TAIC) of New Zealand. Every aircraft accident in New Zealand must be reported to the CAA who may conduct an investigation into the circumstances of the accident. Some of these accidents (normally the more serious ones involving passenger transport aircraft) are also investigated by the TAIC. We were able to obtain information on a variety of environmental factors which might contribute to the injury outcome of an aircraft crash. The variables shown in Table 11.1 all had significant odds ratios for both the risk of fatal injury and the overall risk of injury.

Table 11.1   Environmental factors which increase the chances of injury in an air crash

Variable	Definition	Odds Ratios (Fatal/Overall)
Off Airport	Outside the boundaries of an airfield	9.42 / 5.44
Water	Sea, lake, swamp	3.52 / 5.25
Steep	Incline > 20 degrees	3.34 / 3.96
Bush	Crash into trees/forest	4.21 / 4.69
Altitude	Crash site above 2000ft amsl	4.27 / 3.06
Wind	Strong wind at time of crash	4.28 / 4.15
Rain	Rain at time of crash	2.59 / 2.03

Source: O'Hare, Chalmers, and Scuffham (In preparation)

Li and Baker (1993) also found that the risk of fatal injury was approximately four times greater when the crash occurred under actual Instrument Meteorological Conditions (IMC) compared to Visual Meteorological Conditions (VMC). Our data showed a similar tendency, with 29% of flights crashing in IMC resulting in a fatality compared to 7.5% for crashes in VMC. Flights that were operating under IFR had an 18.75% fatality rate compared to 7.5% for VFR flights.

Unfortunately, these data do not allow us to conclusively determine if the factors identified above contribute to the accident severity (in-crash phase) or whether they act at the post-crash stage to reduce the chances of surviving a crash. Factors which were not significantly associated with the risk of either fatal or overall injury included darkness, time of year, cloud, visibility or temperature.

Since a number of these factors would be expected to reduce the chances of post-crash survivability, we might hypothesise that the factors identified as affecting the likelihood of surviving a crash are most likely to do so at the in-crash phase. That is to say, crashing on a steeply sloping, high altitude location, or in rainy and windy conditions probably results in a more severe crash impact, and thus greater likelihood of injury. Undoubtedly, these same factors must also reduce the likelihood of subsequent survival.

Post-crash fire There is little doubt that the single most important factor in determining pilot survival in a crash is the presence of a post-crash fire. Li and Baker (1993) reported an Odds Ratio of 8.2 for survival in fixed-wing commuter and air taxi crashes. Krebs, Li and Baker (1995) reported an Odds Ratio of 20 in their study of helicopter commuter and air taxi crashes. In both studies, post-crash fire was the most important determinant of pilot mortality. Post-crash fire was also one of the most significant risk factors for pilot survival in the New Zealand study (O'Hare et al., in preparation) with an Odds Ratio of 16.58 for the risk of fatal injury. Whilst steps have been taken to reduce the risk of post-crash fires in automobiles and military helicopters (Li & Baker, 1993), much more could be done for civilian aircraft, particularly those of the single and light twin-engined varieties. One possibility involves the use of polyurethane reticulated foam in GA aircraft fuel tanks as a means of preventing rupture and fire (AvFlash, November 23 1998). The high survival rate in glider/sailplane crashes is probably connected with the absence of post-crash fire risk in non-motorised gliders and sailplanes.

Operational Factors

Phase of flight Although the number of crashes which occur when aircraft are performing aerobatic manoeuvres is not great, they have a very high likelihood of being fatal. In our New Zealand study, the proportion of aerobatic crashes resulting in a fatal injury was 71% with an Odds Ratio for 'aerobatic phase' of 46.88. Both the cruise and circuit phases of flight were risk factors for fatal injury (Odds Ratios of 5.63 and 3.65 respectively) whilst crashes during the landing phase were at significantly less risk of fatal injury (Odds Ratio = 0.14). Exactly the same pattern of findings was evident when looking at the overall risk of injury (O'Hare et al., in preparation).

Instruction An unexpected finding in the New Zealand study was that pilots who were 'under instruction' at the time of the crash were more likely to receive a fatal injury. Whilst 22% of pilots under instruction were fatally injured, only 8% of those not under instruction received fatal injuries. The Odds Ratio was 3.14 (O'Hare et al., in preparation). The same was found for overall risk of injury, with an Odds Ratio of 2.25. There is no obvious explanation for this finding.

Presence of others A pilot-in-command is twice as likely to suffer a fatal injury when there are other people in the aircraft. In the New Zealand study, the fatal rate for solo pilots was 5.7% compared to 10% when there were others on board (Odds Ratio 1.84). Much the same was true for overall risk of injury (Odds Ratio = 1.93). Similar findings have been reported in connection with the risk of fatal car crashes involving younger drivers (Doherty, Andrey, & MacGregor, 1998; Preusser, Ferguson, & Williams, 1998).

There are several possible explanations for such a finding. The presence of others on board may encourage greater risk-taking by the pilot-in-command in response to actual or perceived pressures from those present, thus leading to more severe types of crashes. This is the most likely explanation for the effects of passengers on teenage car drivers (Preusser et al., 1998). Alternatively, the hazard may be primarily at the in-crash stage due to the effects of an increased number of occupants being flung about in the enclosed confines of an aircraft cabin.

Summary

The research described above presents a fairly consistent picture of the risk factors for injury in aircrashes with similar findings from studies of U.S. commuter aircraft and a wide range of New Zealand aircrashes. Clearly, the location of the crash—off airport, particularly into water or steep, mountainous terrain, and the occurrence of post-crash fire are key factors in determining the survivability and injury outcome of a crash. Failing to use a safety. belt, and crashing whilst performing aerobatics or during the circuit or cruise phase of flight are also likely to lead to more significant injury. A crash in a twin-engined aircraft is more likely to be fatal. Two factors have been shown to be risk factors for both the likelihood of a crash and the injury outcome of a crash: one is being a full-time pilot, and the other is the nature of the aircraft.

Microlights and experimental/homebuilt aircraft have higher than average chances of crashing and are then also associated with higher than average chances of injury. The glider/sailplane category presents an interesting case, since they have an above average risk of crashing but a below average risk of injury. The much reduced chance of post-crash fire (possible only in motorised sailplanes) and the semi-reclining seating position used in many designs may both be contributory factors to the good injury risk of these aircraft. This example shows that safety is not simply about accident reduction. Safety levels in microlights and in experimental/homebuilt aircraft could be enhanced either by reducing the risks of accident involvement, or by reducing the risks of occupant injury, or both. In aviation, as in other systems, a multi-faceted intervention strategy is likely to be more effective than measures aimed exclusively at one of the three phases of loss (pre-crash, in-crash, post-crash).

Operational Implications

It is generally accepted that good decisions require good information. Since aviation is all about risk management then the information outlined above should be valuable in making informed decisions about risks. With its culture of individual responsibility, the faulty judgement and decisions of those at the 'sharp-end' tend to attract a disproportionate amount of attention. Whilst the pilot must bear responsibility for the management of certain risks, the control of many of the risks discussed above is in the hands of others—aircraft designers, government regulators, aircraft operators and so forth.

Three significant risk factors for death and injury in aircraft crashes are the category and type of aircraft, the crash site, and the occurrence of post-crash fire. As in automobile crashes, larger heavier aircraft offer better levels of performance and enhanced occupant protection. Where pilots are able to exercise some choice in what they fly, then they should use the risk information provided as part of their decision making. The tendency for some categories and types of aircraft to experience certain kinds of accidents should be carefully noted and steps taken to counteract these possibilities.

Pilots cannot control the landscape beneath them, but they may have some discretion over the routes they fly. The significantly increased chances of death and injury due to crashing into rugged terrain should be borne in mind when planning a route. If circumstances permit, it might be better to choose a route over less inhospitable terrain, viewing the added distance and time as a form of insurance policy against disaster. The human mind takes more kindly to the prospect of a certain loss or inconvenience when this can be viewed as insurance against a greater loss.

Safety in all areas of aviation would be increased if there were less chances of fire occurring during or after a crash. Fire has been shown again and again to be one of the most significant risk factors in aviation. Crashworthy fuel systems that would all but eliminate this risk are perfectly feasible (Baker et al., 1992). Pilots and operators can do little about this, but manufacturers and regulators can take steps to modify this risk factor. Zotov (Chapter 12) also emphasises this point.

Research Agenda

There are many opportunities for increasing our knowledge of the risk factors for accidents and injury in aviation. As noted in the introduction, this is an area which has not been much explored by epidemiologists and public health researchers despite very high rates of death and injury in some areas of aviation. Some opportunities for further research have been noted above, for example it would be of interest to determine if other countries have the same kind of regional variations in accident risk which have been found in the U.S. It is important however, to move beyond descriptive differences to establish the underlying reasons for variation in accident risk due to age, flight experience etc. Insufficient attention has been given to the social, environmental and organisational constraints within which pilots operate. The main limiting factor in drawing conclusions about risk factors for crashes is the absence of good data on flights which don't crash! What is needed is a large, well designed case-control study involving the collection of extensive information from flights which have not been involved in a crash. Such studies have been carried out in connection with automobile driving, but not as yet in the field of aviation.

Acknowledgements

The studies of New Zealand pilots were carried out in collaboration with Dr David Chalmers of the Injury Prevention Research Unit of the University of Otago. The study of risk factors for air crashes was supported by the Civil Aviation

Authority of New Zealand and the Injury Prevention Research Unit. I am very grateful for the support and cooperation of the Civil Aviation Authority of New Zealand and wish to thank the Director, Mr Kevin Ward, and Mr Richard White in particular. The case-control study of air crash injuries was supported by the Health Research Council of New Zealand. I am extremely grateful to Paul Scuffham for all his hard work. I would like to acknowledge the assistance we received from both the Civil Aviation Authority of New Zealand and from Mr Ron Chippindale, the Chief Inspector of Air Accidents at the Transport Accident Investigation Commission. I would like to thank Dr Ross St. George for his helpful comments on the chapter.

References

Air Accidents Investigation Branch (1989). *Report on the accident to Boeing 737-400 G-OBME near Kegworth, Leicestershire on 8 January 1989. Aircraft accident report 4/90.* London: HMSO.

AvFlash (1998, November 23). The EAA initiates a firefight (AvFlash@avweb.com).

Aviation Co-Operating Underwriters Ltd. (1997). How to stop crashing. *Insurance Flight Report, No 7, 1.*

Baker, S. P., & Lamb, M. W. (1989). Hazards of mountain flying: crashes in the Colorado Rockies. *Aviation, Space, and Environmental Medicine, 60,* 531-536.

Baker, S. P., lamb, M. W., Li, G., & Dodd, R. S. (1993). Human factors in crashes of commuter airplanes. *Aviation, Space, and Environmental Medicine, 64,* 63-68.

Baker, S. P., O'Neill, B., Ginsberg, M. J., & Li, G. (1992). *The injury fact book* (2nd ed.). New York: Oxford University Press.

Barancik, J. I., Kramer, C. F., Thode, H. C., Kahn, C. J., Greensher, J., & Schechter, S. (1992). *Epidemiology of fatal and non-fatal injuries in the Avianca plane crash, Avianca flight 052 – January 25, 1990. Department of Applied Science, Brookhaven National Laboratory.* Springfield, VA: National Technical Information Service.

Berlin, J. L., Gruber, E. V., Jensen, P. K., Holmes, C. W., Lau, J. R., Mills, J. W., & O'Kane, J. M. (1982). *Pilot judgment training and evaluation: Volume 1.* (Report No. DOT/FAA/CT-82/56.) Atlantic City, NJ: FAA Technical Center.

Booze, C. F. (1977). Epidemiologic investigation of occupation, age, and exposure in general aviation accidents. *Aviation, Space, and Environmental Medicine, 48,* 1081-1091.

Bruckart, J. E. (1992). Analysis of changes in the pilot population and general aviation accidents. *Aviation, Space, and Environmental Medicine, 63,* 75-79.

Civil Aviation Authority. (1988). *CAP542 – General aviation accident review 1987.* Cheltenham: Civil Aviation Authority.

Civil Aviation Authority. (1997). *CAP667 – Review of general aviation fatal accidents 1985-1994.* Cheltenham: Civil Aviation Authority.

Cryer, P. C., & Fleming, C. (1987). A review of work-related fatal injuries in New Zealand – numbers, rates and trends. *New Zealand Medical Journal, 100,* 1-6.

Doherty, S. T., Andrey, J. C., & MacGregor, C. (1998). The situational risks of young drivers: the influence of passengers, time of day and day of week on accident rates. *Accident Analysis and Prevention, 30,* 45-52.

Evans, L. (1991). *Traffic safety and the driver.* New York: Van Nostrand Reinhold

Friedman, G. D. (1980). *Primer of epidemiology.* New York: McGraw-Hill.

Guide, P. C., & Gibson, R. S. (1991). An analytical study of the effects of age and experience on flight safety. *Proceedings of the Human Factors Society 35th Annual Meeting* (pp. 180-184). Santa Monica, CA: Human Factors Society.

Haddon, W. (1980). Options for the prevention of motor vehicle crash injury. *Israeli Journal of Medical Science, 15,* 45-68.

Hall, C. (1991). Agricultural pilot safety in Australia: a survey. *Aviation, Space, and Environmental Medicine, 62,* 258-260.

Jensen, R. S. (1995*). Pilot judgment and crew resource management.* Aldershot: Ashgate.

Jensen, R. S., & Benel, R. A. (1977*). Judgment and evaluation and instruction in civil pilot training* (Report No. FAA-RD-78-24). Savoy, IL: University of Illinois, Aviation Research Laboratory.

Krebs, M. B., Li, G., & Baker, S. P. (1995). Factors related to pilot survival in helicopter commuter and air taxi crashes. *Aviation, Space, and Environmental Medicine, 66,* 99-103.

Lane, J. C. (1973). Safety in airline and general aviation. *Proceedings of the 10th Annual Conference of the Ergonomics Society of Australia and New Zealand* (pp. 1-16).

Li, G. (1994). Pilot-related factors in aircraft crashes: a review of epidemiologic studies. *Aviation, Space, and Environmental Medicine, 65,* 944-952.

Li, G., & Baker, S. P. (1993). Crashes of commuter aircraft and air taxis – what determines pilot survival? *Journal of Occupational Medicine, 35,* 1244-1249.

Li, G., Hoote, E. G., Baker, S. P., & Butts, J. D. (1998). Alcohol in aviation-related fatalities: North Carolina, 1985-1994. *Aviation, Space, and Environmental Medicine, 69,* 755-760.

Lillehei, K. O., & Robinson, M. N. (1994). A critical analysis of the fatal injuries resulting from the Continental flight 1713 airline disaster: evidence in favor of improved passenger restraint systems. *The Journal of Trauma, 37,* 826-830.

Lubner, M. E., Markowitz, J. S., & Isherwood, D. A. (1991). Rates and risk factors for accidents and incidents versus violations for U.S. airmen. *The International Journal of Aviation Psychology, 1,* 231-243.

McFadden, K. L. (1996). Comparing pilot-error accident rates of male and female airline pilots. *Omega, International Journal of Management Science, 24,* 443-450.

McFadden, K. L. (1997). Policy improvements for prevention of alcohol misuse by airline pilots. *Human Factors, 39,* 1-8.

Modell, J. G., & Mountz, J. M. (1990). Drinking and flying – the problem of alcohol use by pilots. *New England Journal of Medicine, 323,* 455-461.

Mohler, S. R. (1969). Aircraft accidents by older persons. *Aerospace Medicine, 40,* 554-556.

Nagel, D. C. (1988). Human error in aviation operations. In E. L. Wiener & D. C. Nagel (Eds.), *Human factors in aviation* (pp. 263-303). San Diego: Academic Press.

National Transportation Safety Board (1979). *Aircraft accident report – Japan Air Lines Co., Ltd. McDonnell-Douglas DC-8-62F, JA8054, Anchorage, Alaska, January 13 1977* (Report NTSB-AAR-78-7). Springfield, VA: National Technical Information Service.

National Transportation Safety Board (1987). *Annual review of aircraft accident data U.S. general aviation calendar year 1983* (NTSB/ARG-87/01). Washington, DC: National Transportation Safety Board.

O'Hare, D. (1990). Pilots' perceptions of risks and hazards in general aviation. *Aviation, Space, and Environmental Medicine, 61,* 599-603.

O'Hare, D., & Chalmers, D. (1999). The incidence of incidents: a nationwide study of flight experience and exposure to accidents and incidents. *The International Journal of Aviation Psychology, 9,* 1-18.

O'Hare, D., Chalmers, D., & Bagnall, P. (1996). *A preliminary study of risk factors for fatal and non-fatal injuries in New Zealand aircraft accidents. Final report to the Civil Aviation Authority of New Zealand.* Dunedin: Department of Psychology and Injury Prevention Research Unit, University of Otago.

O'Hare, D., Chalmers, D., & Scuffham, P. (in preparation). *A case-control study of risk factors for injury in New Zealand aircrashes.*

Olsen, S. O., & Rasmussen, J. (1989). The reflective expert and the prenovice: notes on skill-, rule-, and knowledge-based performance in the setting of instruction and training. In L. Bainbridge & S. A. R. Quintilla (Eds.), *Developing skills with information technology* (pp. 9-33). London: Wiley.

Preusser, D. F., Ferguson, S. A., & Williams, A. F. (1998). The effect of teenage passengers on the fatal crash risk of teenage drivers. *Accident Analysis and Prevention, 30,* 217-222.

Richter, E. D., Gordon, M., Halamish, M., & Gribetz, B. (1981). Death and injury in aerial spraying: pre-crash, crash, and post-crash prevention strategies. *Aviation, Space, and Environmental Medicine, 52,* 53-56.

Salvatore, S., Stearns, M. D., Huntley, M. S., & Mengert, P. (1986). Air transport pilot involvement in general aviation accidents. *Ergonomics, 29,* 1455-1467.

Vail, G. J., & Ekman, L. G. (1986). Pilot-error accidents: male vs female. *Applied Ergonomics, 17.4,* 297-303.

Wichman, H., & Ball, J. (1983). Locus of control, self-serving biases, and attitudes towards safety in general aviation pilots. *Aviation, Space, and Environmental Medicine, 54,* 507-510.

Wilde, G. J. S. (1982). The theory of risk homeostasis: implications for safety and health. *Risk Analysis, 2,* 209-225.

Yacavone, D.W., Borowsky, M. S., Bason, R., & Alkov, R. A. (1992). Flight experience and the likelihood of U.S. Navy aircraft mishaps. *Aviation, Space, and Environmental Medicine, 63,* 72-74.

12 The Role of Accident Investigation in General Aviation

Dmitri Zotov

Why Investigate General Aviation Accidents?

Of course, we all know that accident investigation is a 'Good Thing'. By investigating accidents we are able to reduce their numbers in the future, together with the associated pain and suffering. However, in these days of financial stringency and 'user pays', a rather more pragmatic approach is needed to persuade governments to finance General Aviation accident investigation.

One of the most persuasive arguments is the actuarial cost of accidents to the community. This cost includes such items as the cost of educating the victims, which cannot now produce a return, the tax which is no longer collectable from them, the cost of rescue services and hospital treatment for the survivors, and so on. The figure adopted in New Zealand is $2 million per fatality; in the U.S. it is $2.7 million. If we can prevent just one or two fatalities per year, we shall more than cover the costs of investigation.

There is a third reason, which is just as compelling. Major air transport accidents are a fact of life, and when they occur, the Government expects that they will be investigated thoroughly, if only to allay public concern. The investigation of minor accidents provides the essential training which will enable investigators to handle the 'big one'. The investigation of a General Aviation accident is the investigation of an airline disaster writ small: similar techniques are used to investigate the mid-air break-up of a homebuilt aircraft as were used to investigate the loss of the Pan American Boeing 747 at Lockerbie, for example.

There is still an inclination, on the part of those who hold the purse strings, to suggest that those who incur the risks of pleasure or sporting activity should bear the costs of investigation. The difficulty of this approach is that the various sporting bodies—gliding, microlight, hang-gliding and so on—are essentially amateur. They lack the financial resources to meet the expenses which can be involved, and even if their own members had the expertise to conduct the investigation and analyse the evidence, they are most unlikely to have the time to do so.

The costs of the on-site phase of the investigation can mount rapidly if work in mountainous terrain or underwater recovery is involved, and people are seldom kind enough to have their accidents in accessible places. One underwater recovery of a light helicopter cost NZ$100,000—not the sort of money that the average aeroclub could find—and that was just to get to the stage where the investigation could start. And the time taken to investigate and analyse the evidence typically runs into months of full-time work

Investigating the Accident

Any investigation will involve collecting the available information, and analysing that information to understand how the accident came about. In turn, collecting the information can be separated into on-site work, and subsequent detailed investigation. It is often said that all the information must be gathered before the investigator starts to analyse it. This is fine for getting the news media off our backs, and avoids committing ourselves to tentative explanations which we may subsequently disprove. However, the reality is that we start to formulate possible explanations as the data comes to hand; this may direct our attention to various lines of inquiry which will either tend to confirm, or may eliminate, those explanations. Naturally, we must record all the 'volatile' information which will disappear if we do not gather it right away, so this is the first priority.

Let us look at an actual investigation (Transport Accident Investigation Commission, 1990), to see how the various parts come together. The accident involved a light homebuilt aircraft, which went missing on a flight over rugged bush-clad terrain in the central North Island of New Zealand. The weather at the time had been severe, with thunderstorms reported in the area. Two days later, a pile of wreckage was spotted by a searching helicopter. The investigators were flown to the site by helicopter, and on their way in were shown a piece of white fabric lodged in the top of a tree, some two miles away.

The first task on arrival at the site was to liase with the Police who were guarding it, to ensure that visitors did not unwittingly destroy evidence. It is easy enough to do so, if you do not understand the possible importance of scuffmarks in the ground, or seemingly unimportant fragments of wreckage. The Search and Rescue unit had set up base in a large caravan nearby, and some 50 forestry workers and SAR personnel were on hand. The distant piece of fabric suggested that the aircraft might have broken up in flight, so there was the possibility that a ground search might be needed: if so, the availability of the searchers could save much time.

At any accident site, the first thing to do is nothing. The investigators will take stock of the scene, wandering about and starting to take photographs to document it before anything is disturbed. If the terrain permits (and it did in this case) it is conventional to photograph the wreckage from eight points around it, every 45 degrees. Not only does this process give the investigator time to gather his thoughts, but also the pictures can sometimes contain information whose importance is not realised until later.

The next thing is to ensure that all the aircraft is there. If both wingtips, the spinner and the empennage are all there, it is reasonable to suppose, for a first look, that the parts in the middle are there too. In this case, even at a cursory inspection there were no wingtips, and no empennage. Since the aircraft was predominantly white, the piece of white fabric became important. The helicopter was dispatched to retrieve it, and to note the exact co-ordinates for future reference. Meanwhile, further examination showed that large parts of the aircraft were missing, and what remained had struck the ground in a vertical dive.

While the wreckage examination was in progress, the helicopter returned. Not only had it recovered the fabric, but also the fin and rudder, which by good fortune had landed on a forestry road and so been visible from the air. It was on a direct line between the piece of fabric and the main wreckage site. A ground search was set up along this axis, with the object of recovering as much of the wreckage as possible. A grid pattern was marked out on large-scale maps, and a system of annotation was devised so that each item could be identified by its location on the map. The searchers were split into teams of five, equipped with hand-held radios, and instructed to space themselves about 5 metres between individuals; each group was to follow a path parallel to the search axis. For the nearest part of the search the terrain was undulating, with recently planted pines. Further on the searchers entered a mature pine forest, with trees about 75 feet tall; further still the pines were bounded by native beech forest, which might present difficulties.

Work on site continued, with successive layers of wreckage being peeled away and laid out for inspection. Before each layer was moved the whole assembly was again photographed, and features of the wreckage were documented and photographed as they were revealed—downward failures of the wing spars, flogging of wing fabric, the positions of switches and controls, and so on. The bodies of the occupants were removed, and it was then necessary for aviation pathology to be arranged and the bodies to be flown to a suitable venue.

Aircrew who die in an accident are always examined by pathologists specially qualified in aviation pathology, because important data might be overlooked in normal procedures. For example, routine pathology would not require X-rays of the hands and feet, but these can show who was holding the controls at impact, and this can be important. Likewise, if we find that the pilot was likely to have been dead before impact, this can save a lot of effort in trying to explain the accident.

One of the first finds by the searchers was the complete tailplane and elevators. The loss of the tailplane almost always results in failure of the wings in downward bending: the tailplane is under a download in normal flight, and sudden release of this load lets the aircraft bunt violently, the negative G load causing the wings to break near the roots. The immediate question, therefore, was what caused the tailplane to come off? A preliminary inspection showed that the elevator trim tab had travelled beyond its normal limits, and had distorted the hinge; the paint on the tailplane fabric showed diagonal crazing. Both of these features are indicative of flutter, a destructive oscillation of the control surfaces something like a flag flapping in the wind. A look at the rudder showed that it, too, had overtravelled, with strike marks on the bump stops, and much damage to the internal structure.

The damage to the root of the fin showed that it had been broken while being bent to the left, perhaps under the force from the fluttering rudder, and the load from the fin and rudder might have torn off the tailplane. This line of inquiry looked promising, but needed to be followed up in a workshop. In the meantime the search continued, as did the documentation of the wreckage and ground marks.

The instrument panel did not reveal much of interest, but the two fuel tanks had different characteristics. The main tank showed characteristic hydraulic bursting, while the auxiliary tank was little damaged, indicating that it was empty at the time. The main tank was in use at the time of the accident, but might a crisis have arisen if the auxiliary tank had run dry and the engine stopped for a time during the changeover? The fuel load at take off would have to be ascertained, as would the pilot's fuel management practices. The propeller had been rotating at impact, but there was no real indication that the engine had been delivering power.

As is usual, the engine was removed from the wreckage for strip inspection, and the auxiliary components were examined. One peculiarity was that the control to the carburettor hot air selector (which can protect against carburettor icing) appeared to have failed before impact. The implication of this was uncertain, but this assembly was set aside for workshop examination.

The ground search went on for two days. Bits as small as an inch square were recovered, the lighter pieces being further along the wreckage trail from the main wreckage. This is characteristic of an in-flight break-up at a considerable altitude, the lightest pieces being borne furthest by the wind. The search had to be terminated at the beech forest, which proved virtually impenetrable, so we had to make do with what we had.

While all this was going on, other routine matters were being attended to, some in the nearby town, and some by colleagues at Head Office. Witnesses were sought and interviewed, meteorological records and aftercasts were obtained, ATC tapes impounded and played. Files held by the Civil Aviation Division were impounded—records of the aircraft's construction and maintenance, the pilot's file and medical record, a copy of the aircraft plans. The pilot's flying logbook was requested from his family. All of these would be reviewed subsequently, to build up a picture of aircraft and pilot which might help in understanding the accident.

Two sets of witnesses were particularly helpful: two shooters who had been sheltering a mile or so from the accident site; and the farmer with whom the pilot and his passenger had spent the night before the flight. The shooters had heard a burst of engine noise followed by a brief silence and then a thump. One of them was an ex-policeman, and noted both the bearing of the sound and the time. The bearing pointed to the main wreckage site, and the time was consistent with the time it would have taken the aircraft to fly to the point, from the time of take-off.

The farmer's information included the fuel state: the main tank was filled; the auxiliary was empty (so there was no question of fuel mismanagement). Also, the pilot had obtained the weather at his destination on the other side of the ranges, which was fine, but had been unable to get an official forecast. He had discussed flying the much longer route via the East Coast, but could see a large break in the cloud and planned to climb through it and fly above cloud on the direct route. (Flying below cloud on the direct route was unlikely to be feasible, because a moist unstable westerly airflow was causing a build-up of cloud on the mountain ranges between departure point and destination. Cloud base was likely to be below the mountaintops; in such conditions the cloud tops would typically be nine to ten thousand feet).

After take-off, the aircraft was seen to climb through the gap in the cloud, whose base was estimated to have been about 4,000 feet. What the investigators now knew was that the pilot had intended to fly above the cloud layer which covered the ranges between his departure point and destination; that the aircraft had broken up in the air following loss of the tailplane; and that there appeared to have been flutter of the elevators and rudder. Once medical incapacitation was ruled out by the aviation pathologist, the next question was, what caused the flutter?

To answer this, it was necessary to examine the wreckage in detail. The hundreds of pieces recovered from the wreckage trail, together with the principal wreckage, were moved to a workshop, and painstakingly reassembled. The jigsaw puzzle was greatly assisted by the construction drawings, because it was possible, by measuring the wood sizes, to establish whereabouts on the framework it had been used. For example, spruce strips ¾ inch square came from the fuselage main framework. The debris was sorted into pieces of the same wood size, and fractured pieces were then matched and bound together with tape so that the framework could be built up.

Figure 12.1   ZK-DAG reconstruction

As the framework was assembled, It became apparent that there was a lot wrong with the construction. Glue joints had failed in several different ways, none of which should have happened. A sound glue joint is stronger than the wood, and if a test joint is broken, at least 75% of the failure should be the wood, not the glue. Were these defective joints the cause of the flutter? It was a possible explanation; because failed joints will cause the structure to be more flexible than it should have been, and so prone to flutter at speeds lower than would normally be safe. However, in this case it was concluded that the fuselage joints failed in consequence of the flutter, rather than the other way around. The wings had evidently been essentially intact at the time the tailplane departed, because they had been able to generate the force required to fracture the main spars, so it was unlikely that flutter had started there. Attention was therefore focussed on the elevator trim tab, as being a likely source.

The tab had certainly been fluttering. When the elevators were opened up, it was found that the tab had generated sufficient force to fracture the spar on which it was mounted. The tab had driven the half of the elevator that it was attached to, and this had in turn driven the other half, via the carry-through spar. This had the potential to get the two halves out of phase with each other, putting a twisting load on the fuselage. Cracking of glue joints in the fuselage bracing would then allow twisting, which could set up sympathetic flutter in the rudder. That rudder flutter occurred was demonstrated by the two strikes on the rudder stops, and by evidence that the rudder had overtravelled and struck one elevator corner when the elevator was itself at full deflection. It was not surprising that the fin and rudder could generate sufficient force to twist the tailplane off the already weakened fuselage.

Why had the tab fluttered? It was intended to be connected to the lever in the cockpit by two thin piano wires in tension. One of these remained attached to the tailplane, but the other had come detached from its termination, and microscopic examination showed that there had been no grip between termination and cable for a long time, if ever. So the tab was controlled by a single length of thin piano wire. In effect, it was connected to the aircraft by a spring, and was almost guaranteed to flutter if the speed was sufficient.

It is a good principle to ask 'Why?' at least five times. So, why did the tab flutter? We knew that quite recently, the aircraft had been involved in an air race, which had involved flying as fast as was possible in level flight, in quite turbulent conditions. Clearly, it had not fluttered then. In the circumstances of the flight as they were intended to be, the aircraft should have been flying in smooth air above the clouds. Because the engine would have been delivering less power at altitude than it would at sea level, it should have been flying more slowly than it had in the air race. Was it possible that the aircraft had inadvertently entered cloud, and subsequent loss of control led to the high speed that set the flutter sequence in motion?

The pilot had intended to divert to the coast (where the weather had remained clear) if his first plan of a direct flight above cloud had not been feasible. Might he have been trapped in some way? If the aircraft was already at its operational ceiling close above the cloud tops, a partial power loss would cause it to subside into cloud. It was time to look at the engine in more detail.

The strip inspection disclosed nothing amiss, but the disconnected carburettor heat control cable provided the next clue. This was not the first time it had failed. The lever had become detached from the valve spindle previously, and had been repaired by placing it in position, and brazing over the joint. It was the braze metal, less than 1/16 inch of it, which had failed in torsion. Failure in the 'Hot air' position would deprive the engine of a significant amount of power. The time of the accident was consistent with the aircraft having climbed to its ceiling, and it would probably have needed to do so to cross the ranges above cloud. The explanation for the cable failure could have been that the pilot had applied hot air briefly, as a routine precaution, and the valve had stuck. A forceful attempt to move it had then caused the joint to fail.

Dissecting the hot air box, which contained the valve, showed the reason for the difficulty in moving the valve. The box was home made from sheet aluminium. This had been welded together, and excessive heat had caused the sides of the box to buckle. The valve plate was rubbing against the sides, and jamming before it quite reached the fully 'Hot' position—it had never been able to travel to full deflection. This was the explanation of the force which had twice broken the joint between lever and spindle. There was no question that the valve was in this position at impact, and that it was not deflected there during the impact sequence: a clear witness mark[1] showed that it was stationary at impact.

There was, however, another possible explanation for the aircraft entering cloud, namely that it had been trapped by a build-up of cloud and had been unable to return to clear air. The application of hot air might then have been an instinctive reaction to entering cloud, and the subsequent failure would then have been purely incidental. This was a real possibility, because it had been a day of quite exceptional weather. Several tornadoes had caused a great deal of damage in the central North Island, and it was quite possible that the pilot could have found cumulonimbus boiling up all around him. Could the Weather Office help? Unfortunately, not. Satellite technology has its limitations, and there had been a layer of cirrus over the area of the accident. There was no way of knowing whether the tops of the clouds were ten thousand feet or thirty thousand.

In the end, we were left with two feasible explanations. A cockpit voice recorder or flight data recorder might have been able to resolve the dilemma, but in their present form they are far too big and heavy—to say nothing of expensive—to be fitted to General Aviation aircraft. In this age of microminiaturisation anything seems to be possible, so perhaps we can look forward to developments in the future.

What did we gain from this Investigation?

Firstly, the whole thing was triggered off by a human error—the choice of a faulty strategy. Flying above cloud in a single engined aircraft, when plane and pilot are not fit for instrument flight, is a gamble on the continued working of the engine. Even if a gliding descent on instruments is possible, where the terrain is unsuitable for landing in the event of engine failure, the whole thing is not a good idea; if the cloud base is below the mountaintops, it is a recipe for disaster in the event of any malfunction. This is not to say that the pilot was at fault in making this decision. At the time, flying above cloud when the destination was clear was sometimes advocated, in preference to trying to get through below cloud. After this accident, and one more where much the same thing happened, this idea was seen to be unsound, and there have been no subsequent accidents from this cause. (This is not to say that we do not get accidents from pilots trying to get through below the weather, unfortunately.)

Secondly, the faulty glue joints came about from defective construction techniques. At the time of its construction, the surveyors who had inspected the aircraft wrote of it in glowing terms, advocating it as an example of how such construction should be done. Clearly, there was a need for education in this area. There were very few craftsmen left who had been trained in the fabrication of wooden aircraft, and the information was close to being lost. One of those few craftsmen was consulted during the investigation of the structural defects, and in order to make this information available to amateur constructors, a monograph was written. This was sent to the Amateur Aircraft Constructors Association, and was published as an Annex to the Report. The New Zealand Gliding Association asked for sufficient copies to send them to all their authorised repairers.

Better understanding of the risks involved in single-engine flight above cloud, and of the techniques required for the construction of wooden structures, should reduce the incidence of similar accidents in future.

One thing that should be evident is that, as the ICAO Accident Investigation Manual (International Civil Aviation Organization, 1970) points out, the investigation of a light aircraft accident must be done with the same thoroughness as is an airline disaster. An arbitrary deadline, or a 'once over lightly' approach is not an option. If the investigation is not followed through to completion, valid conclusions cannot be drawn, and the whole affair would be a waste of time and resources.

Human Errors in General Aviation Accidents

We have looked at an accident where the error that set the sequence in action was faulty strategy. This is rather a novel way to look at pilot errors. Until recently, most systems of classifying errors listed faulty performance of the task: 'Failed to maintain airspeed', or 'Did not avoid obstacle', or 'Misjudged altitude'. These classification schemes were very cumbersome, often comprising hundreds of categories. Indeed, they were so complex that coding errors were commonplace, and so the validity of the database as a whole was open to question. Besides, these classification schemes (or taxonomies) were really only lists of what happened. They were of little value in trying to decide what remedial action should be taken.

Rasmussen (1980, 1982) argued that what was needed was not a classification by tasks, but by the task as performed by the human. He devised such a scheme after studying errors made by operators of nuclear power plants. On first sight, you might think that this had little to do with aircraft accidents, but this work was adapted by O'Hare, Wiggins, Batt, and Morrison (1994), who devised a six-element taxonomy to classify pilot errors. (Figure 12.2). This taxonomy has been extensively tested (Wiegmann & Shappell, 1997; Zotov, 1997). Zotov (1997) found that every reported accident in the New Zealand database between 1965 and 1990, which had been subject to official investigation, was able to be classified by this system.

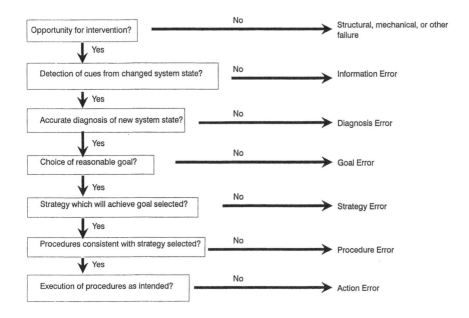

Figure 12.2   Taxonomic algorithm for pilot errors in aircraft accidents
Source: O'Hare et al., 1994

To classify the pilot errors which led to an accident, we start by examining whether the necessary information was available to the pilot, so that he could have taken corrective action. If not, there was nothing he could have done. However, information may have been available, but he may not have sought or detected it. For example, an airliner with a 'glass cockpit' display was being flown by a crew who were used to conventional instruments. A partial engine failure led to high vibration. The engine instruments correctly displayed which engine was vibrating, but the inconspicuous display was not noticed by the crew, who shut down the wrong engine. (Air Accident Investigation Branch, 1990). Such an error is coded as an Information Error.

Next, the crew may receive information, but their appreciation of the situation may be at variance with reality. A Boeing 737 had a major fire on take-off as a result of uncontained turbine disk failure (Air Accident Investigation Branch, 1988). The passengers, and everyone at the airport, saw the flame streaming behind the aircraft like the tail of a comet. The pilot's subsequent RT call:

"We appear to have a fire in the port engine"

has to have been the understatement of the year. Such faulty appreciation is classified as a Diagnosis Error.

Having correctly diagnosed the state of play, the pilot may have to choose between competing goals, for example, to divert to an alternate, or to hold until the weather improves. A frequent Goal Error in General Aviation is deciding that it will be possible to proceed towards the destination in the face of deteriorating weather.

There may be a number of ways to achieve a goal. These are termed 'Strategies'. For example, to achieve the goal of 'Diverting to an Alternate', the pilot might decide to climb only to minimum safe altitude, perhaps minimising traffic delays, or to climb until the descent path is intercepted, often the most economical way. The choice of strategy may be less than optimum.

Procedures are groups of tasks by means of which strategies are achieved. Thus, the strategy of diverting to an alternate may commence with an overshoot: raise the nose, apply climb power, raise the undercarriage, accelerate to flap raising speed and raise the flaps, and so on. Performing the wrong procedure, or the correct procedure wrongly, is classified as a Procedural Error.

Of course, any action in a procedure can be performed incorrectly. An example of such an Action Error would be raising the undercarriage instead of the flaps, at the end of the landing roll.

We can analyse what we might term the 'triggering' error: the first departure from accepted and prudent practice. Of course, there are likely to be other errors after the initial event, as the pressure builds up. Take as an example an accident to a motor-glider which suffered an engine failure at 400 feet in the circuit after take-off (Office of Air Accidents Investigation, 1986b). Prima facie, an engine failure at 400 feet in the circuit should not result in an accident. However, the pilot flew round the circuit with the engine extended, and at an excessive speed, which resulted in rapid loss of height. In pulling up over a tree on the approach, the pilot stalled the aircraft, and it dived into a swamp short of the airfield. Stalling the aircraft was an Action Error, but was preceded by the Procedural Errors of not retracting the engine, and flying with excessive speed.

But there was more to it than that. The engine stopped because the fuel was exhausted. The pilot was aware that the contents shown by the fuel gauge indicated a lower consumption during the previous flight than he had expected. It would have been possible (though not easy) to have dipped the tank before continuing the flight. Had the pilot been aware of the fuel state he could have topped up the fuel, so the engine would not have stopped from fuel exhaustion, and the pilot would not then have been put in the position which led to the subsequent errors. There was therefore an Information Error, in that the pilot did not seek available information which would have averted the accident.

When we look at the triggering errors that have set in motion fatal accidents to powered fixed-wing General Aviation aircraft, we find the distribution shown in

Figure 12.3 (helicopter accidents have a slightly different distribution—see Zotov, 1997). As you can see, by far the commonest triggering event is faulty goal selection. This is understandable. If you make a bad choice of goal, such as deciding to press on into adverse weather because you 'have to' get home, the pressure will build up as the weather gets worse. You will be less able to think clearly, you may make poor choices of strategies, you may perform procedures poorly, and your aircraft handling will suffer. The aircraft is likely to strike the terrain with considerable energy, with undesirable consequences. Remember the old saying:

"The superior pilot uses his superior airmanship to avoid situations requiring him to demonstrate his superior skills".

By contrast, if you make an action error—say, applying insufficient power to counter a wind-gradient on approach—the result is likely to be a red face, rather than a disaster.

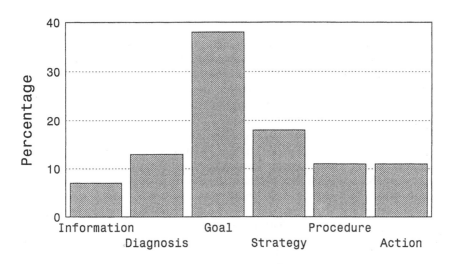

Error Type

Figure 12.3   Distribution of triggering errors for fatal light aircraft accidents
Source: O'Hare et al., 1994

How can we use this error taxonomy? It is helpful to accident investigators, because while we can reconstruct the wreckage to determine what failed mechanically, it is not helpful to reconstruct the pilot, except in some cases of medical incapacitation. Classifying the errors in this way helps us to understand what the pilot was trying to do, rather than what the investigator thinks the pilot ought to have been trying to do—not at all the same thing. It has been used for such diverse purposes as deciding which of two possible flight paths constructed from the witness observations was the more likely, and determining if the workload built up so much that the pilot became overloaded.

For you, the average (er, sorry, *superior*) pilot, the most important lesson is that Aeronautical Decision Making, dry-as-dust though the subject may be, is *important*. It is the single thing most likely to keep your little pink body in one piece, and accident investigators out of a job.

Survivability in General Aviation Accidents

Survivability—the ability to have an accident and walk away from it—is a specialised aspect of human factors. If we cannot eliminate all accidents, perhaps we can make them less dangerous? Light aircraft generally fly at much lower speeds than do airliners; they weigh less. The energy involved in an accident is therefore much lower. You might think that light aircraft accidents would be much more survivable, but you would be wrong. There are several differences which do not help the light aircraft occupants.

An airline accident usually occurs during the take-off or landing phase. It is thus likely to be in the aerodrome environment, where there is relative freedom from obstacles, the ground is likely to be reasonably level, and rescue services are on hand. The aircraft is bigger and stronger, there is more disposable structure to absorb the energy during the ground slide, and the occupiable space is likely to be better preserved. Over the years, the interior has been progressively 'de-lethalised' by removing sharp projections which could damage the occupants.

By contrast, light aircraft accidents can happen in any phase of flight, usually away from an aerodrome. The aircraft is necessarily light and is readily distorted, and there is not much structure between the occupants and whatever the aircraft strikes. The pilots, at least, must be in reach of a variety of knobs and levers in order to fly the aircraft. Even the volume control knob on a radio has been known to cause a fatal injury. Since light aircraft occupants are at greater risk—and the majority of aviation casualties involve the occupants of light aircraft—it is evidently worth considerable effort trying to make light aircraft accidents more survivable.

Some measures which would undoubtedly succeed, seem to be simply unacceptable. There is no doubt that wearing helmets and flying clothing (boots, gloves and flying suits) would save many lives. Air forces throughout the world equip their aircrews with them. Why should they be so unpalatable to General Aviation? Perhaps wearing survival equipment might be seen as an admission of vulnerability? Whatever the reason, it would be worth overcoming this resistance.

Other measures require only knowledge. For example, a seat cushion in a light aircraft, if made from upholstery foam, can double the impact force experienced by the occupant. This is because, while the rest of the structure is slowing progressively, the occupant continues downward with unabated speed until his posterior contacts the now stationary structure. High hysteresis cushion materials such as 'Temperfoam' are available; while more expensive (say $70 for a cushion) the expense is trivial compared with the cost of operating an aircraft. Another area where improvement would be simple is shoulder harness: if it leads back behind the pilot at a level below his shoulders, it may feel secure but will compress his torso vertically during deceleration, allowing him to be flung forward. This in turn can cause spine and lung damage, and perhaps let him strike an object otherwise out of range. Both of these factors (foam cushion and low harness attachment points) conspired against the pilot of an aerobatic aircraft who made a late recovery from a spin. The aircraft struck the ground in a level attitude and not especially hard (11G peak vertical deceleration), and there was a clear ground slide with steady horizontal deceleration, but the combination of spinal and lung damage proved fatal (Office of Air Accidents Investigation, 1986a).

Fires are not uncommon in accidents to light aircraft. The fuel is highly inflammable, and is in close proximity to sources of ignition such as broken wires and hot engine parts. However, the risk could be greatly reduced by adoption of available devices. It would be possible to have tanks which did not spill fuel when ruptured—Indy race cars have them—and disconnect fittings with shut-off valves where fuel lines are liable to be pulled apart in an accident. By emphasising the preventive capabilities of such devices when fire has occurred in accidents, it may be possible at least to get owners to retrofit them, though it seems that legislative action would be necessary to have them fitted to new aircraft.

Yet another possible lifesaver is the fitting of a space frame to protect the occupiable space, so that the occupants are not crushed during the deceleration sequence. Some agricultural aircraft have them; the pilot of ZK-CSC probably owed his life to the structural design of his Agwagon, after it plunged 300 feet into a forest (Office of Air Accidents Investigation, 1989). A space frame has been designed for one series of light aircraft, but is not available commercially because of American product liability considerations. Here, it seems, would be a useful area for lobbying by aviation groups.

In all of these survivability areas, we are not looking at anything new: the problems are well known, and the potential for saving lives is considerable. What is required is education, both of pilots and of the Authorities with power to get something done.

Investigating Repeated Accidents

So far, we have worked on the implicit assumption that investigating accidents will result in recommendations for avoiding their recurrence, and so accidents from identifiable sources will be eliminated. Regrettably, the review of the entire New Zealand accident database since 1965 (Zotov, 1997) showed that there were still many accidents occurring now which are essentially the same as those which happened twenty years or more ago. This gives weight to the view of those who say that accidents are inevitable, and we should not waste effort in investigating them. Certainly, there is a need to be selective. As Dr Lee of the Australian Bureau of Air

Safety Investigation has put it, "There is no point in investigating the 980th occasion on which a student pilot has broken the nosewheel leg of a Cessna 152".

Some accidents, such as the C152 nosewheel leg, are not really accidents at all, being entirely predictable. They are normal incidents of training—"learning by error", as Reason put it (Reason, 1990). Apart from the expense they cause, they are not really a matter for concern[2.]

Other accidents seem to happen time after time. Pressing on into adverse weather claims as many lives now as ever. Is there any merit in investigating those accidents? Perhaps not, if we just continue to investigate them in the same way as in the past. There is no point in determining that the airframe and engine were functioning properly, and that the pilot tried to do something beyond his capabilities—"Pilot Error". If we are to do something to stop the steady drizzle of aluminium from the skies, we need to understand *why* the pilot made such a faulty decision. This probably requires investigation of the human factors in greater depth than is at present done for General Aviation accidents. The need to understand 'why' is beginning to be appreciated in connection with air transport accidents (Helmreich, 1990; Reason, 1990; Reason, 1991; Zotov, 1996). There is no point in saying 'Pilot failed to configure flaps correctly'; we need to know *why* he did not set them properly, so that others can avoid the same trap.

Air Transport accidents can often be traced back to organisational failures, which resulted in pressure being put on the pilot to such an extent that he was unable to make sound decisions. Could this also be a factor in accidents to private pilots? At first sight this looks unlikely. Apart from flying training, General Aviation operations are essentially unstructured. Other than a few supervisory failures, there have been so few serious training accidents in New Zealand that they can almost be neglected. So where do the pressures come from?

At the moment we can only speculate, based perhaps on our own experiences. For example, aeroclubs can put pressure on pilots. If a pilot who has (sensibly) diverted because of weather is then berated because the aircraft is needed for another flight, he may be less likely to make the best decision next time. Or the pressure may be subtler. Suppose the pilot diverted on Sunday and came back by road, and the Club then asked him how he proposed to get the aircraft back on Monday (when the pilot had to be at work): would the pilot make the same decision next time? The reasonable approach would be for the club to use its professional staff to get the aircraft back, without additional cost to the individual member, but not all clubs would see it this way!

The pilot can also generate pressure on himself. If he *has* to complete the flight because he has an important appointment to make, he may press on to such a point that turning back or making a precautionary landing may be beyond his skill. Perhaps we should remind pilots of the old saying: "Time to spare—go by air".

There again, perhaps the pilot was deceived by some effect that suggested to him that the weather was going to be satisfactory when in fact it was not? None of these questions, or others like them, is likely to be answered by current investigations, but if we are to improve matters, they need to be.

Sometimes, however, examining a series of accidents can lead us to a solution that might not have been apparent from one case taken in isolation. Some series of accidents have been attributed to a variety of factors, yet when the reports are grouped together the common features are striking. Such a series was the group of accidents to Pterodactyl canard microlights.

The Pterodactyl was, at one time, the commonest training microlight in New Zealand. A great many of them had been built in the USA, where they were designed, and still others in Australia. In New Zealand they were built under licence by a flourishing concern. As we have seen, with a basic training aircraft some accidents were to be expected, but these aircraft seemed to be prone to breaking up in the air—not at all the typical mishandling failure. The accidents—generally fatal—were attributed to a variety of causes: pilot-induced oscillation, pre-existing damage, flying too fast, turbulence. But, when the wreckage plots recorded at each accident site were placed side by side, the structural damage in each case was virtually identical. Time to look for a common factor.

What transpired was that the failures in the tubular wing spars were in compression, like a yacht's mast when it is overloaded. Such compression forces could only be generated by the wire bracing, in the presence of wing twisting. No static load could possibly generate enough force, so the cause was likely to be torsional oscillation. Examination of video recordings of two such break-ups showed that indeed there was such oscillation, and that it was preceded by marked nose-down pitch excursions. These in turn were traced to insufficient foreplane power to handle the centre of pressure shift which could occur at higher speeds or in gusty conditions: the resulting excessive speed caused torsional flutter of the wings. Unfortunately the design was not susceptible to improvement, and the type was, effectively, grounded. At least this prevented the fatal accidents which were recurring every year or so (Transport Accident Investigation Commission, 1993).

There is, therefore, merit in investigating repeated accidents, even if the earlier efforts are not successful in preventing the accidents recurring. Information gathered during the earlier investigations can provide the clue which subsequently allows the accidents to be 'solved'. It is important, however, that investigators should review past accidents for common features, in the same way that any scientist will review the literature before conducting an experiment. This might seem self-evident, but the 'Bible' of accident investigation, the ICAO Accident Investigation Manual (International Civil Aviation Organization, 1970) says that each accident must be investigated on its own merits, without being prejudiced by comparison with earlier accidents.

What the Manual is saying is that the investigation should be approached without preconceptions: superficially similar accidents may stem from quite different causes. This is indeed true. For example, two Convair airliners dived into the sea near airports in Norway and New Zealand, within a short time of each other, both having recently been refurbished by the same agency. The immediate causal factor in one case was a tailplane hinge failure; in the other, somatogravic illusion trapped an inexperienced pilot (Transport Accident Investigation Commission, 1991).

But the Manual does *not* say that investigators should not examine past accidents for common factors, the basic investigation having been completed.

Some investigators have interpreted it this way, but to do so is to refuse to benefit from past investigations.

The Accident Database

This raises several questions: Who should maintain (and pay for) the database? Some investigation authorities would argue that they have neither the time nor the resources to do so. What should be in the database? Should it be restricted to local accidents, or extended to overseas accidents? What about incidents? How should the information be organised, especially the human factors information? Multiple, overlapping error-coding schemes may obscure commonalities.

At present, some investigation authorities rely on 'corporate memory', i.e., the recollections of individuals, passed on to others during informal discussion, and to their successors during their apprenticeship training. This has worked surprisingly well, but requires stability in the investigating organisation, and people who stay in the job for a long time. It would be desirable to have a system so that the organisation was less dependent on individual memory. The complexity of database systems has been a severe handicap in the past, but fortunately technology has come to our aid. There are now systems which require only narrative entries, and are capable of grouping accidents by whatever words we want to key in, without even the need for predetermined keywords.

As to who should maintain the database, it may seem that this should be done by the users. Investigating authorities jealously preserve their independence, both real and perceived, from regulatory authorities, and should have completely independent systems. However, there could be merit in having the database maintained by a research institution like a University, which will have the facilities and expertise, and may well use the information for research projects.

It makes no sense whatever to have only fatal accidents on the database: many fatal accidents have minor accidents as precursors, and the information from these can be of value in solving a subsequent fatal accident. By the same token, information from a minor accident has enabled earlier fatal accidents to be solved.

Take, for example, a series of accidents in which hang gliders soaring close to a slope had turned downwind and flown straight into the hill, with detrimental results. These accidents were attributed to tip stalling, and pilots were exhorted to maintain flying speed, but the accidents continued. Ultimately a pilot survived with only minor injuries, and was able to tell us that he had had adequate speed for normal control, but although he had applied maximum roll control away from the hill, the aircraft had turned and flown into it. This prompted a study of how a hang glider turns, the available roll authority, and the wind shear in close proximity to the upwind side of a hill. It became evident that it was possible for the roll induced by the wind shear to be greater than the available control to counteract it. The remedy was simple: don't try to fly close to the hill in an attempt to soar in weak lift conditions. There have been no further accidents of this sort in New Zealand. (Office of Air Accidents Investigation, 1987) There can be a real advantage in having a live pilot to talk to.

For small countries, there is a considerable advantage in having access to the larger databases of larger countries. This is a further argument for the database to be maintained by the investigating authority, since overseas investigating authorities may be more willing to share information with another such, rather than with a regulatory authority

The question of how incidents (where no accident occurred, but there was the potential for one) should be recorded, is controversial. Unless the incident is independently investigated, the report may be coloured by the pilot's perception, and is subject to all the usual problems of self-reporting. Increasingly, voluntary (confidential) incident reporting is being put under the aegis of separate, independent organisations - the US ASRS and UK CHIRP schemes, for example. These reports are not investigated in depth, but collectively may show areas where action can be taken to pre-empt an accident.

Accident Reports

Most countries write accident reports on major accidents in the format recommended by the International Civil Aviation Organisation (ICAO). Minor accidents will often be reported in 'Brief' format, a narrative of a page or two. The ICAO format is shown in Figure 12.4. The division between 'Factual Information' and 'Analysis' is logical enough, but the order within the Factual Information is somewhat arbitrary. The idea of the set headings in this section is that if the investigators have gathered all the information under each heading, they are unlikely to have overlooked anything important which might be needed to write the Analysis. Also, researchers looking for information across a number of accidents will know exactly where to find it.

The format was originally devised to handle reports of investigations into mechanical failures, and for these it generally serves well. However, information on human factors is scattered over a number of separate sections—medical, communications, personnel, and so on. There is no set format for Analysis, and the various strands of a human factors investigation are often presented individually, as the investigator worked on them. It can be difficult to put in writing the way in which a complex web of interacting factors can lead to an accident.

Alternative formats have been tried (Bureau of Air Safety Investigation, 1994; Zotov, 1996), based on what we know about how human factors accidents happen. There is no impediment to any country using such a format if it chooses, and the Australian Bureau of Air Safety Investigation, for example, has attempted to use the Reason model (Reason, 1991) as a template. In time, a better standard format can be expected to emerge, and it may then be adopted by ICAO.

ICAO REPORT FORMAT

SECTION 1 - FACTUAL INFORMATION

1.1	History of the Flight
1.2	Injuries to Persons
1.3	Damage to Aircraft
1.4	Other Damage
1.5	Personnel Information
1.6	Aircraft Information
1.7	Meteorological Information
1.8	Aids to Navigation
1.9	Communications
1.10	Aerodrome Information
1.11	Flight Recorders
1.12	Wreckage and Impact Information
1.13	Medical and Pathological Information
1.14	Fire
1.15	Survival Aspects
1.16	Tests and Research
1.17	Organisational and Management Information
1.18	Additional Information
1.19	Useful or Effective Investigative Techniques

SECTION 2 - ANALYSIS

SECTION 3 - CONCLUSIONS

SECTION 4 - SAFETY RECOMMENDATIONS

Figure 12.4   The ICAO report format

Writing Recommendations

There is a general conception that the ultimate product of an investigation is the Accident Report. Nothing could be further from the truth; the Report is simply a summary of the investigation and its findings. What really matter are the Safety Recommendations, where the lessons learnt from the investigation are put into concrete form so that (we hope) that particular accident will never occur again. You might well wonder to whom the investigators are making the Recommendations. As a safety authority, can't the investigating organisation just say what needs to be done?

In most countries there is a strict separation between the investigating authority, and the Civil Aviation Authority. There are two reasons for this. Firstly, because the Authority is responsible for enforcement—a policing role, if you like—there is an understandable reluctance for people to come forward with information which might show themselves in a less-than-perfect light. However, the information they have might be essential for the understanding of an accident, and so save lives in future. Having an independent investigation agency makes it much more likely that information will be volunteered. Secondly, the investigation of an accident may put the CAA in the embarrassing position of having to investigate itself. Defective regulation, inadequate surveillance and organisational deficiencies may have contributed to an accident; even the Authority's staffing policy has been called into question in investigations.

With the separation of responsibility between the investigators and the Civil Aviation Authority, executive power rests with the Authority. It alone can make new regulations, or change policies, or issue Airworthiness Directives. The investigators will recommend (usually) to the Authority, but the Authority will have the final say. Generally, the Investigators will welcome this division of duties. They are probably less well informed of the difficulties of making changes than is the Authority.

An ill-advised change can give rise to unforeseen problems just as bad as those which it seeks to avert. Some years ago, one type of glider proved unstable in pitch. The solution found was to introduce a mass-balance to the all-flying tail. Naturally, if you put a substantial piece of lead in the rear of a lightweight aeroplane, you will move the centre of gravity aft considerably, so the cockpit placard was amended to show a higher minimum pilot weight. A pilot who was familiar with the aircraft, but not the modification, flew the glider. When he turned steeply the aircraft flicked into a spin and would not recover. Subsequently the mass balance modification was rescinded, and friction damping was increased.

It is not generally a good idea, therefore, for the investigators to say *how* the improvement should be effected: it is better to say *what* needs to be done, but leave it up to the CAA to decide on the practical implementation. Unfortunately, this assumes both goodwill and competence on the part of the CAA. There have been some spectacular examples where these assumptions have proved invalid. Perhaps the worst was the Orly disaster (Air Accident Investigation Branch, 1976), when a fully laden DC-10 lost a cargo door shortly after take-off. The loss of pressure from the baggage compartment resulted in the floor of the cabin bowing down, and jamming the elevator controls; the aircraft flew into a forest in a shallow dive. What made this accident inexcusable was that there had already been almost a replica of the accident, except that in the precursor the crew managed to regain control of the aircraft and landed safely (National Transportation Safety Board, 1973). The US National Transportation Safety Board, the Federal Aviation Authority and the manufacturer were all perfectly aware of the problem, but no effective action was taken. The subsequent Congressional Inquiry was most scathing of those whose inaction led to the disaster. Today, the NTSB publicises its Recommendations and the reasons for them, and follows up their implementation by the FAA.

If there are at times difficulties in getting straightforward mechanical problems fixed, you can imagine that getting human factors problems addressed can seem well nigh impossible. Part of the problem seems to be that physical causes of accidents are seen as being capable of proof, while human factors causes have to be inferred. Physics is seen as 'hard science', whereas psychology is a 'soft science'. This distinction is not altogether valid. For example, transient phenomena such as wind-shear and icing can rarely be 'proved' by physical exhibits. If we have an engine failure due to carburettor icing, the ice will have melted long before the investigators arrive on the scene. They will consider the combination of humidity and temperature, with the likely power setting, and may infer that carburettor icing was highly probable. If they are able to eliminate fuel supply problems and mechanical failure, then what remains, carburettor icing, is the probable cause.

Conversely, some of the human factors effects are capable of precise demonstration and are quite repeatable in experiments—reaction times in given circumstances, for example, may be more predictable than the time of failure of a spar under fatigue. But it is certainly true that, more often than not, the human factors which led to an accident must be inferred. This does not make them any less real, as anyone who has suffered spatial disorientation in IMC will know!

Here, I think, it may be desirable for the Investigators to go further than merely stating what the problem is, in their Recommendations. They will have been involved in analysing it, and so should have as good an understanding of the problem as anyone.

This is not to say that nothing has been achieved. Some significant improvements have taken place. For example, Crew Resource Management and its General Aviation equivalent, Aeronautical Decision Making, have been introduced into licensing requirements. In the case of CRM, there have been well-documented successes where crews have averted or minimised accidents: there are few sceptics now. It is perhaps too early to claim similar success for ADM, but experimental trials have been encouraging, and the need for the concept has been highlighted by the major part played in GA accidents by faulty decision-making.

The Paradox

The chief obstacles to a safer future are in the realm of policy. Certification standards, 'Grandparenting' clauses, a 'hands-off' approach to General Aviation safety: none of these help to reduce the accident rate.

It is almost standard practice for authorities to accept the standards of the country of manufacture, even where those standards are manifestly inadequate. For example, the pilot of a motor glider was killed in a minor impact when his aircraft nosed into a swamp short of the runway (Office of Air Accidents Investigation, 1986b) because the extended engine, mounted behind his head, hinged forward and struck him. The airworthiness authority refused to do anything about the light restraining cable which was supposed to prevent this happening, because it (just) met the certification standard. A small increase in wire size would have doubled the breaking strain, and prevented such injury in any likely survivable impact. That the problem is not confined to General Aviation was demonstrated after the Leeds-Bradford over-run, in which an aircraft that was flown within the prescribed limits for landing never-the-less ran off the end of the runway. (Air Accident Investigation Branch, 1987).

Grandparenting clauses permit aircraft certified before a change in the airworthiness regulations to continue in production unmodified, even though they have been constructed after the change. This results in even new aircraft having inadequate restraints, or defective seats. (Transport Accident Investigation Commission, 1992).

The 'hands-off' approach is now fashionable: "Organisations are responsible for their own safety". This can work, if there is a suitable organisation: gliding associations, world-wide, have a very good record. Perhaps this is because of the collaborative nature of gliding, with the combined efforts of a number of people being needed for a successful flight. Whatever the reason, the concept has not worked so well in other areas of General Aviation. In particular, there is no-one looking after the interests of the typical aeroclub pilot. He may encounter unprofessional maintenance (Office of Air Accidents Investigation, 1990), pressure to make poor decisions (as discussed above), and aircraft with longstanding defects. In a perfect world, the regulatory authorities might be able to rely on 'organisations', but the world is imperfect.

We investigate General Aviation accidents because, over time, we hope to reduce the accident rate. By applying the lessons from past accidents, we could also reduce the number of lives lost when accidents do occur. But the fundamental paradox is that, to be credible, the investigation process has to be outside the mainstream of regulators, manufacturers and operators; yet by being outside the action, the process may have little chance of influencing those inside to change anything.

The raison d'être for accident investigation tends to fall apart if there are immovable objects between the investigation and recommendations, and corrective actions. In New Zealand, there is little evidence of any reduction in the accident rate for general aviation in recent decades (see Figure 1.3 in the Introductory chapter to this volume).

Who are the Investigators?

Most accident investigators are either pilots or engineers with a wide aviation background. In New Zealand, the minimum requirement for pilots is an Air Transport Pilot Licence and 3000 hours command time. Engineers must have, in addition to a Bachelor of Engineering degree or equivalent, a Commercial Pilot Licence with instructor and instrument ratings. In most countries the requirements are similar.

Their training consists partly of academic instruction—everything from structures and strength of materials, to what makes some accidents survivable rather than others. Mostly, however, the training is a lengthy apprenticeship. The

International Society of Air Safety Investigators prescribes a minimum of five years experience before an investigator can be admitted as a full member, and this figure seems to be generally accepted as the experience required before an investigator can be considered fully qualified.

In the past, investigators have become knowledgeable about mechanical and operational problems, but there was undoubtedly a weakness in the training in human factors. This is now being addressed, either by stressing the human factors aspects during case studies, or by specific training.

However, there is far too broad a scope of knowledge needed for any individual ever to master it all, whether in aerodynamics and structures, or in human factors. The real trick is to recognise what sort of problem you have found, when to call in experts, and what questions to ask them. Since experts invariably talk a language of their own, it also helps if you know enough to understand what they are telling you—the first time a psychologist mentioned 'cognitive dysfunction' to me, I had to call for time out while he explained what he meant.

Areas for Research

Generally, Safety Recommendations are made on the 'creeping tide' principle: if we can improve many small points, then over a period of time, flying should become safer. This approach certainly has its merits. If we find something unsafe during the course of our investigations, it makes sense to remove the hazard if we can. Many of the improvements in crashworthiness and survivability have come about in this way.

However, this approach does not cope with those stubborn accidents that seem to recur no matter what we do. Take weather-related accidents: we exhort, train, cajole—and still we find VFR pilots flying into IMC. When we get such recurring accidents, it may be that we need to seek a 'silver bullet', the neat solution that will eliminate a group of accidents altogether. It can be done, sometimes, for example the 'downwind turn' accidents with hang gliders. (Office of Air Accidents Investigation, 1987). Generally, such repeated accidents come about as a consequence of lack of knowledge. With the hang gliders, there was a lack of understanding of the limitations of roll authority, and of the wind shear close to the hill when slope soaring in light airs, for example. They may therefore be a profitable subject for research.

One of the most perplexing areas today is the VFR into IMC accident. The conventional wisdom is that these are due to 'bad attitudes' on the part of the pilots, but this does not seem to tie in with experience, in every case. Students on dual cross-country flights have blundered into cloud despite all sorts of prompting. Perhaps unawareness might be nearer the mark? Weather related accidents to students flying solo seem to have this hallmark, too. Other accidents seem hard to explain, without postulating that a critical situation had arisen without the pilot having been aware that it was becoming critical.

In a number of cases, the accidents were so precisely replicated that it was almost possible to walk up to the site on dead reckoning. It is hard to argue that all the pilots suddenly had bad thoughts at just that point! Could there have been some sort of visual effect that led the pilots to believe that the situation was still under control?

More generally, the discovery of a process for deciding when and how to find 'silver bullets' would be of great benefit to investigators.

Summary

The purpose of investigating accidents is to reduce the numbers of accidents. By applying the lessons from past accidents, we can also reduce the numbers of lives lost when accidents do occur. Governments see this as a useful activity, because lives lost in accidents represent a considerable cost to the community. There is the added benefit that the investigation of light aircraft accidents is valuable training for the time when the investigators are called upon to inquire into (fortunately rare) airline accidents.

The product from investigations is the Safety Recommendation. Since investigators are usually separated from the Civil Aviation Authority which has the executive power, the recommendations advise what the problem is, but not (generally) how one should go about fixing it. However, in the relatively new field of human factors it may be that more specific information could be helpful.

In the area of survivability, many of the solutions to making aircraft safer are well known. By highlighting deficiencies that have led to avoidable injury or loss of life, the investigators may be able to bring about needed changes.

Investigating General Aviation accidents may seem a costly exercise, but those costs could be redeemed many times over in a safer future.

Notes

1   A witness mark is an indentation, scratch or smear, made by another part when distortion occurs during the impact sequence. It shows the relative position of the two objects at that moment.

2   An accident is, by its nature, unexpected, whereas we expect students to bend nosegear. We know that students learn most when they are on their own, and to an extent they are experimenting to find the limits for themselves. They are sent off on their own before they can do things perfectly, because to hold them back until perfection was reached would be counterproductive. While it would be nice if they 'learnt from error' without actually bending the aircraft, we accept that from time to time this ideal state of affairs won't happen. Where the consequences of error would be catastrophic, we adjust matters so that there is more of a buffer in the system. For example, stall-spin accidents on approach have been virtually eliminated by teaching the approach at 1.3 Vs, and using circuit patterns that do not require steep turns onto final approach.

References

Air Accident Investigation Branch. (1976). *Aircraft Accident Report 8/76 Douglas DC 10-10 TG-JAV.* London: HMSO.

Air Accident Investigation Branch. (1987). *Aircraft Accident Report 2/87 Lockheed Tri Star G-BBAI.* London: HMSO.

Air Accident Investigation Branch. (1988). *Aircraft Accident Report 8/88. Report on the accident to Boeing 737-236 G-BGJL at Manchester Airport on 22 August 1985.* London: HMSO.

Air Accident Investigation Branch. (1990). *Aircraft Accident Report 4/90. Report on the accident to Boeing 737-400 G-OBME at Kegworth on 8 January 1989.* London: HMSO.

Bureau of Air Safety Investigation. (1994). *Piper PA 31-350 Chieftain VH-NDU* (Report No. 9301743). Canberra: BASI.

Helmreich, R. L. (1990). Human factors aspects of the Air Ontario crash at Dryden, Ontario. In V. P. Moshansky, *Final Report of the Commission of Inquiry into the Air Ontario Crash at Dryden, Ontario* (Technical appendices, pp. 319-348). Ottawa: Minister of Supply and Services.

International Civil Aviation Organization. (1970). *Manual of aircraft accident investigation.* (4th ed.). Ontario: Author.

National Transportation Safety Board. (1973). *Aircraft Accident Report AAR-73-2. Douglas DC 10-10 N103AA.* Washington, DC: Author.

Office of Air Accidents Investigation. (1986a). *Aircraft Accident Report No. 85-039 Pitts Special S-1E ZK-ECO.* Wellington: Government Printer.

Office of Air Accidents Investigation. (1986b). *Siren Pik 30 motor glider ZK-GSG. (Report No. 86-096).* Wellington: Government Printer.

Office of Air Accidents Investigation. (1987). *Lancer 4L hang glider. (Report No. 87-078).* Wellington: Government Printer.

Office of Air Accidents Investigation. (1989a). *Cessna C188b ZK-CSC. (Report No. 89-088).* Wellington: Government Printer.

Office of Air Accidents Investigation. (1990). *Aircraft Accident Report No 90-038 Cessna A150L Aerobat ZK-DPK.* Wellington: Government Printer.

O'Hare, D., Wiggins, M., Batt, R., & Morrison, D. (1994). Cognitive failure analysis for aircraft accident investigation. *Ergonomics, 37*(11). 1855-1870.

Rasmussen, J. (1980). What can be learned from human error reports? In K. D. Duncan, M. Gruneberg & D. Wallis (Eds.), *Changes in working life* (pp. 97-113). New York: Wiley.

Rasmussen, J. (1982). Human errors: a taxonomy for describing human malfunction in industrial installations. *Journal of Occupational Accidents, 4,* 311-335.

Reason, J. (1990). *Human error.* Cambridge: Cambridge University Press.

Reason, J. (1991). Identifying the latent causes of aircraft accidents before and after the event. *Proceedings of the 22nd International Seminar of the International Society of Air Safety Investigators* (pp 39 - 45). Sterling, Virginia, USA: ISASI.

Transport Accident Investigation Commission. (1990). *Aircraft Accident Report 91-012 Gardan Minicab ZK-DAG.* Wellington: Author.

Transport Accident Investigation Commission. (1991). *Aircraft Accident Report No 89-064 Convair 340/580 ZK-FTB Manukau Harbour*. Wellington: Author.

Transport Accident Investigation Commission. (1992). *Aircraft Accident Report No 91-004 Robinson R22 Beta ZK-HDD*. Wellington: Author.

Transport Accident Investigation Commission. (1993). *Pterodactyl Ascender 11 + 2 ZK-FKF* (Report No. 92-007). Wellington: Author.

Wiegmann, D. A., & Shappell, S. A. (1997). Human factors analysis of post-accident data: applying theoretical taxonomies of human error. *The International Journal of Aviation Psychology, 7*, 67-81.

Zotov, D. V. (1996). Reporting human factors accidents. *The Journal of the International Society of Air Safety Investigators, 29*(3), 4-20.

Zotov, D.V. (1997). *Pilot error: cognitive failure analysis*. Unpublished master's thesis, Massey University, Palmerston North, New Zealand.

Index